T0332011

COVERED WITH DEEP MIST

Covered with Deep Mist

The Development of Quantum Gravity (1916–1956)

Dean Rickles

University of Sydney

OXFORD

UNIVERSITY PRESS

OXFORD

UNIVERSITY PRESS

Great Clarendon Street, Oxford, OX2 6DP,
United Kingdom

Oxford University Press is a department of the University of Oxford.
It furthers the University's objective of excellence in research, scholarship,
and education by publishing worldwide. Oxford is a registered trade mark of
Oxford University Press in the UK and in certain other countries

First Edition published in 2020

Impression: 1

Published in the United States of America by Oxford University Press
198 Madison Avenue, New York, NY 10016, United States of America

British Library Cataloguing in Publication Data
Data available

Library of Congress Control Number: 2019955139

ISBN 978–0–19–960295–7

DOI: 10.1093/oso/9780199602957.001.0001

Printed and bound by
CPI Group (UK) Ltd, Croydon, CR0 4YY

To the memory of my father, Peter Rickles (1950–2010)

Preface

In view of the great difficulties of this program, I consider it a very positive thing that so many different approaches are being brought to bear on the problem. To be sure, the approaches, we hope, will converge to one goal.

Peter Bergmann[1]

Quantum gravity [QG] is the attempt to draw together in the same framework general relativity and quantum theory, often (incorrectly) described as the 'physics of the large' and 'physics of the small' respectively. Though the problem of quantum gravity (that is: of simultaneously giving due consideration to both theories, and the phenomena they describe) has been known for almost as long as either of these two 'ingredient theories,' progress has been very slow and we still have no such theory to hand. Instead, we are faced with a veritable profusion of approaches, each dedicated to solving the basic problem. A recent book on *Approaches to Quantum Gravity* (D. Oriti, ed.) displays a wide representative sample of this array of distinct and diverse ways of solving the problem of quantum gravity, many of which are but fledglings.[2] This situation of a very large number of quite distinct quantum gravity proposals existing in parallel seems to be a relatively recent development[3]: the failure to produce a workable theory after almost a century of hard labour by the very best physicists of this century and the last has led to a somewhat desperate struggle to try anything!

In this monograph I trace the historical and conceptual development of QG for the first forty years of its existence, up until the mid-1950s. This is in fact a fairly natural stopping-off point: the late 50s and the 60s see the flowering not only of new experimental possibilities for gravity, but of radically new approaches, such as discrete gravity, string theories, semiclassical gravity, and also far more detailed work on the canonical approach

---

[1] Closing remarks about the problem of quantum gravity, at *Conférence Internationale sur les Théories Relativistes de la Gravitation*, held in Royaumont, June 21st–27th, 1959.

[2] That book also draws attention to a feature that will become apparent in the present book: namely, that very often the different approaches correspond to quite different conceptions of what the problem of quantum gravity actually is, and to different underlying motivations—I call this the problem of *polysemicity*. Because of these differences, I shall adopt an extremely liberal view with respect to the question of what constitutes 'quantum gravity.' I discuss this broadly methodological issue in the opening chapter.

[3] Earlier on, attempted solutions were often simply bound to the various methods of quantization which were at that stage not usually considered to have strong conceptual implications of the kind that would warrant divisions into anything like todays 'camps.' We find in this book, then, also the pre-history of the present day schisms between approaches to the quantum gravity problem, locating the origins (or the glimmerings of such) of the now firmly established quantum gravity camps (e.g. between strings and loops) only in the later parts of the selected period.

(e.g. the Wheeler–DeWitt equation) and the covariant approach (e.g. the computation of the Feynman rules for general relativity). This book attempts to trace the origins of this explosion of approaches—itself, partially an aftereffect of the so-called 'renaissance' of general relativity; though one might trace some of this very renaissance to work on quantum gravity—along with the reasons behind the fairly sudden explosion. As we will see, institutional factors have a central role to play in this story, and much of the explosion comes about because of the establishment of new funding sources, institutes, conferences, summer schools (and other such colloquia), networks, and so on.[4]

This book is in many ways, then, an attempt to find the *roots* of these later proposals and institutional changes. We find them very often bound up with the *internal* problems faced by these earlier approaches, and by *external* features such as the success of some particular quantization approach applied to another interaction (such as Yang–Mills theory and the strong force), or else some new development in a *prima facie* disparate field, such as cosmology, computation, or statistical physics. Characteristic of the earliest work on QG is a notable lack of an 'inner life' of its own guiding the development of the field. Instead we find a field that is subject to exogenous buffeting and influences, or a desire to *use* QG to resolve problems that are strictly separate from QG per se (a crucial example being the possibility that gravity might act as a 'natural regulator' curing or taming the ultraviolet divergences of quantum field theories). This book might in fact be read as charting the transformation of quantum gravity into a research field in its own right, which I believe can be positioned at my chosen endpoint.

In writing this book I was very much inspired by Sam Schweber's masterly study of the creation of quantum electrodynamics *QED and the Men Who Made It* (Princeton University Press, 1994). I look upon his book as a model of good history of physics writing. If my own book were to have a subtitle like his, it might be: *QG and the Men Who Failed To Make It*! This is not meant to be in any way disparaging to those currently working in the field of quantum gravity, nor those who tilled its forbidding soil in the past, for whom I have the very highest respect.[5] The problem of quantum gravity is

---

[4] Of particular importance is the Chapel Hill Conference on the Role of Gravitation in Physics (see C. DeWitt and D. Rickles, eds., *The Role of Gravitation in Physics: Report from the 1957 Chapel Hill Conference*, Berlin, Edition Open Access, 2011: http://edition-open-access.de/sources/5/index.html.). It is here, at the inaugural conference of the Institute for Field Physics, directed by Bryce and Cécile DeWitt, in January 1957, that we find extensive discussions of the measurability of the quantum gravitational field, the divergence problem in the context of quantum general relativity, the canonical approach, Feynman (path-integral) quantization (and, in the comments and *published* papers related to the conference, Everett's 'relative state' interpretation of quantum mechanics, initially devised in part to provide an interpretation for quantum gravitational contexts). This conference (and the papers that were written up for *Reviews of Modern Physics*) displays a distinct advance in the conception of the problem of quantum gravity. In many ways, the problem of quantum gravity becomes an independent area of research at this point, and so I have quite self-consciously chosen to stop the discussion immediately prior to this conference. I will, however, chart the events leading up to it, including an earlier preliminary 'warm up' workshop held at Roaring Gap in 1956, and including the establishment of the Institute of Field Physics. A recent paper by Alexander Blum and Thiago Hartz investigates a followup meeting in July 1957 focusing solely on quantum gravity ("The 1957 quantum gravity meeting in Copenhagen: An analysis of Bryce S. DeWitt's report," *The European Physical Journal H* (2017), 42(2): 107–57).

[5] I note that at least one of those involved in the earliest phases of quantum gravity, Wolfgang Pauli, explicitly referred to this failure to solve the quantum gravity problem in just these terms. In a letter to Erwin Schrödinger, on the latter's 70th birthday, Pauli pointed out that they are connected through their focus on the "two main themes of the general theory of relativity and the quantum theory." He goes on: "Also, our age difference of 13

clearly one of enormous complexity. However, the development of QG research is, in a very real sense, the history of *failed* attempts and dead ends. There is nonetheless historical interest in QG despite this. One has to wonder *why* physicists have persisted with the problem, why it has been pursued for so long without closure, and (also of methodological interest) what factors have governed the many casualties. The historical study of QG is also of more direct historical interest since the 'side-effects' of work on quantum gravity have themselves been of vital importance to the development of other areas of physics, notably renormalizable quantum gauge field theories (a side-effect of the search for the Feynman rules of quantum gravity), statistical physics, lattice methods, numerical relativity, conformal field theory, and more—including, though of a different character, new interpretations and constraints on old interpretations of quantum mechanics. Classical general relativity has also benefited greatly from the pursuit of quantum gravity, and for a long time the study of gravitational radiation was viewed as a mere prelude to quantum gravity and thus its underlying motivation. All the more reason to write quantum gravity's history despite the glaring absence of a distinguished paradigmatic theory of quantum gravity.

I expect this book to appeal to several audiences because of the nature of this complexity in content and context. On the one hand it will naturally be of interest to quantum gravity researchers: these are, of course, the readers this book is primarily directed at, and I hope they will find many surprises in the following pages. Historians of physics too ought to find plenty that is new and interesting in this book, simply because there is to date no large-scale historical study of quantum gravity.[6] However, the book should also be of the utmost interest to philosophers and sociologists of physics and science in general. Quantum gravity is very curious from a methodological point of view; it seems to fall outside of the usual methodologies of science (i.e. those models attempting to explain 'how science works'). There are—or at least *were* in the proposals mentioned in this book—no direct experiments to guide theory selection and rejection.[7]

years will soon appear insignificant, and we will be counted in the same generation of physicists: to the one who, e.g. failed in synthesizing the two mentioned topics - general theory of relativity and quantum theory" ["Auch unser Altersunterschied von 13 Jahren wird bald als unwesentlich erscheinen, und man wird uns zur selben Physiker-Generation zählen: zu denjenigen, der z.B. eine Synthese der beiden genannten Themen – allgemeine Relativitätstheorie und Quantentheorie – nicht gelungen ist"] (letter from Pauli to Schrödinger, August 9th, 1957; in Karl von Meyenn, ed., *Scientific Correspondence with Bohr, Einstein, Heisenberg, a.o. Volume IV, Part IV, A: 1957*, Springer, 2005: p. 519—note, in what follows, all translations are my own, unless otherwise specified). I might also note that I find evidence here, in the quantum gravity context, of Schweber's argument that the development of the field is a story of a relatively small number of exceptional individuals.

[6] One can find references to it in passing in historical books on, e.g., unified field theories, classical general relativity, and quantum field theory. However, it is usually glossed over in a manner that often suggests some embarrassment with the subject and its still incomplete state. My view is that almost 100 years of rich history ought to trump this 'hands off until complete' position seemingly adopted by many historians of physics. I should, however, point out that there is a kind of companion to this book in the form of a sourcebook I co-edited with Alexander Blum: *Quantum Gravity in the First Half of the 20th Century: A Sourcebook* (Berlin: Edition Open Access, 2018). I have also studied the history of superstring theory, from roughly the endpoint of this present book until the present day, in a separate monograph: *A Brief History of String Theory. From Dual Models to M-Theory* (Springer, 2014)—I leave the tracing of the further, post-1956 development of quantum gravity more generally to those with more stamina than I.

[7] Though we will find that there were some very early proposals (from 1919) to test the gravitational behavior of quantum systems and look for conflicts between the two. These had the effect of highlighting why further tests of QG would be practically impossible (on account of gravity's weakness), thus quickly ushering in the

The proposals must clearly be guided by other factors, such as old evidence, old and new puzzles (including external puzzles, as mentioned above), unifying power (and a sense that one *needs* a unified worldview), and mathematical consistency. I shall attempt to probe this issue, of locating the source of decisions in the construction, selection, and rejection of approaches, throughout the book. In this sense, this book is as much an account of the *methodology* of quantum gravity as it as about its history. In this sense, the book also provides useful support to the burgeoning field of philosophy of quantum gravity, especially for those philosophers of science that favour an 'integrative' approach, blending historical and philosophical aspects.

Finally, though likely wishful thinking, this book might be read by that growing breed: the interested non-academic. Though the book is often dealing with highly technical material—a full understanding certainly calls for considerable mathematical sophistication on the part of the reader—if the reader were to skim the more formal parts they should be able to gain a fairly good picture of the origins and early development of one of the most interesting and profound, and still unresolved puzzles of modern physics. To this end, I have tried to avoid regurgitating formalism from the source papers discussed unless it seems necessary for the overall narrative—as mentioned, many of these papers are reproduced (in open access form) in the volume cited in footnote 6. The social context involved behind the scenes, especially as revealed in the later chapters, is, I think, of sufficient independent interest to offer plenty of stimulation to those without a physics or mathematics background.

Let me say in closing that when I began this project, I expected somewhat meagre pickings in terms of historical fruit from my chosen period. The consensus was that hardly anything of interest happened in quantum gravity before the 1950s, save for a couple of papers by Rosenfeld that in any case fell more or less deadborn from the press. This was far from the case, and indeed I could have continued the book for longer in various directions (e.g. the full extent of the electrodynamical analogy; the role of early cosmology; the bearing of quantum gravity considerations on the history of renormalization; the relationship with the study of gravitational radiation;…). At some stage, a desire the keep this book manageable in length and scope overtook, for my own sake as much as the reader, the quest for an all-encompassing history. As such, there remain some incompletenesses in this account that I hope others will be motivated to fill in.

D.R., Bowral, NSW: May, 2019.

---

long-persisting pessimistic attitude with respect to any hopes of QG phenomenology (only very recently dislodged). This had the effect of rendering the problem of quantum gravity primarily one of 'aesthetic distaste' at the prospect of a 'half and half' world (i.e. half quantum and half classical)—though with a debate about whether this went beyond the merely aesthetic and in fact constituted an inconsistent situation.

Acknowledgments

This book features research that has already or will appear in the following articles by the author:

- "Patronage of Gravitational Physics and the Relativity Community in the USA (1949–1959)." To appear in A. Blum, R. Lalli, and J. Renn (eds.), *Einstein Studies*. Birkhaüser.
- "Discreteness and Divergences". In *Quantum Gravity in the First Half of the 20th Century: A Sourcebook* (co-authored with Alex Blum). Edition Open Access: Max Planck Research Library for the History and Development of Knowledge, Volume 10, 2018.
- "Geon Wheeler: From Nuclear to Spacetime Physicist." *European Journal of Physics H* **43**(3) (2018): 243–65.
- "The Price of Gravity: Private Patronage and the Transformation of Gravitational Physics after World War II." (Co-authored with David Kaiser). *Historical Studies of the Natural Sciences* **48**(3) (2018): 338-79.
- "Paul Weiss and the Origins of Canonical Quantisation." (Co-authored with Alexander Blum). *European Journal of Physics H* **40**(4) (2015): 469-87.
- "Pourparlers for Quantum Gravity: Some Early Sources of Quantum Gravity Research." In S. Katzir, C. Lehner, and J. Renn (eds.), *Traditions and Transformations in the History of Quantum Physics* (pp. 149–80). Max Planck Research Library for the History and Development of Knowledge, 2013.

I have generated an enormous number of debts in the writing of this book, from colleagues and readers, to interviewees and administrators. I will not be able to remember all who have contributed, but I'll try.

Firstly, a big thank you to my editor at OUP, Ania Wronski, for persistently wrangling this book out of me over many years with many delays and as many projected deadlines 'whooshed' by. I wonder if it would ever have been completed without her. The American Institute of Physics (and Greg Good in particular) supported many of the 'data collection' parts of this book, including many interviews. I also acknowledge the American Institute of Physics' Emilio Segrè Visual Archives, Lande Collection, from which the very fine portrait of Erwin Schrödinger was taken.

I am grateful to the the Dolph Briscoe Center for American History at the University of Texas at Austin for allowing me to access and reproduce materials from the Bryce S. DeWitt Papers. I thank the late Cecilé DeWitt for allowing me access to her own personal archive of material on the Institute of Field Physics, amongst other things, which really kick-started this entire project—and her daughter Chris, for supplying me with excellent photographs. I would also like to thank CARA (The Council for Assisting Refugee Academics) for permission to reproduce material from the Archives of the Society for the Protection of Science and Learning [Folios: MS. SPSL 286/2 and MS. SPSL 444/1]; the Bodleian library for reproducing the documents; Churchill College, Cambridge for allowing me to reproduce papers from the Born and Dirac archives (and Florida State University for permission to reproduce documents from the Dirac archive). Many of these trips were generously aided by the FQXi. This research was supported by grant number FQXi-RFP-1817 from the Foundational Questions Institute and Fetzer Franklin Fund, a donor advised fund of Silicon Valley Community Foundation. The Australian Research Council funded two research fellowships (DP0984930 and FT130100466), without which this book would not have been possible: for this I thank the Australian tax payers!

My thanks to Paul Weiss's family for their help with this project, and also to Agnew Bahnson's family. The team involved in the *Roots of Quantum Gravity* workshops, at Caltech and Berlin—I thank Caltech also for the Maurice Biot award that enabled me to visit the Feynman archives. Jürgen Renn, and the Max Planck Institute, were pivotal in many ways in terms of the cultivation of many ideas in this book, providing much intellectual (and financial) support. I thank Donald Salisbury for joining me on an excellent and enjoyable set of interviews of many of the old guard of quantum gravity research: James Anderson, Dieter Brill, Stanley Deser, Charles Misner, Felix Pirani, Josh Goldberg, and Louis Witten. Specific people that have aided me directly or indirectly (via their works) are: Alex Blum, Gennady Gorelik, David Kaiser, Carlo Rovelli, Silvan Schweber, John Stachel, and Karim Thébault.

* * *

Finally, I thank my beautiful muse Mira for her abundant love and support and my beautiful daughter Gaia ("noot noot"), for always keeping me well grounded.

Contents

At the back of our striving for a unitary field theory, the great problem awaits us of bringing it into line with quantum theory. This point is still covered with deep mist.

Erwin Schrödinger, 1944

1

The Problem of Quantum Gravity: From Feelings to Phenomena

No question about quantum gravity is more difficult than the question, 'What is the question?'

John Wheeler ([Wheeler (1984)], p. 224)

Programs of quantization derive their motivation both from a general philosophical desire (just as [with] the unitary field theories) not to permit a compartmental approach to the theories of physics, and from the more specific 'feeling' that quantized sources of a field require also a quantized field.

Peter Bergmann ([Bergmann (1962)], p. 466).

1.1   The Quantum Gravity Paradox

The problem of quantum gravity is perhaps the most stubbornly persistent problem in modern physics. In a nutshell, the problem involves the incompatibility between our physics of large, massive systems (general relativity) and small, light systems (quantum mechanics). One has the "feeling," as Bergmann puts it, that there shouldn't be a division in nature according to which gravity stands separate (as a fundamentally *classical* interaction) from the rest of the world's phenomena, which are described by quantum theory. It seems intuitively sensible that given the universal nature of gravity (i.e. the fact that any and all sources of mass-energy gravitate), when we have quantized sources of gravity the gravitational field itself must be "caught up" in this quantization like a quantum mechanical infection.[1] To deny this would appear to imply that we inhabit a curious schizophrenic world.[2]

---

[1] This infection process is modelled after the famous Bohr–Rosenfeld analysis of 1933 ("On the Question of the Measurability of the Electromagnetic Field Strengths," published in the *Proceedings of the Danish Academy of Sciences*), where it was shown to be inconsistent to have a classical electromagnetic field coupled to a quantum mechanical object such as a measurement apparatus. The "feeling" that Peter Bergmann refers to in the opening quote is based on broadly analogous reasoning—though there are important *dis-analogies* due to the unavailability of gravitationally neutral charges. It is fair to say that much of the earliest work on quantum gravity was based on the attempt to treat gravity analogously to electromagnetism, and it was supposed by most that gravity would succumb to quantization with just a few modifications to the electromagnetic case.

[2] The earliest history of this issue is covered in Chapter 5.6.

Covered with Deep Mist: The Development of Quantum Gravity (1916–1956). Dean Rickles, Oxford University Press (2020). © Dean Rickles. DOI: 10.1093/oso/9780199602957.001.0001

Fig. 1.1 *Penrose's impossible triangle as a model for the problem of quantum gravity: locally (e.g. focusing on general relativity or quantum mechanics in isolation), we can sense of the picture, but not so globally, in which case the situation becomes paradoxical.*

And yet, sensible though this seems, and despite 100 years of struggle, we still have no consistent theory capable of describing quantum gravity in such a way that a picture of our world emerges—though several proposals are currently viewed as strong contenders. We only appear to be able to make sense of general relativity independently of quantum theory and quantum theory independently of general relativity. If we try to gain a unified picture, things break down, much like Penrose's impossible triangle, (see fig. 1.1). Experimentally the situation has matched this impossible triangle-like configuration: if we try to get objects massive enough to enable measurement of gravitational strengths, then the objects' quantum mechanical aspects effectively disappear thanks to the inevitable environmental decoherence that occurs. But if we try to get objects small enough to retain their quantum coherence, the gravitational aspects are too weak to be measured.[3] This has led some, most notably Roger Penrose (see, e.g., [Penrose (1986)]), to argue that gravity is inextricably bound up with the quantum measurement problem: gravity *causes* the collapse of wavefunctions that become too macroscopic since superpositions of spacetime geometries (that would be associated with the macroscopic superpositions, e.g. of position states) are unstable.

There is an inescapably conceptual aspect to the problem of quantum gravity having to do with the very distinct ways that space and time are treated within the two ingredient theoretical frameworks. This is really at the heart of the paradox. In the case of general relativity the geometrical structure of spacetime (manifesting as a gravitational field) has to be solved for, like any other field. However, in quantum mechanics the geometry

---

[3] See, e.g. [Schmöle (2016)], for a good description of this idea. Though as we see in the final section, this experimental roadblock is finally beginning to look more like a merely practical problem that might yield within a matter of years. There have also been proposals that attempt to draw experimental results applicable to quantum gravity, albeit in a more indirect way (known as "quantum gravity phenomenology": [Amelino-Camelia (2013)]) by using astrophysical and cosmological features (such as patterns in the cosmic microwave background radiation), rather than terrestrial experiments.

is fixed at the outset, and never varied once set. In general relativity, then, space and time are dynamical variables, with the geometrical structure appearing as one of many other degrees of freedom. In quantum mechanics, they are instead parameters against which the evolution of other quantities occurs. One needs the fixed causal structure thus provided in order to preserve causality, unitarity, and other seemingly essential features of quantum theory. Though there is ongoing controversy about how best to define and understand it, this cluster of features is known as background independence and includes the so-called 'problem of time'.[4]

There are a host of related apparent inconsistencies feeding into the quantum gravity problem, as Rovelli and Vidotto colourfully explain:

> A good student following a general relativity class in the morning and a quantum-field-theory class in the afternoon must think her teachers are chumps, or haven't been talking to one another for decades. They teach two totally different worlds. In the morning, spacetime is curved and everything is smooth and deterministic. In the afternoon, the world is formed by discrete quanta jumping over a flat spacetime...that the morning teacher has carefully explained not to be features of our world.
>
> ([Rovelli and Vidotto (2015)], p. 5)

Yet there is only one world, and so solving the problem of quantum gravity then involves devising a framework in which such conflicts are resolved.[5] As we shall see, there are a variety of ways that one might tackle this, treating one or the other of quantum mechanics or general relativity as fundamental, or treating neither as fundamental, or showing how there can be peaceful coexistence despite the apparent conflict.

By way of preparation for a variety of reader backgrounds, in this opening chapter we briefly review, in extremely broad brushstrokes, the problem of quantum gravity and its overarching history in highly schematic form. In particular, we draw attention to two features that render the problem unusually difficult as a scientific problem, beyond the obvious technical hurdles, both pertaining to historical features. We will briefly depart here from our 1956 cutoff point in order to see how the earlier work we then go on to examine fits into the wider picture that exists at present. It is important to note that this chapter couches the problem of quantum gravity in modern day terms, and therefore much of what we have to say here cannot readily be transplanted onto the earlier research without anachronism. The next chapter will then provide a corrective for this, outlining some of the challenges facing historical studies of quantum gravity, as well as some novel prospects for such studies.

---

[4] See [Pooley (2017)] for a thorough recent review of background independence. See [Thébault (2019)] for a recent review of the problem of time.

[5] There are, however, well-understood schemes in which we can treat quantum field theory over curved spacetimes, though these are viewed rather as approximations—we discuss the origins of these approaches in Chapter 4.

1.2   The Dimensions of Quantum Gravity

Research in quantum gravity can very simply be defined as any attempt to construct a theoretical scheme in which ideas from general relativity and quantum theory appear together in some way. For example, Ashtekar and Geroch, in their review of quantum gravity, characterize quantum gravity as "some physical theory which encompasses the principles of both quantum mechanics and general relativity" (Ashtekar and Geroch, 1974, p. 1213). This leaves a fair amount of elbowroom for the form such a theory might take, and is clearly too broad to pin down specific theoretical schemes: we need to be clear on what the essential principles are that ought to be brought together and the exact manner in which they are thus brought together. We can make things a little more specific by noting that a universal property of any such scheme is the existence of units with dimensions of (*L*)ength, (*M*)ass, and (*T*)ime formed from combinations of the characteristic constants of the ingredient theories: Newton's constant *G*, Planck's constant $\hbar$, and the speed of light *c*. We naturally expect a quantum theory of gravity to include all three constants, and the construction of these so-called "Planck units" enables us to see the fundamental limits of the theory, demarcating its domains of validity/applicability just as the constants do separately with respect to the ingredient theories. While the speed of light sets a limit on the speed of information transmission in relativistic contexts, so the Planck units set limits on the smallest possible lengths, areas, volumes, time intervals, and masses, in quantum gravitational contexts. Armed with this knowledge, we are able to say with certainty at what scales quantum gravitational effects would be expected to manifest themselves directly:[6]

$$L_P = \sqrt{\frac{\hbar G_N{}^3}{c}} = 1.616 \times 10^{-35} m \tag{1.1}$$

$$T_P = \sqrt{\frac{\hbar G^5}{c}} = 5.59 \times 10^{-44} \sec \tag{1.2}$$

$$M_P = \sqrt{\frac{\hbar c}{G_N}} = 2.177 \times 10^{-5} g \tag{1.3}$$

At these scales, both quantum mechanics and general relativity are expected to play a non-negligible role, and (if we accept that general relativity is a theory of spacetime geometry) it is this scale that we expect quantum geometry to dominate.[7] It is hoped

---

[6] Though as we see in §1.5, it is a mistake to think that quantum gravity is relevant *only* at these scales, and therefore, given their extreme magnitudes, out of bounds for the foreseeable future—note that the Planck mass, M_P, which might not look so extreme as the Planck time and length, must be localized within a region of space L_P^3, to generate directly observable quantum gravitational effects.

[7] Curiously, these units were discovered by Planck almost three decades before quantum field theory was discovered, and almost two decades before general relativity was completed (and six years before special relativity): [Planck (1899)]. Planck was interested in producing *universal* descriptions of the world, that could

that features of spacetime geometry (and perhaps topology) at these minuscule scales (such as "graininess" of spacetime) will nonetheless develop into features discernible at presently (or soon-to-be) detectable scales (see §1.5).

One of the basic questions to be faced by any proposed solution to the problem of quantum gravity is what role is played by the Planck units. Whatever response one gives to this question, the small scales point to the kinds of issue that quantum gravity proposals might tackle: especially interiors of black holes (black hole singularities), the conditions close to the big bang (the big bang singularity), the high-frequency behaviour of quantum fields (light cone singularities). It is thought that the fundamental limits on length scales imposed in quantum gravity (i.e. the discrete quantum geometry) might serve to "smear out" singularities of each of these kinds—most of the major approaches achieve exactly this.

1.3   The Ways of Quantum Gravity

We can describe the various solutions to the problem in a fairly simple way. Let us return to the simplistic definition of quantum gravity given by Ashtekar and Geroch:

> [QG is] some physical theory which encompasses the principles of both quantum mechanics and general relativity.

We can simply write this in terms of the schema: GR + QM = QG (where QM usually amounts to some quantization method). A common way of reading this schema is in terms of the fundamental constants characterizing the theoretical frameworks of GR and QM, as we intimated above, so that GR + QM = QG can be read as $cG + \hbar = cG\hbar$. That is, one must demonstrate how both Planck's action constant and gravitational fields can both exist in one and the same world. However, there are contexts (such as semiclassical gravity or quantum fields on curved spacetime) in which we have the presence of all three constants that would not *strictly speaking* qualify as quantum gravity proper (i.e. in which gravity itself is quantized)—they still qualify as responses to the basic problem, however. Rather, these approaches consider the effects of strong (classical) gravitational fields on quantum phenomena (such as the behaviour of quantum fields near black holes).[8]

---

even be understood by extraterrestrial civilisations! For this reason he pursued a set of natural scales that would make no reference to such local circumstances as the size of the Earth or aspects of its orbit and rotation. See [Gorelik (1992)] for more on the curious discovery of these units and their subsequent propagation into early quantum gravity research. Note that by the mid-1950s the notion of the Planck length was understood by those working on the so-called canonical approach as a measure of the fluctuations of spatial geometry, leading to topological features as multiple-connectedness known as "spacetime foam." For those working along spacetime covariant approaches, the Planck length marked a natural boundary to the wavelengths of quantum fields and held the promise of taming the ultraviolet divergences that result from considering fields at ever small distances. See §1.3 for more on these two approaches.

[8] In fact, there has been at least some experimental work on the behaviour of quantum systems with respect to classical gravity, e.g. in testing whether quantum particles obey the equivalence principle, as well as tests to determine quantum mechanical phase shifts induced by the Earth's gravitational field (this is the famous

In the following very general partitioning of approaches we split in terms of which of the two ingredient theories (general relativity [GR] and quantum mechanics [QM]) is more fundamental. "Fundamental" here does not necessarily mean ontologically fundamental,[9] but instead just the principles of one are more significant in terms of determining the final features of the approach to quantum gravity that results, so that one *modifies* features of the other for example.

1.3.1 QM as Fundamental

Perhaps the most common approach to resolving the problem of quantum gravity involves treating quantum mechanics as the most basic element of the world (see fig. 1.3.1). According to such approaches, quantum theory *contains* classical general relativity in some sense, or less radically modifies (or "corrects") classical general relativity.

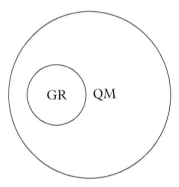

Fig. 1.3.1

This can take a variety of forms, however. Many of the earliest attempts involved the direct quantization of general relativity as a classical field theory, so that the classical theory emerged in the appropriate limit. The main lines of research are of this type, depending on which of the various quantization methods one employs: covariant, canonical, or path-integral. The most common approach along the lines of covariant quantization is based on the idea that gravitational interactions are mediated by a particle,

"COW" experiment, where "COW" stands for "Colella, Overhauser, and Werner"—[Colella, Overhauser, and Werner (1975)]).

[9] Though it sometimes does, e.g. in the case of unified field theories in which classical general relativity is the *sole* source of ontology, with discrete quantum-like particles emerging from it. We might also note that the AdS/CFT duality, linking a string theory on a 5-dimensional anti-de Sitter spacetime with a gauge theory on the 4-dimensional boundary of the spacetime of the string theory, does not fit comfortably within this framework, since *qua* duality neither side is fundamental, and both sides are in fact supposed to be physically identical in some sense. Of course, this duality is still somewhat controversial, given the ill-defined string theoretic side; but even despite this, the duality would seem to fit within the "neither is fundamental" approach below, pointing instead to some deeper underlying theory.

the graviton. To get this picture, however, requires some violence to a certain view of general relativity as a "distinctive" theory in that it is a theory of spacetime as much as a theory of the gravitational interaction. Covariant quantization (i.e. Lorentz covariant) approaches involve a four-dimensional formulation in which the symmetries of flat, Minkowski spacetime are preserved intact, though the metric is compartmentalized into a flat part and a perturbation (which is then quantized). Canonical approaches break covariance, spliting this spacetime into space and time and have to demonstrate preservation of symmetries via more indirect means. The path-integral approach, due to Feynman, involves summing over all four-dimensional metrics, all without splitting the spacetime metric or spacetime, though one must specify initial and final hypersurfaces. There are, of course, interrelations between these formulations, but in the case of general relativity they are far from simple. Also of importance is the fact that the covariant (perturbative) approach takes place using the linear approximation to the full Einstein equations—the "graviton" concept itself only really makes sense in this approximation.

Since the middle of the last century there has been a more or less philosophical clash of ideas with respect to whether the canonical or covariant approaches are the way to go. The canonical approach involves expressing general relativity in Hamiltonian form and then applied the standard prescription for Schrödinger quantization. This involves the singling out of a time (with a slicing into hypersurfaces representing instants), which seems to run counter to the spirit of general relativity.[10] However, in a similar way, the covariant approach, while keeping space and time together as one, splits apart the metric into background terms and perturbations (in at least one approach[11]). This is a different kind of violence to general relativity, attacking its so-called "background independence." However, if background independence is retained, as in the canonical approach, then a serious problem must be faced regarding the fact that since time reparametrization is a gauge transformation in general relativity (being a diffeomorphism), the Hamiltonian vanishes, which in turn implies that the quantum theory is "timeless."

Superstring theory is the most popular contrast to the canonical approach[12] (and still the most widely pursued approach to the problem of quantum gravity). Gravitation (and the results of general relativity) arise in string theory as a result of the presence of a massless spin-2 particle (the graviton) in the spectrum of a closed string (which all string theories contain). String theory itself is a sort of quantum field theory of strings, with no gravitation at a fundamental level, emerging instead, as with the other forces, from the

---

[10] Or rather a *class* of times is singled out, since any such foliation is not unique in the context of general relativity.

[11] The path-integral approach does not face these issues, and works by computing the amplitude to go from one 3-geometry to another 3-geometry by summing over all possible interpolating 4-geometries (the "histories," each weighted by a complex number amplitude). However, it faces its own issue in the form of a "measure problem" in which it isn't clear how to assign probabilities to outcomes in the path integral—in a nutshell, since the domain space is a space of all 4-manifolds (and, e.g., state which manifolds are equivalent and inequivalent) one needs to solve one of the most pressing problems in topology to resolve this measure problem.

[12] Of course, string theory can be given a canonical formulation, but it is not the spatial metric of general relativity that is quantized but the string degrees of freedom.

string dynamics. It is for this reason that some advocates of string theory often refer to string theory "predicting the existence of gravity." In a certain sense this is perfectly true: it is a logical consequence of the theory. But, of course, in another sense it seems somewhat contrived since its presence constituted part of the reason for pursuing the theory in its current form after it had originally been employed as a model for the strong interaction.[13]

1.3.2   GR as Fundamental

Though he initially believed that the quantum theory must modify general relativity, Einstein's later work on his "unified field theory" adopted the view that general relativity was the fundamental theory from which quantum phenomena (such as discreteness and the existence of particles) were to be derived (fig. 1.3.2). Here we find the converse of the problem above: getting a discrete theory/ontology from continuum theory/ontology.

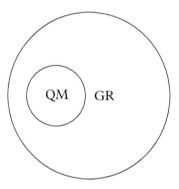

Fig. 1.3.2

Such an apparently paradoxical problem was resolved to some extent by John Wheeler, through his enigmatic "charge without charge" ideas (in which the flow of charge in and out of a wormhole mimics the positive and negative poles of a discrete elementary charge).[14] Together with his students he got remarkably far with getting the structure of particles from gravity treated via the path-integral approach (and classical gravity), including the existence of spin—we consider Wheeler's work in Chapter 8.[15]

More recent proposals along lines that prioritize general relativity come from the notion that gravitation might induce modifications of more basic aspects of quantum

---

[13] For a philosophically inclined history of string theory, see [Rickles (2014)].

[14] Wheeler later dropped this interpretation of wormholes as elementary particles and viewed them instead as a consequence of a path-integral formulation of general relativity in which the fluctuations of the gravitational field induce a topologically multiply connected manifold. An earlier version of pulling quantum-like effects from topological trickery was Oskar Klein's five-dimensional theory of relativity, in which the fifth dimension was compactified down to a small circle, giving charge quantization through the circle's closure and periodicity [Klein (1927)].

[15] Spin was long believed to be the sticking point for such unified field theory approaches since it seems to have no classical counterpart (the gravitational field appears to be purely bosonic: integer spin).

mechanics. For example, Roger Penrose [Penrose (2014)] speaks of "gravitizing" quantum mechanics so that the usual quantum mechanical principles (such as superposition) have exceptions in cases where significant masses in such superpositions are concerned (such superpositions have a "lifetime" due to instabilities in the spacetime geometry associated with superpositions of locations of masses). As we see in §1.5, there are recent almost-performable (conceptual) experiments that seek to examine such conditions.

1.3.3 Both as Fundamental

If both (classical) general relativity and quantum theory are equal partners in the world (fig. 1.3.3), then we must live in a hybrid world. In this case there is no such thing as quantum gravity: it is not the kind of thing that needs to be quantized. Yet we still face the same puzzle (Bergmann's "feeling") that we started with: how can we have a world in which quantum mechanical systems interact with classical gravitational systems without the quantum mechanical properties seeping into the classical gravitational field? A world dappled in this way seems to be problematic.

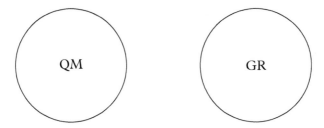

Fig. 1.3.3

The problem also has a mathematical aspect too since, if matter is treated quantum mechanically while gravity is treated classically, the two sides of the Einstein field equation appear to involve different structures: a c-number term and a q-number term. One of the earliest, fully worked out suggestions of this type, that offered a solution of the mathematical problem, was that of Christian Møller [Møller (1962)], who argued that one could convert the right-hand (matter) side to a c-number by using the energy-momentum's expectation value instead.[16] The problem with this is that it faces the measurement problem with renewed force: either there is a discontinuity in the gravitational field on measurement, or the spacetime geometry is put into a superposition of geometries in the style of the many-worlds interpretation. There are also arguments that suggest a potential violation of the uncertainty relations in such a world (see [Mattingly (2009)] for a review).

---

[16] In fact, the first expression of this idea (replacing the energy-momentum tensor by its expectation value) was in 1948 in Iwao Sato's "An Attempt to Unite the Quantum Theory of Wave Field with the Theory of General Relativity" (*Science Reports of the Tohoku University, Series 1* 33: 30–37)—see §5.6 below.

1.3.4   Neither as Fundamental

A new class of approaches having their roots in "effective field theory" views both general relativity (and so spacetime) and quantum field theory as valid within a certain range of (low) energy values, below the Planck energy (or, equivalently, for long wavelengths). These are known collectively as "emergent gravity" approaches (fig. 1.3.4). One way of conceptualizing such approaches is in terms of the search for the microscopic structure of spacetime. The idea is that the metric is a collective variable of some microscopic theory that does not possess the notion of a metric in its own fundamental ontology. This simply means that as one probes higher energies the metric does not appear, so that rather than viewing classical general relativity as simply one of the classical limits of a quantum theory of gravity (e.g. understood via quantization), the theory (and its features) are instead emergent much as hydrodynamics (a continuum theory) is emergent from a particle theory.

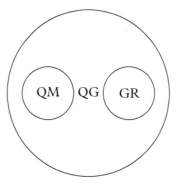

Fig. 1.3.4

Although the fundamentality of general relativity is not asserted in this effective field theory approach, the low energy part works perfectly well as a quantum field theory. What this shows us, rather than being a solution to the problem of quantum gravity (in the sense of an account of what happens when both gravitational and quantum effects are strong) is that at everyday distances there is no real conflict between quantum field theory and general relativity. This is important since it at least reveals that there is no problem with the ingredient theories at currently observable energies. However, it leaves us none the wiser about what happens at the more extreme energies that are expected to involve new physics with conceptual revisions.

1.4   Why is the Problem of Quantum Gravity so Hard?

> You do not win battles by debating exactly what is meant by the world "battle". You need to have good troops, good weapons, a good strategy, and then hit the enemy hard. The same applies to solving a difficult scientific problem.
>
> Francis Crick ([Crick (1995)], p. XI)

Wolfgang Pauli once remarked to Bryce DeWitt, on hearing that the latter was working on the problem of quantum gravity: "That is a very important problem—but it will take someone really smart!" [DeWitt (2009), p. 414]. The efforts of very many smart people have not yet resulted in a resolution. Crick's notion of solving a difficult scientific problem as a battle might seem apt then: as scientific problems go, the problem of quantum gravity has to be one of the most difficult in the entire history of human thought. But *why* is it so hard? Part of the issue is that the shape of the problem ("the enemy") changes depending on time and context. We are, in fact, not even fighting the same battle that we were once fighting in the earliest days of quantum gravity: developments in the ingredient theories and background theories have revised the problem again and again. If the enemy keeps shifting, then it doesn't matter how good our troops, weapons, and strategies are: they simply might not be well-matched anymore. This isn't just an evolutionary arms race: the battlefield itself changes, as well as the enemy:[17]

> Today's theoretical physics is largely built on two giant conceptual structures: quantum theory and general relativity. As the former governs primarily the atomic and subatomic worlds, whereas the latter's principal applications so far have been in astronomy and cosmology, *our failure to harmonize quanta and gravitation has not yet stifled progress in either front.* Nevertheless, the possibility that there might be some deep dissonance has caused physicists an esthetic unease, and it has caused a number of people to explore avenues that might lead to a quantum theory of gravitation, no matter how many decades away the observations.
>
> [Bergmann (1992), p. 24, my emphasis]

That progress has not been stifled in either of the ingredient theories (GR and QM) is precisely a key part of the curious nature of the problem of quantum gravity—there has often been *too much* progress which has the effect of modifying what is required in a solution. I expect that one can find similar instances in the history of physics, in which a unification is required between two theories while changes are going on in either or both,[18] but the timescales involved in QG (a century of struggle) and the pace of progress might make it qualitatively different in this sense.

So there is something peculiar about the problem, I think, making an already hard problem, a *really* hard one. Referring to the Pauli quote: smartness might not be enough, if we are faced with a hydra that sprouts more heads every time one is lopped off! Precisely because we do not have this direct experimental grasp, or specific phenomena to match up with, the problem itself is subject to fluctuations with respect to the ingredients of the problem: QM and GR (and background knowledge relevant to our understandings of these). This is precisely why we see analogies playing such a rich role in the history

---

[17] This shifty nature is especially evident in the very earliest work that we cover in Chapters 3 and 4.

[18] Certainly, the business of a scientific problem being buffeted around by external influences can be found in other areas (the early days of cosmology springs to mind, in which we have an initial phase of untestability)—it is part and parcel of the growth of scientific knowledge to update as new information comes in. However, observations entered early on in cosmology, whereas quantum gravity has never settled enough to have a life of its own.

of quantum gravity: because we need to look for something concrete to latch onto. This much captures the non-stationarity of the problem, and it highlights the extent to which the problem is an *historical* one.

There is a further somewhat related problem, which I shall call the problem of *polysemicity*, whereby the ingredient theories each allow for different formulations and interpretations which impacts on the possible solutions to the problem of quantum gravity. For example, if we agree with Steven Weinberg that the geometrical (gravity as genuine spacetime curvature) approach drove a "wedge" through physics, separating gravitation (general relativity) and elementary particle physics, then we face two very distinct paths: either treat gravity like a standard field theory, or treat gravity as something special. We face the same issue with respect to how we view quantum theory. As we have seen, the various quantization methods suggest very different solutions to the problem of quantum gravity. Other developments such as the discovery of spin, or the discovery of Yang–Mills theory, radically altered the terms of the problem, offering up new (and very difficult) mathematical structures that now had to be incorporated in the theory, or offering up a host of new analogies, respectively. We can further envisage that different interpretations (e.g. Bohmian versus Copenhagen) would suggest very different solutions and even outlaw certain approaches.[19]

1.5   To Phenomena and Experiment

Léon Rosenfeld was the first to attempt a direct quantization of the gravitational field (using both then-available methods: covariant and canonical) in 1930. In his later life he was adamant that it was not proven that quantum gravity was even necessary since there were no "empirical clues" available (see [Rosenfeld (1966)]). All that existed were some (difficult) theoretical problems—such as the final stages of gravitational collapse and the physics near the big bang—and the feeling of unease described by Bergmann. Data has remained a problem throughout quantum gravity's life, and this has led to a proliferation of approaches, exacerbated by the shifts in the ingredient theories that we saw in the last section. Thus, a kind of (quite understandable) pragmatism guided the early neglect of quantum gravity research; the scales at which phenomena would be apparent were known then to be well out of reach of direct tests.[20] Dirac expressed this standpoint particularly clearly:

---

[19] An additional, distinct aspect to the polysemicity problem concerns the *aims* of a quantum theory of gravity. The most obvious example which I won't discuss directly is the differing points of view of string theory and others on this point: does the QG problem involve unifying gravity with the other interactions or not? In this sense, string theory and, e.g., loop quantum gravity, though both claiming to offer *solutions* to the problem of quantum gravity, are really tackling different problems: they are fighting different battles.

[20] The characteristic "Planck length" is computed by dimensional analysis by combining the constants that would control the theory of quantum gravity into a unique length. As shown above, this is $l_p = \sqrt{\hbar G/c^3} = 1.6 \times 10^{-33}$ cm: a minuscule value, making gravity (effectively) a "collective phenomenon" requiring lots of interacting masses. That quantum gravitational effects will not be measurable on individual elementary particles is, therefore, quite clear. Bryce DeWitt devised rigorous arguments to show this to be the case: the gravitational field itself does not make sense at such scales. He showed that the static field from such a particle (with a mass of

There is no need to make the theory conform to general relativity, since general relativity is required only when one is dealing with gravitation, and gravitational forces are quite unimportant in atomic phenomena.

[Dirac (1966), p. 66]

There is a very simple way of exposing the problems that direct measurement of quantum gravitational effects might face. It is a basic fact that for any distance we wish to measure, the measurement probe must be at least as small as that distance. In the case of quantum gravitational measurements, the distances are absolutely tiny (and the energies huge), as we have seen. As we probe smaller distances, the probe's particles must be of ever higher momentum values (thanks to the uncertainty relations). But this higher momentum is a source of stress-energy and so is responsible for generating large spacetime curvature, so much so that the device will generate a black hole and be rendered unobservable—or, at the very least, will distort spacetime to such a degree that measurement becomes a practical impossibility.[21]

However, a common confusion still persists which is no doubt a hangover from this early pragmatic stance and such arguments as these. This is the idea that in order to generate quantum gravity phenomenology (e.g. to detect the quantum properties of the gravitational field) one must be able to detect *individual* gravitons, and this therefore would require probing the Planck scale directly. For example, Freeman Dyson recently wrote:

This talk is concerned with a different question, whether it is in principle possible to detect individual gravitons, or in other words, whether it is possible to detect the quantization of the gravitational field.

[Dyson (2012), p. 1]

But detecting individual gravitons and detecting the quantization of the gravitational field are two quite different things.[22] An analogy would be the conflation of the direct and indirect detection methods of gravitational waves: the decay rates of pulsars offer a means of detecting gravitational radiation beyond the kinds of direct detections one finds in the interferometer observations at LIGO and elsewhere. The pulsar data also counts as evidence of gravitational radiation despite the indirectness. There are two recent avenues that rectify the parallel quantum gravity situation, by "going large" (using the heavens as an observatory) or by "going small".

---

the order 10^{-20} in dimensionless units) would not exceed the quantum fluctuations. The static field dominates for systems with masses greater than 3.07×10^{-6}. The gravitational field is from this viewpoint an "emergent" "statistical phenomenon of bulk matter" ([DeWitt (1962)] p. 372).

[21] Note that this same reasoning is often used to show that the notion of a spacetime point will no longer make physical sense in quantum gravity, since no physical significance can be attached to such a concept.

[22] Cf. Sabine Hossenfelder's blog post "Quantum gravity phenomenology $\neq$ detecting gravitons" for a useful discussion on this point: http://backreaction.blogspot.com.au/2013/06/quantum-gravity-phenomenology-neq.html.

Obviously the major problem with the latter approach is precisely the seemingly killer combination Dyson is referring to: the extreme tinyness of the Planck scale, the weakness of gravity (certainly well known from the earliest days, as the next chapter shows), and the presence of decoherence, making it very hard to generate quantum effects on objects that are nonetheless big enough to measure gravitational effects. However, technological rather than theoretical/conceptual advances look set to dislodge the century-old embargo on quantum gravitational data. What is required is simply enough control of quantum coherence (for masses in location superpositions) combined with enough sensitivity to measure the gravitational field for such (still very small) masses. Then it suffices to check whether the gravitational field is itself in a quantum superposition. Almost performable experiments have now been planned that make use of massive quantum oscillators capable of generating such location superpositions for objects as weighty as one nanogram, together with micromechanical machines that can radically reduce the size of experimental equipment thus improving sensitivity to the required degree for gravitational field measurements at one nanogram (see, e.g., [Schmöle et al. (2016)]).

The alternative ("going large") approach attempts to access the large energies by utilizing the large scale of the universe: that is, one can probe the minute Planck length by accessing the vastness of the universe. Such observations would be more along the lines of the measurements of binary pulsar decay. An example is the potential for Lorentz invariance violation as a result of photons travelling over "grainy" spacetime (which is treated as a lattice mathematically, and therefore possesses different symmetries to a continuum spacetime). Observationally this would be measured as an energy–dependence on the time of arrival of the photons from very distant sources. Since some approaches to quantum gravity *predicted* such features, such measurements have the power to confirm or rule out such approaches, thus bringing quantum gravity research in line with more standard scientific methodologies (featuring selection and rejection of theories on the basis of empirical evidence). In fact, observations *were* made with no such effects discernible (a null result), thus reducing the space of possible theories (see [Albert et al. (2008)]: the MAGIC telescope collaboration). Though such data do not allow for a direct testing of specific approaches to quantum gravity (such as string theory), they do constrain along the lines of perfectly standard empirical results. They already demonstrate that quantum gravity research has progressed from Bergmann's vague feelings about how the world ought to be.

1.6   Conclusion

The problem of quantum gravity spent much of its earliest history at the mercy of wider changes with respect to the ingredient theories, general relativity and quantum theory. Even once those theories settled down, quantum gravity remained firmly detached from experiments. This situation has only recently changed and promises to offer new phenomena to test proposed solutions to the problem which will enable us to make firmer statements about the more physical implications of these proposed solutions. However, we may still face the problem of polysemicity stemming from the very differing interpretations and formulations that the ingredient theories allow, as well as differing motivations for pursuing quantum gravity. We will see this in action in Chapter 3, dealing with the earliest attempts to unite gravitational and quantum phenomena in which the instability of the ingredient theories (more so in the case of quantum theory) is at its most extreme. Subsequent chapters reveal how the problem of quantum gravity shifts its scope and definition depending on external developments (such as the discovery of spin and the Dirac equation; cosmology; new methods of quantization; new understandings of gauge symmetry; new problems in physics, such as the resolution of singularities, and so on). However, we first look in more detail at the specialness of quantum gravity from the point of view of historiography and history of physics.

....

REFERENCES

Albert, J. et al. (2008) Probing Quantum Gravity using Photons from a Flare of the Active Galactic Nucleus Markarian 501 Observed by the MAGIC Telescope. *Physics Letters B* **668**(4): 253–7.

Amelino-Camelia, G. (2013) Quantum-Spacetime Phenomenology. *Living Reviews in Relativity* **16**(5).

Ashtekar, A. and R. Geroch (1974) Quantum Theory of Gravitation. *Reports on Progress in Physics* **37**: 1211–56.

Bergmann, P. G. (1992) Quantization of the Gravitational Field, 1930–1988. In J. Eisenstaedt and A.J. Kox (eds.), *Studies in the History of General Relativity* (Einstein Studies, Vol. 3, pp. 364–6). Boston: Birkhäuser.

Bergmann, P. G. (1962) Summary of the Colloque Internationale de Royaumont. In A. Lichnerowicz and M.-A. Tonnelat (eds.), *Les Théories Relativistes de la Gravitation* (pp. 464–72). Centre national de la recherche scientifique.

Colella, R., A.W. Overhauser and S.A. Werner (1975) Observation of Gravitationally Induced Quantum Interference. *Physical Review Letters* **34**: 1472–4.

Crick, F. (1995) *The Astonishing Hypothesis: The Scientific Search for the Soul*. New York: Scribner.

DeWitt, B. S. (2009) Quantum Gravity: Yesterday and Today. *General Relativity and Gravitation* **41**: 413–19.

DeWitt, B. (1962) Quantization of Geometry. In L. Witten (ed.), *Gravitation: An Introduction to Current Research* (pp. 266–381). New York: John Wiley and Sons.

Dirac, P. (1966) *Principles of Quantum Mechanics*. Oxford: Clarendon Press.

Dyson, F. (2012) Is a Graviton Detectable? Poincare Prize Lecture. URL: https://publications.ias.edu/sites/default/files/poincare2012.pdf.

Gorelik, G. (1992) First Steps of Quantum Gravity and the Planck Values. In J. Eisenstaedt and A. J. Kox. (eds.), *Studies in the History of General Relativity* (Einstein Studies, Vol. 3, pp. 364–79). Boston: Birkhäuser.

Klein, O. (1927) Zur Fünfdimensionalen Darstellung der Relativitätstheorie. *Zeitschrift für Physik* **46**: 188–208.

Mattingly, J. (2009) Mongrel Gravity. *Erkenntnis* **70**(3): 379–95.

Møller, C. (1962) The Energy-Momentum Complex in General Relativity and Related Problems. In A. Lichnerowicz and M.-A. Tonnelat (eds.), *Les theories relativistes de la gravitation* (pp. 15–29). Du Centre National de la Recherche Scientifique, Paris.

Penrose, R. (2014) On the Gravitization of Quantum Mechanics 1: Quantum State Reduction. *Foundations of Physics* **44**(5): 557–75.

Penrose, R. (1986) Gravity and State Vector Reduction. In R. Penrose and C. J. Isham (eds.), *Quantum Concepts in Space and Time* (pp. 129–46). Oxford: Clarendon Press.

Pooley, O. (2017) Background Independence, Diffeomorphism Invariance and the Meaning of Coordinates. In D. Lehmkuhl, G. Schiemann, E. Scholz (eds.), *Towards a Theory of Spacetime Theories* (pp. 105–43). Basel: Birkhäuser.

Planck, M. (1899) Über Irreversible Strahlungsvorgänge. *Sitzungsberichte der Königlich Preußischen Akademie der Wissenschaften zu Berlin.* **5**: 440–80.

Rickles, D. (2014) *A Brief History of String Theory: From Dual Models to M-Theory.* Berlin–Heidelberg: Springer-Verlag.

Rosenfeld, L. (1966) Quantum Theory and Gravitation. In R. S. Cohen and J. Stachel (eds.), *Selected Papers of Léon Rosenfeld* (pp. 599–608). Dordrecht: D. Reidel Publishing Company.

Rovelli, C. and F. Vidotto (2015) *Covariant Loop Quantum Gravity.* Cambridge: Cambridge University Press.

Schmöle, J., M. Dragosits, H. Hepach, and M. Aspelmeyer. (2016) A Micromechanical Proof-of-Principle Experiment for Measuring the Gravitational Force of Milligram Masses. *Classical and Quantum Gravity* **33**: 1–19.

Thébault, K. (2019) The Problem of Time. Forthcoming in E. Knox and A. Wilson (eds.) *Routledge Companion to the Philosophy of Physics.* London: Routledge.

Wheeler, J. (1984) Quantum Gravity: The Question of Measurement. In S. M. Christensen (ed.), *Quantum Theory of Gravity* (pp. 224–33). Bristol: Hilger.

2

On Writing a History of Quantum Gravity

> *To understand a science it is necessary to know its history.*
>
> Auguste Comte

The history of science is not a straightforward matter. It is certainly not as straightforward as most popular accounts portray. According to Dennis F. Shaw, *quondam* Keeper of Scientific Books at Oxford University, Niels Bohr once claimed that "[t]he history of physics is not a different subject from physics" for "it is in reading about physics that one learns the history of the subject" ([Shaw (1990)], p. 3). This is almost entirely wrong.[1] In reading physics one is inevitably reading it through the lens of the victors in a battle that often included multiple worthy opponents. One does not, in general, learn anything about the theories that fell by the wayside, for whatever reason. Thomas Kuhn [Kuhn (1977)] quite rightly argued that this neglect can perform a positive pedagogical function (if not a creative one): students learning from the books don't get bogged down with details that are largely irrelevant when it comes to mastering and *using* the

---

[1] Though as David Kaiser [Kaiser (2013)] has emphasized, in working through a scientific subject's *textbooks* over time, one can gain a feel for the historical development of that subject—though this is, in itself, an historical exercise (see, e.g., [Bensaude-Vincent (2006)] and [Badino and Navarro (2013)]). Problematically, however, in the case of quantum gravity, textbooks came very late to the party, and much of this has to do precisely with the "specialness" of quantum gravity from an historical point of view: it is an old yet still unresolved problem. As a matter of fact, it does seem that the arrival of textbooks (especially in string theory, for example; but also in loop quantum gravity) plays a not inconsiderable role in generating a community, via the provision of a shared set of concepts, tools, decisions as to what are the important problems, and so on. Hence, the *absence* of textbooks on quantum gravity is intimately entwined with an absence of strong *communities* devoted to some specific research project. Inasmuch as communities existed (and they did, as we will see—albeit *micro*-communities), they were thoroughly local and depended on some leading figure with the community most often formed from their students (e.g. a Peter Bergmann, who had at least his textbook on general relativity that included remarks about quantum gravity, or a John Wheeler, who came to the game late, but quickly made a large impression). Textbooks, on the other hand, allow a community to be more non-locally dispersed yet retaining some unity. Note that while edited collections devoted to quantum gravity have been around since the 1970s, the first textbook on quantum gravity (on string theory, to be exact) did not appear until 1988—this was Green, Schwarz, and Witten's *Superstring Theory* (Cambridge University Press, 1988). It is not out of the question to think that the existence of such a textbook (two very systematic and attractive volumes) played a role in string theory's early dominion over matters quantum gravitational—see [Rickles (2014), p. 172] for more on this point.

Covered with Deep Mist: The Development of Quantum Gravity (1916–1956). Dean Rickles, Oxford University Press (2020). © Dean Rickles.
DOI: 10.1093/oso/9780199602957.001.0001

theory. Yet those failed theories, and the reasons behind their failure, often contain vital clues about the structure of the physical theories that would ultimately prove victorious. Knowing something about them has the potential to provide much greater insight into the "winning theory." Moreover, it simply gives a more accurate representation of the process of science.

The historical image that one gets from reading about physics, at least physics textbooks, is of a pristine, perfectly rational progression of theories that were victorious *because they were right*. The errors that led up to the theory (if in fact they were errors), and the factors that led to the selection of the consensus theory, evaporate. If there is some token historical background chapter, it will invariably involve a perfunctory, stock description of the development of the subject that shows that the approach discussed in the book succeeded simply because it was true, while the others were wrong-headed. This so-called Whiggish view (involving the writing of history from the vantage point of the *present*[2]) isn't quite right, and indeed one of the things this book will show is that there is a great deal amount of contingency in the development of physics.[3]

Given the enormous complications of writing history of any kind—let alone of a theory that doesn't yet exist—before turning to the history itself, we first cover some of these basic issues, and will explain and justify the methods and concepts used in writing this work—including the selection procedure governing what falls under the "Quantum Gravity" label (itself a highly non-trivial historiographical decision). I also discuss some of the *unique* historical challenges (and opportunities) thrown up by quantum gravity in view of its still incomplete status (so that no "victor" yet exists that might enable one to construct narratives that point towards it as if by necessity): here, the rug is seemingly pulled out from under the Whiggish historian's feet. Quantum gravity remains a story of parallel, sometimes intersecting paths, with no *terminus ad quem* to glorify, à la Butterfield's villain, in sight.

---

[2] In the classic text on the subject of the Whig interpretation of history, Herbert Butterfield defines the approach as the production of a "story which is the ratification if not the glorification of the present" ([Butterfield (1965)], p. v).

[3] In this I am very much indebted to the late James Cushing, especially through his book *Theory Construction and Selection in Modern Physics*. In one sense, the extent to which contingency is in operation is considerably less than Cushing has argued, perhaps, due partially to the fact that so much hinges on mathematical consistency in the context of quantum gravity, so the elbow room for contingencies to emerge is radically narrowed. In another sense, however, there is even more scope for contingency since, in quantum gravity, there are no experimental guideposts constraining approaches. A different set of epistemological strategies must therefore be defined in order to provide some kinds of constraints to assess what makes a good resolution of the basic problem (we return to these aspects in §2.2). In the earliest phases, in addition to consistency already mentioned, there are other constraints provided by known low energy results or "principles" (such as general covariance) in the ingredient theories, which provide some semblance of *reliability* through a kind of "theoretical calibration." However, it must be emphasized that there was little sense of *rivalry* amongst competing approaches in the earliest research; this came much later once the approaches were aligned with particular worldviews. New constraints came still later in the form of quasi-experimental checks based on expected predictions of the semiclassical theory (e.g. in the form of Hawking radiation and the numerical value for the Bekenstein–Hawking entropy), though there is much controversy over whether these ought to be used as guideposts in this sense (see, e.g., [Curiel (2001)] for an argument against, and [Unruh (2014)] for an argument for treating results in black hole thermodynamics as empirical checks; [Wüthrich (2019)] argues that the entire framework of black hole thermodynamics should, in terms of its evidential role, be taken *cum grano salis*).

2.1   Carving out a History

The idea of a *definitive* history is nonsensical. There are many histories to be told from many perspectives, any number of such can be equally authoritative. Even the idea that there is one set of *events* underlying all of these many histories itself presupposes some prior individuation of historical facts. This does not imply woolly-headed relativism nor social constructivism: there isn't unlimited freedom in the perspective one can choose. The world kicks back if the perspective doesn't fit. This book is such a perspective, filtered through a largely "conceptual" lens. I am interested in how quantum gravity and its guiding concepts developed from the very earliest work that warrants, however loosely, the "quantum gravity" title. Yet, there is a very great deal of social history also covered here. Given the period of history covered (spanning world wars and cold wars), the intrusion of the wider social and cultural climate can scarcely be avoided, nor can socially and economically influenced curiosities in the changing fortunes of quantum gravity research be ignored.[4] Hence, while being undoubtedly a theorist's dream, the history of quantum gravity is also bound up with wider and more mundane historical events, such as the emergence of new sources of private philanthropy, which we shall briefly touch upon.

I aim, then, to go beyond a merely "technical" (internal) history of ideas, and try to shed some light on the contextual factors that were in place when key ideas were being developed. But I do not make the general claim that these contextual factors played a *causal* role in the ideas (though in some cases we can in fact see evidence of ideas being thus directly determined). Again: this is *not* a work in the "social constructivist" mould. I simply wish to show the surprising richness of the history of quantum gravity, even in a period in which many believe there was no such thing as quantum gravity research. Quantum gravity's history has, thus far, been amputated from the histories of (classical) general relativity and quantum field theory (the two ingredients of quantum gravity).[5] This book also aims to show that quantum gravity is, however, an integral part of the history of both ingredient theories and should no longer be neglected in such accounts; to ignore its role renders them radically incomplete.

The course of quantum gravity research has been carved out already in several ways, including in terms of the now orthodox "triplet," based on the specific quantization

---

[4] Many of quantum gravity's key players (e.g. Bryce DeWitt, Richard Feynman, and John Wheeler), for example, were heavily involved in war work, and there are certainly areas in which the influence works in both directions: from physics to war work and vice versa—[Galison (2012)] provides an account of the impact of the Los Alamos/atomic bomb work (and other more practical influences) on Wheeler's special "machine-like" conception of mathematics; a similar study, involving Feynman's work (with its distaste for mathematical rigor and penchant for visual methods) is [Galison (1993)] (cf. also [Kaiser (2005)], pp. 35–6).

[5] We can speculate as to the reasons for this neglect. Perhaps there is a prevailing "hands off until complete" attitude amongst historians, waiting for the dust to settle in order that a coherent narrative might emerge? Certainly, the ship is not yet in the bottle, but all the more reason to probe a theory under construction, especially one whose underlying problem has so long a life. Or perhaps it might be supposed that quantum gravity's history is not interesting enough, since it does not occupy a central role in the development of recent physics, unlike relativity and quantum theory themselves. This is a mistake: there are very many points of overlap with more standard sectors of physics, but even irrespective of this, the history is rich and rewarding.

method employed: (Canonical, Covariant, Path-Integral[6]). Carlo Rovelli [Rovelli (2002)] employs just such a carving in his historical reconstruction. In what he labels "the prehistory" period (1920–1957: largely corresponding to the entire temporal segment tackled in this book), we find the following diagram (fig. 2.1):

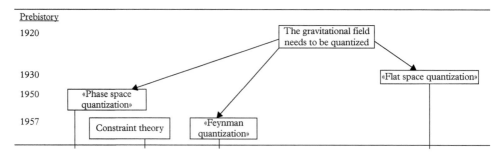

Fig. 2.1 *Carlo Rovelli's reconstruction of the quantum gravity research landscape prior to 1957 (source: [Rovelli (2002)], p. 746).*

This can certainly be useful in terms of seeing the present-day lay of the land, but it is a little more difficult to see things as so clear cut in the early research: to project this newer partitioning of approaches onto the older work is rather anachronistic—not least since in the very earliest period, these methods (and the notion of "field quantization") simply did not yet exist; nor could it really be said that it was established that the gravitational field *needs* to be quantized in a phase of development as early as the 1920s.[7] More so than quantization considerations, it was the *harmonization* of the curved world of general relativity and the discrete, probabilistic (and eventually spin-laden) world of quantum theory that was at stake. This was often conducted without any consideration of quantum properties of such curved spaces, as for example in the study of the wave equation and the Dirac equation in Riemann spaces, from the mid- to late 1920s—later morphing into the study of quantum field theory on curved spacetime resulting in the first "test" of quantum gravity in the form of Hawking radiation and the Bekenstein–Hawking entropy for black holes. Once specific field quantization methods did appear, the approaches

---

[6] This latter class of approaches is also sometimes labelled "sum-over-histories" or "Feynman quantization." The first explicit version of this tripartite carving appears to be in Charles Misner's paper (essentially just his PhD thesis, written under Wheeler's supervision at Princeton: see Chapter 8 for a discussion of Wheeler's work, including Misner's work as a doctoral student) on the Feynman quantization of gravity [Misner (1957)], published exactly at the conclusion of this book's historical period, in 1957—Peter Bergmann did mention the Feynman/Schwinger approach and the linearized spin-2 theory *very* briefly as alternatives to his preferred canonical approach in his review of progress in the quantization of the gravitational field from the Einstein Jubilee conference in Berne, in 1955: [Bergmann (1956)]. (Though I won't pursue or defend the claim here, it seems that the emergence of review papers, summarizing the state of the art in a systematic fashion, has the effect of entrenching and polarizing positions in quantum gravity and seems to be another strong way of building divisions in the possible approaches and those pursuing them—a good recent study of the review paper genre is [Lalli (2014)].)

[7] Also absent is an indication of proposals, starting with Einstein's, involving the idea that general relativity might in fact have a role to play in quantum theory.

adopted were highly overlapping, resulting a tangled web, often on account of the fact that researchers were not so strongly bound to some specific approach, instead exploring multiple avenues, but also because approaches were simply more multifarious at that time. Moreover, the specific difficulties facing each approach, and an understanding of their various conceptual implications (e.g., concerning the kind of picture of spacetime one is required to uphold, given some quantization method), took considerable time to emerge—again, largely because the ingredient theories were in such a state of uncertainty and flux. However, this discrepancy in itself reveals another way in which multiple historical accounts are possible: such divisions of approaches suggest underlying guiding research agendas offering different lenses through which to view the development of the subject.

John Stachel ([Stachel (1995)], p. 318), for example, has suggested a simpler split in terms of "conservative" and "radical," which he claims characterizes research since the mid-1930s.[8] Conservative approaches are simply understood to be those relying on well-established quantization methods, and bring with them the mindset there is no "special" problem of quantum gravity (i.e. that a technical assault is all that is required). By contrast, the radicals feel that general relativity has certain features (pertaining to space, time, and symmetry) that simply render these conservative approaches inapplicable, so that deeper issues (and most likely significant changes to our worldview) need to be considered.[9] Though not so fine-grained, this still serves as a fairly reasonable partition of approaches of the present day, and we shall certainly find elements of this Stachelian division; but as we see, there is a definite consensual trend towards the view that general relativity is really not like any other field. This manifests itself in the dawning recognition that the electrodynamical analogy, which was often called upon, breaks down in several distinct ways.[10]

---

[8] Asher Peres too suggested a very simple binary split: "those which are based on canonical quantization, and those which are not" ([Peres (1968)], p. 1335). Of course, the *possibilities* for partitioning are themselves time-variant. In the early period we consider, the existence of discrete approaches split into canonical and otherwise does not quite make sense: the theoretical frameworks simply had not yet been established and "covariant" *principles* often infiltrated what would become the canonical approach. Even once they had been initially formulated, it is not straightforward to make such simple splits.

[9] This "conservative–radical" distinction maps quite closely onto both Chris Isham's "conservative versus iconoclastic" split [Isham (1991)] (where "conservative" means that nothing special is expected to happen at the Planck scale and "iconoclastic" means that fundamental revisions are required) and also onto Lee Smolin's "craftspeople" versus "seers" division [Smolin (2007)] (in which the former are akin to Kuhn's normal science folk, probing existing concepts and applying known techniques, and the latter to revolutionaries, questioning those same concepts and techniques). Note that Stachel ([Stachel (1998)], p. 532) later added a third grouping involving the claim that general relativity is a purely macroscopic theory that applies to bulk masses only (and so would not involve quantum field theoretic techniques), in which case "quantum gravity," strictly speaking, is a non-referring term. Of course, this has *radical* implications since one must then consider from what the classical spacetime of general relativity *emerges*.

[10] The analogy makes good intuitive sense, since both gravitation and electromagnetism are long-range interactions, and there is a striking similarity between Coulomb's and Newton's laws. Its breakdown was connected to external developments in areas such as cosmology in which high energy sources were discovered rendering the neglect of the non-linearities (involving self-interactions not present in electromagnetism with its Abelian group structure) of Einstein's equations (and the focus on *weak field* approximations) less palatable. Of course, it is often the case that some new development in physics (or indeed any science) infuses the rest

Also related to the electrodynamical analogy, is a distinction (found in DeWitt [DeWitt (1962)] for example) between "flat spacetime" and "geometrical" approaches—these can also be parsed in terms of the method of incorporating non-linear features into the theory. However, the idea that conservative approaches apply to "well-established" quantization methods breaks down if we transport ourselves back to the 1930s, where Stachel sees his view as legitimately generalizable too. Clearly, the methods had only just been devised: *everything* was radical back then! But perhaps a simple modification to render the labeling also time-variant resolves the problem, so that this is not an "eternal" carving. While covariant and canonical methods might well have been radical at the time of their creation, they are no more.

One might also track the history of quantum gravity in terms familiar from elementary particle physics in the earliest phases, which Sam Schweber characterizes in terms of "oscillations between the particle and field viewpoints" of Dirac and Jordan, viewing matter particles as fundamental (with radiation alone given the field treatment) versus matter as field quanta ("second quantization") respectively [Schweber (2003), pp. 377–8].[11] These oscillations become rather more entrenched after 1957, ultimately reflecting fairly deep divisions, disciplinary and otherwise, leading to the emergence of distinct *communities* within quantum gravity.[12] Carlo Rovelli puts the conflict as follows:

> Partially, the divide reflects the different understanding of the world that the particle physics community on the one hand and the relativity community on the other hand, have. The two communities have made repeated and sincere efforts to talk to each other and understanding each other. But the divide remains, and, with the divide, the feeling, on both sides, that the other side is incapable of appreciating something basic and essential: On the one side, the structure of quantum field theory as it has been understood in half a century of investigation; on the other side, the novel physical understanding of space and time that has appeared with general relativity. Both sides expect that the point of

of the scientific field. For example, as Oliver Darrigol [Darrigol (2009)] has so nicely shown, Lord Kelvin was keen to apply hydrodynamics to all of the world's phenomena, making use of "flow" analogies wherever he could. Similarly, electrodynamics, being the first classical field to be quantized, quickly became the model for quantization of the gravitational field and suggested to many the possibility of harmony between field quantization and gravitation. But the crucial disanalogies that cropped up led many to the view that gravity was something entirely apart from ordinary field theories. Towards the later part of the present book's story, we find that (non-Abelian) Yang–Mills theories take centre stage (with the 1955 discovery, by Ryoyu Utiyama, that general relativity can be put into this form), with the electrodynamical analogy breaking down and the Yang–Mills analogy taking off, restoring at least the possibility of peaceful coexistence, structurally speaking, once more (cf. [Trautman (1982)]).

[11] Chris Isham briefly suggests this kind of division in the context of quantum gravity in [Isham (1975)], pp. 10–11; cf. also [Jacobsen (2012), pp. 32–35].

[12] Rudolf Haag points out the creation of a similar separation, occurring 1945–1975, into distinct quantum gravity communities (separated "sociologically and perhaps even ideologically," [Haag (1986), p. 258]) and caused by the influx of elementary particle physicists into research on gravitational and cosmological phenomena and, inversely, the influx of gravitational physicists (trained in differential geometric methods) into elementary particle physics, each with their distinct disciplinary backgrounds. David Kaiser [Kaiser (1998)] has focused attention on the influx of particle physics into relativity research in order to explain some of the renaissance behavior, accounting for the mid-century rise of general relativity research, taking as his time-span: 1942–1975.

the other will turn out, at the end of the day, to be not very relevant. One side because all the experience with quantum field theory is on a fixed metric spacetime, and thus is irrelevant in a genuinely background independent context. The other because GR is only a low energy limit of a much more complex theory, and thus cannot be taken too seriously as an indication on the deep structure of Nature.

([Rovelli (2002)], p. 761)

This "worldview"-based sense of division, though fleetingly explored by a few visionaries (such as Arthur Eddington, Oskar Klein, and Matvei Bronstein), is precisely what took time to emerge, and precisely why such splits are not quite so legitimately transplanted onto the older research landscape. In other words, while we have a general tendency to assume that quantum gravity must involve some kind of revolutionary revisions, the earliest practitioners had no such viewpoint in mind, especially since they were still coming to terms with the individual quantum and relativistic revolutions which had yet to crystallize. Moreover, while we now view theories involving the quantization of the metric tensor as tantamount to theories of quantum spacetime (given a direct spacetime interpretation of the tensor), the very earliest attempts established no such spacetime interpretation, and, as we will see in Chapter 6, were rather more ambivalent about the matter.

In terms of carving approaches, I shall prefer to let the presentation reflect the absence of very firm lines prior to 1957.[13] But beyond 1957 we certainly find the orthodox partition explicitly laid out in several texts, and indeed, as mentioned above, the more it is laid out as a triple of distinct positions, the stronger the sense of division becomes.[14] To a certain extent, it had to await a more thorough understanding of quantum field theory and renormalization, which leads us into the issue of a different kind of carving up: periodization. However, before closing off this discussion, inasmuch as a "natural carving" does exist, as reflected in this early period, it is one of external *motivation* for

---

[13] A related issue concerns *where to begin* one's history. In a recent review, Carlip et al. ([Carlip et al. (2015)], p. 326) trace the beginnings of "modern quantum gravity" to the development of the canonical formalism in 1959–1961 by Arnowitt, Deser, and Misner—clearly lying beyond the entire historical period covered here! There is some sense to this if we remember that this was the stage at which the structure of classical general relativity was finally fully worked out—and so the problem of quantum gravity was put into a more stable form. The orthodox view is that Léon Rosenfeld's 1930 papers constituted the "origination" event—which is true only if we are talking about *field quantization* approaches (cf. [Bergmann (1962)]). I prefer to think in broader terms, focusing on the recognition of the *problem* of quantum gravity, rather than its attempted resolutions, as the starting point. In these terms, there was plenty of activity prior to Rosenfeld's papers—and indeed, Rosenfeld's papers themselves issued from the earlier thoughts of Wolfgang Pauli (then his postdoctoral supervisor). I will note, however, that it was only with these 1930s papers on field quantization that the conflict between quantum theory and gravitation was really able to be properly laid bare.

[14] Though I have not seen it laid out as such, the spread of quantum gravity options (including my liberal inclusion of quantum fields in curved spacetime) seems to reduce to options concerning how much of the metric tensor $g_{\mu\nu}$ we consider to be subject to quantization: *none* of it (as in quantum field theory in curved spaces, in which $g_{\mu\nu}$ remains fully classical); *some* of it (as in covariant perturbative approaches, in which $g_{\mu\nu} = \eta_{\mu\nu} + h_{\mu\nu}$, and only the perturbation $h_{\mu\nu}$, corresponding to the deviation from flatness, is quantized, with $\eta_{\mu\nu}$, the background part, remaining classical—recall that "covariant" refers to Lorentz covariance); or *all* of it (as in canonical approaches, and also Feynman quantization, in which $g_{\mu\nu}$ is replaced by an operator $\hat{g}_{\mu\nu}$). Hence, the triple ⟨None, Some, All⟩ adequately exhausts the options across a range of time periods.

pursuing quantum gravity, rather than any internal features of the resultant approaches. In particular, there is the issue (persisting throughout) of whether gravity (classical and quantum) might somehow be used as a *resource* to resolve issues in particle physics or not. The vast majority of early approaches had their eye on something other than gravitation, and there were very few studying quantum gravity purely for its own sake. Those that were were motivated by a harmonization of the foundations of physical theories, as mentioned above.

My preferred periodization focuses on the relative status of the ingredient theories of quantum gravity (quantum theory and general relativity) and largely matches this suggested partitioning of approaches. The various stages of quantum gravity's development are highly attuned to shifts in the understanding of these input theories, and the entire field of research changes abruptly given such changes (e.g. the introduction of spin in quantum theory). That is, the very nature of the problem (what is required to resolve it and how it is conceived) is dependent on external factors that, in this early period, are highly volatile. My 1956 cutoff year was chosen precisely because this appears to be when quantum gravity research starts to have a life of its own independently of the changing fortunes of its two parents.[15]

We can find three broad classes of response to the problem in the timeframe this book focuses on: (1) conflict; (2) peaceful coexistence; (3) utility or "resource"-based— note that (3) is not mutually exclusive with respect to (1) and (2), though obviously (1) and (2) are mutually exclusive. The conflict approaches insist that there is indeed a genuine conflict to be resolved, between the ingredient theories, and that one or the other theory must be modified or at the very least *something* must be said about the problem. This was Einstein's earliest thought on the matter, and he would eventually adopt the radical step of dropping quantum theory altogether as a fundamental description of the universe, though general relativity itself would also require a successor to explain what quantum theory had explained. On the other hand, there were early conflict proposals to drop general relativity as a fundamental theory, in favour of the priority of quantum theory (of gravitons on a flat background), with general relativity emerging in some classical limit. As already mentioned, the peaceful coexistence route tended to focus on structural harmonization between the ingredient theories and, in the earliest work focused on the study of the incorporation of quantum systems in a classical general relativistic framework. Features that emerged from the study of these various lines of attack were recognized to be relevant to more "serious" problems in making sense of quantum field theory (especially, the UV divergences that emerged when one probed short wavelengths of quantum fields, which were seen to be impacted on in quantizations of $g_{\mu\nu}$).[16] These

---

[15] Of course, this does not mean that there are no longer influences from changes in the ingredient theories, for there were many (e.g. advances in gauge theory, the understanding of ghosts, loop computations, renormalization advances, developments in cosmology, and so on). Only that from this point "quantum gravity" constituted a field in its own right, e.g. with conference sessions and workshops devoted to it alone.

[16] The treatment of quantum fields in curved space was initially of less utility in this respect, since the gravitational side was described by a c-number throughout, and so could not deliver resources for providing, e.g., cutoffs of troublesome high-frequency field values. For this reason, this approach tended to be discussed more as a mathematical exercise, revealing the structural compatibility of the formal frameworks of quantum

ramifications helped lift quantum gravity research into the more serious business of "real physics," namely elementary particle physics.

2.2   Convergence, Constraints, and Counterfactuals

Philosopher of science Ian Hacking notes that the convergence on some feature or result can prove very convincing (in evaluative terms concerning, e.g., the pursuit of a theory) in cases where the convergence comes about through quite different instruments and experiments, using quite distinct physical principles. He notes that,

> we are convinced because instruments using entirely different physical principles lead us to observe pretty much the same structures in the same specimen.
>
> ([Hacking (1983)], p. 209)

Hacking had microscopy in mind, but we can extrapolate to purely theoretical scenarios too: If we keep finding some similar behavior or feature in a wide variety of conditions/contexts, then we are the more prone to believe that it is a "real," universal feature, not some artifact of our model. Certain results crop up in multiple formalisms and in the context of quite distinct investigations in quantum gravity research. For example, the relevance of the Planck scale, or the existence of quantum geometry, or the divergences in the perturbative computation of relevant quantities. This ability to isolate general features is especially important in the absence of experiments. Naturally, when we are forced to rely on the theoretical constructions themselves (independently of their empirical consequences, beyond known, old evidence), one needs to consider different ways to guide the direction of the research. It should be clear that one can still probe a purely theoretical construction without considering its worldly implications; theoretical physics often follows this path. Given some such theoretical construction, one can certainly ask if it is mathematically consistent, for example. One can investigate its space of solutions; assess whether it is deterministic, or local, finite, and so on.

One can usefully view this through the lens of Peter Galison's notion of "constraints" [Galison (1987)]. Constraints are very much the life-blood of science. They minimize the latitude one has in theory construction. The satisfaction of constraints can in itself

theory and general relativity, than as a part of physics—indeed, quantum field theory in curved spaces was eventually given a mathematically rigorous formulation (in the context of algebraic quantum field theory: [Dimock (1980)]) that had direct and profound implications on the notion of particle in orthodox quantum field theory, as well as improving understanding of many other structural properties of flat-spacetime quantum field theory (cf. [Wald (1994)], [Ashtekar and Magnon (1975)]). Of course, the extension of this work to quantum fields on black hole backgrounds with back-reaction (of particle creation processes) on the metric [Hawking (1975)] (which naturally had to wait until the black hole concept was understood, in the late 1950s and 1960s), an arena thought to provide a good approximation (for strong gravitational fields, but negligible quantum gravitational effects) to full quantum gravity (i.e. to 'All'), would radically transform this area of research, bringing it back in line with physics (and cosmology), not least because it led to the puzzle of explaining black hole evaporation and the potential information loss it involves, as well as providing a model of the early universe—[DeWitt (1975)] offers an excellent early survey of this work.

act as an evaluative measure. Constraints mark a kind of limit on what is possible (or reasonable) in a theory, such that failure to satisfy some particular constraint usually means the end of that particular approach. They precisely reduce the elbow room one has in making theories, and this is of course crucial given the obvious infinity of "free theories" that we can imagine existing in their absence.[17] What one ideally desires, from constraints, is that they ultimately furnish enough filtering power to uniquely select one theory.[18] Of course, experiment forms the most important constraint as far as physical theories go. Yet quantum gravity has no experiments—certainly not in the period under investigation here.

In fact, Galison has considered the impact that quantum gravity has on scientific methodology in the specific context of superstring theory [Galison (1995)]. However, whereas Galison argues that the constraint structure has "shifted" in the context of quantum gravity (he has string theory in mind, but it is readily generalizable since it is the remoteness of the Planck length that is motivating this view), I would argue that much the same constraint structure was present before string theory (and, by extension quantum gravity), only it was working in the background, swamped by the more powerful experimental structure. When this is taken away, however, naturally the non-empirical constraints come to the fore. The unificatory constraint is a good example of this, readily apparent in the (experimentally accessible) gauge theoretic unifications of the standard model, for example. Likewise, anomaly cancelation was a strong constraint before and during string theory; there was no shift, only an alteration in relative importance. In fact, quantum gravity research gives us a unique probe into the wider constraint structure in physics in this way since it removes the usually dominant constraint of experiment allowing the full gamut of other constraints to come to the surface.

In the absence of experiments and observation, then, non-experimental constraints must come to the fore, to guide theorizing.[19] The black hole entropy value I mentioned earlier functions, in some sense, as a constraint, almost (but not quite) like an experimental constraint so that the new approaches are tested against it for support.[20] It thus provides new material for constructing theories, or working out the possibilities of old theories. Let us briefly return to Erik Curiel's objection from footnote 3 (his "plea for modesty") to viewing the entropy as possessing any evidential warrant whatsoever:

---

[17] Gerardus 't Hooft and Martinus Veltman's ['t Hooft and Veltman (1974)] proof of the finiteness of pure (matter-free) Einstein gravity involved the demonstration that a single scalar coupled to gravity would destroy the finiteness. However, to theorists, this then served up a constraint, not only forbidding, but also guiding the selection of other non-scalar couplings (cf. [Goroff and Sagnotti (1986)], pp. 710–11).

[18] This is equivalent, of course, to Einstein's dream of a uniquely determined theory, which he claimed to derive from Leibniz's doctrine of pre-established harmony (see [Einstein (1918)], cited in [Vizgin (1994)]).

[19] Note: the three classical tests of general relativity (perihelion of Mercury, deflection of light, and gravitational red shift) functioned as constraints on the early approaches in a more or less standard way, i.e. as *empirical* constraints, much like a correspondence principle. What is more interesting is the cases in which we have approaches that match up on all pre-existing knowledge, and don't make any novel testable predictions either. Given only empirical constraints, researchers should be in the position of Buridan's Ass, indifferent to the various proposals. Yet they are not, pointing to the existence of more structure at work.

[20] David Wallace has argued that black hole entropy is a necessary feature of low-energy quantum gravity that *any* plausible theory of quantum gravity must reproduce (see [Wallace (2019), p. 110]).

The derivation of the Bekenstein–Hawking entropy formula by counting the microstates in this or that theory of quantum gravity, impressive and intriguing as it may be in many ways, cannot serve as a demonstration of the scientific merit of a theory of quantum gravity, for the Bekenstein–Hawking entropy formula itself has no empirical standing. There is not the slightest shred of empirical evidence that directly supports or controverts it

[Curiel (2001), p. S435].

If we allow this, however, we threaten our practices regarding approximate theories. Much of what we take to be our knowledge of the world has its roots in such approximate theories. Alexander Rüger [Rüger (1989)] has an earlier discussion that bears directly on this point. He first links the problem of unifying quantum field theory and general relativity to an idea of Noretta Koertge's [Koertge (1971)], who considers cases of inconsistent parallel theories, both empirically adequate, and both mutually relevant, and that require a deeper unifying theory to resolve the *impasse* (increasing content in the process). As Rüger points out, such cases clash with the idea of a succession of rival programmes and it seems not to be the case here that either side (quantum field theory or general relativity) will subsume the other in a fundamental sense—covariant approaches notwithstanding. Shifting to Imre Lakatos' idea of "hard cores" of scientific research programmes,[21] he points out that we have in the problem of quantum gravity precisely a pair of incompatible hard cores in search of a new single, hard core. The problem is we do not have any such new hard core to guide us. Rüger rejects Koertge's own idea of giving one of the theories a purely instrumentalist interpretation, to avoid any ontological clash. Instead, he points to what he views as the actual practice of physicists, one that is neither instrumentalist nor that involves true unification leading to a successor. He argues, rather, that the two theories can be "approximately unified," as in the case of semiclassical gravity (from which we have the Bekenstein–Hawking result that concerns Curiel). This means that both hard cores are modified to regain compatibility, by restricting quantization to non-gravitational sources, and by coupling gravity to the expectation value of the stress-energy tensor (thus keeping things purely classical: $G_{\mu\nu} = \kappa \langle T_{\mu\nu} \rangle$). This amounts to suspending universality claims. This is not quantum gravity, of course, but it is supposed to approximate quantum gravity *at scales above the Planck length, in which we can neglect gravitons.*

Rüger then turns to what is essentially Curiel's question: can we take such a theory seriously, ontologically? And here the robustness aspect of Hacking's convergence idea is crucial, since taking the semiclassical theory's claims seriously would involve its claims being robust enough to survive in full quantum gravity. And, Rüger argues, this is indeed the case: the results are highly insensitive to the structure of whatever theory comes along, since the basic semiclassical equation does not prejudge the quantization of gravity, nor does it commit us to viewing gravity as purely classical. I would argue further, that the ability to then actually compute semiclassical results, such as the Bekenstein–Hawking

---

[21] That is, the basic principles defining a theory that are protected (from immediate falsification) by a "belt" of auxiliary hypotheses (see [Lakatos (1978), Chapter 1]).

entropy value, within proposals for full quantum gravity (such as string theory, loop quantum gravity, and casual set theory), provides *evidence* of such insensitivity and so simultaneously demonstrates convergence, which is evidentially significant.[22]

Renormalizability too acted as a crucial constraint in the post-WWII years. Weinberg describes the importance of this constraint:

> [I]t seemed to me to be a wonderful thing that very few quantum field theories are renormalizable. Limitations of this sort are, after all, what we most *want*; not mathematical methods which can make sense out of an infinite variety of physically irrelevant theories, but methods which carry constraints, because these constraints may point the way towards the one true theory. In particular, I was very impressed by the fact that [QED] could in a sense be *derived* from symmetry principles and the constraint of renormalizability; the only Lorentz invariant and gauge invariant renormalizable Lagrangian for photons and electrons is precisely the original Dirac Lagrangian.
>
> ([Weinberg (1972)], p. 1213; cited in [Galison (1995)], p. 22)

The original perturbative approach to quantum gravity was eventually rejected because it conflicted with this constraint.[23] By contrast, string theory was initially given so much credence because its topological propagator offered the prospect of a finite theory. However, it was then found to violate the further consistency constraint that there be no quantum anomalies in the theory (i.e. symmetries that are in the classical theory but broken at the quantum level). The subsequent satisfaction of *this* constraint, using a specific and highly interesting symmetry group, provided almost as significant a degree of motivation for renewed interest in the theory as a successful experiment since the prior probability for finding such anomaly-free cases was so low at the time.[24] However, as with experiments, we shouldn't place full weight on them: they are rarely decisive in the sense of grounding falsification and terminating research. As I mentioned in the

---

[22] I think exactly what is going on in this case is a little more complex. The Bekenstein–Hawking entropy is enveloped in a cycle in which the passing of the test (i.e. of computing microstates) amounts to a theoretical success, and that success *feeds back* into treating the entropy as a robust feature of quantum gravity, precisely since it is calculable in a variety of contexts. That is, when a variety of approaches were able to get the correct value out for the entropy, belief in the entropy as a real feature of the world was thereby strengthened for the reasons just outlined, and so the successful calculation is assigned yet more weight.

[23] As the dates of the papers in footnote 24 indicate, this process in fact took a long time, and while the constraint functioned as an arbiter on the truth of research programs, it wasn't sufficient to see them off entirely, and often served to point out necessary modifications to avoid infinities—unitarity violation was perhaps more problematic than the infinities, which were after all seen to be a fairly general feature of quantum field theories.

[24] See [Rickles (2014), Chapter 8] for a discussion of this episode. A similar scenario occurred for a cancellation of infinities shown by Deser and Zumino and independently by Ferrara, Freedman, van Niewenhuizen in the 1970s: [Deser and Zumino (1976)], [Freedman, van Nieuwenhuizen and Ferrara (1976)]. This idea led to the emergence of "supergravity," in which a spin 3/2 field is coupled to gravity: the supersymmetry between the bosonic and fermionic fields "tames" the ultraviolet properties of the theory. Note that Deser had together with van Niewenhuizen [Deser and Van Nieuwenhuizen (1974)] worked on a demonstration that standard general relativity with matter (for models with electromagnetism, Yang–Mills fields, Dirac fermions, and a cosmological constant) diverges in 1-loop calculations—'t Hooft and Veltman showed this first, though for a narrower range of matter sources: ['t Hooft and Veltman (1974)]. There followed computations for pure gravity, showing 2-loop divergences: [Goroff and Sagnotti (1986)]. Supergravity theories looked so promising because they outlawed divergences at these first two loop orders.

string theory case, these constraints can lead one to drop a theory prematurely, only to be found at a later date to satisfy it.

In the very earliest approaches, there was no quantum field theory available, so what would become an important constraint (renormalizability) was of course absent. Once QED was constructed, however, one could get a handle on *computable* aspects of quantum gravity, and compare them to what had been done in QED. Rosenfeld's computation of the self-energy of the graviton was just such an example. The development of renormalizability led to an easily applicable criterion to decide whether a theory was worth pursuing. Interestingly, the constraint of renormalizability played a lesser role once the tools of renormalization *group* theory had been assimilated.

Another kind of constraint based on analogy can be found in the lessons of the Bohr–Rosenfeld analysis of measurability of the electromagnetic field; these were taken to transfer over to the gravitational case (yet another example of the electrodynamic analogy so prevalent in quantum gravity research).[25] The idea was that the gravitational field would *necessarily* have to be quantized if it were coupled to another quantized field, or to quantized matter. This belief spurred on physicists in the early days. However, in the 1957 Chapel Hill conference (*On the Role of Gravitation in Physics*), Rosenfeld argued that the analysis he performed with Bohr does not translate into the gravitational case. The crucial disanalogy is that one cannot (even theoretically) find a measuring instrument that would not generate perturbations in the measurement result: this is due to the equivalence principle. In the electromagnetic case the fact that there are both positive and negative charges allows one to control the perturbations. The electromagnetic field can be shielded. This supposed necessity (suggested by the thought experiment) previously functioned as a constraint on quantum gravitational theorizing. However, John Wheeler was willing to suggest (following Rosenfeld's remarks during Salecker's talk, [Salecker (1957), p. 179]) that perhaps the measurement problem for quantum gravity could be ignored for the present and that one place more emphasis on "the organic unity of nature" as a key constraint—as we saw in Bergmann's remarks about "aesthetic unease" in the previous chapter, this certainly seemed to function as a guiding principle.[26]

In the case where the scientists have as their goal the "navigation among the potentialities proffered by nature" John Stachel has described the process of convergence as one of "negotiation" ([Stachel (1994)], p. 143). Any account of scientific discoveries must take account of this negotiation, though Stachel is quick to point out that this does *not* imply the neglect of nature. Though there is a certain amount of elbow room in scientific discoveries, and so the evolution of scientific research and the nature of the theories that result, all of this this must be in accord with the "the potentialities proffered

---

[25] Note that such analogies can themselves be interpreted as constraints since one is essentially making a claim that two systems are sufficiently similar so that the (well-known) constraints that apply to one will most likely apply to the other—we return to the issue of analogies in quantum gravity research in Chapter 5.

[26] Likewise, in the editors' introduction to the first published volume devoted to quantum gravity (from the first Oxford Symposium on Quantum Gravity): "there is the intuition that the whole of physics should be described by a single fundamental theory" [Isham, Penrose, and Sciama, eds. (1975), p. i.]

by nature".[27] This is not strong social constructivism, then. The contingency is very heavily constrained by nature, which is paramount.

Stachel goes on to give an alternative possible scientific history, in which a different theory of gravitation was "discovered" that was perfectly in accord with nature's potentialities (since it matches Einstein's version on all relevant observables). He borrows the example from Feynman who asked: "Suppose Einstein never existed, and the theory [of GR] was not available" (cited in [Stachel (1994), p. 146]). Could one replicate "the physics" of Einstein's theory using what other (non-geometrical) tools were available, namely special relativistic quantum particle theory? The answer is Yes,[28] as several people had already suspected before Feynman posed his question. One uses the fact that the gravitational interaction has observed qualitative properties that can be encoded into field quanta (the gravitons) with specific properties. These include (note the three classic tests):

- Obeys inverse-square law—and so is *long range*
- Is always attractive and couples universlly to all massive objects with equal strength independently of their constitution
- Macroscopicly observable
- Causes a red shift
- Bends light around the Sun
- Causes a correction (relative to Newton's theory) in the perihelion of Mercury.

One then assigns properties to the exchange particle in a somewhat bespoke fashion. We can see immediately that the particle must be massless in order to satisfy the long-range requirement (and also to get the right value for the bending of starlight around the Sun). The fact that gravitational effects can be seen at macroscopic scales means that the particle must have integer spin. A more complex argument is required for the attractiveness properties. We will skip this here (but see [Weinberg (1964)] for the full argument), and simply note that a spin-2 particle is demanded in order to have *universal* attraction that couples in the right way to matter whilst also being consistent with the other requirements. The particle must also self-interact by virtue of universality—it is this feature that causes the nasty, untameable divergences in the quantum theory at high frequencies, since it leads to graviton-graviton coupling (i.e. gravity gravitates).

---

[27] Indeed, he gives as a very apt example, the U.S. Air Force's support of "anti-gravity" projects: no amount of support of any calibre could generate such a phenomenon. In this sense, the goals of the U.S. Air Force were not in accord with the potentialities of nature: no amount of coaxing was able to bring it about. I might add that there was even a convergence between government and industry (in the form of Roger Babson, a wealthy businessman who was searching for a gravity shield). We consider the role of this funding in Chapter 9.

[28] Although the matter is not as straightforward as Stachel (and Feynman) suggest. For details on the subtleties involved, see [Wald (1994)]. Note also that Stachel suggests the analysis takes place in the context of quantum field theory; in fact the analysis involves the particle picture only.

Hence, had certain contingent factors been otherwise, we might have had a very different theory of gravitation based on flat rather than curved space. Michael Duff makes a very similar point very clearly in his discussion of the approach to quantum gravity that follows this "alternative path" (covariant quantization): "the historical development of a physical theory and its logical development do not always proceed side by side, and logically, the particle physicist has no strong *a priori* reason for treating gravity as a special case" [Duff (1975), p. 79]. Though it is clear that there isn't an unlimited supply of empirically adequate alternatives, and they are often very hard to construct. Inasmuch as one approach could be rationally (or logically, in Duff's terms) justified, so could the other. In fact, many early approaches were rejected because they did *not* meet the requirements set by these constraints. However, in 1939, Pauli and Fierz, in a general study of the quantization of fields [Fierz and Pauli (1939)], employed the linear approximation of general relativity (another flat space approach), and only considered this linear field in interaction with the electromagnetic field.[29] This approach was important for future developments in quantum gravity research, but it suffered from an inability to recover the perihelion in the classical limit (when coupled to matter).[30] However, despite the clear empirical failure, the desire to have a theory of gravity that was in step with the other forces, the approach was developed further.

Nonetheless, it is very possible that had quantum field theory been to hand earlier, and had quantum gravity been seen as more of a problem, the flat space, special relativistic approach to gravity might have superseded Einstein's on account of its amenability in terms of quantizability. George Birkhoff developed in the early 1940s a theory of gravitation based on flat spacetime [Birkhoff (1944)], precisely in order to bring gravitation more in line with the rest of physics, and this was then quantized by his student Marcos Moshinsky [Moshinsky (1950)]. However, aside from requiring an overly stringent "perfect fluid" as its matter, it suffered from the same empirical problems

---

[29] In this paper, we also find explicitly mentioned the idea that gravity corresponds to a massless, spin-2 field, so that the particle carrying the force would be massless and spin-2 (note that the presence of spin-2 particles implies that a theory containing them would, ceteris paribus, be generally covariant). Thus, they write: "for vanishing rest-mass, our equations for the case of spin 2 go over into those of the relativity theory of weak gravitational fields (i.e. $g_{\mu\nu} = \delta_{\mu\nu} + \gamma_{\mu\nu}$, neglecting terms of order higher than the first in $\gamma_{\mu\nu}$); the 'gauge-transformations' are identical with the changes induced in $\gamma_{\mu\nu}$ by infinitesimal co-ordinate transformations" [Fierz and Pauli (1939)], p. 214. Marcus Fierz had briefly considered the spin-2 case in an earlier paper (which would become his habilitation thesis, supervised by Pauli—cf. [von Meyenn (1985), p. 635]), but the links to gravity are nowhere mentioned [Fierz (1939)]. The first recorded expression of a link appears to be in a letter from Heisenberg to Pauli in setting up the Solvay Congress schedule, dated April 23rd, 1939. Heisenberg [von Meyenn (1985), p. 635] mentions that Fierz had told him about the gravitational quanta with spin 2 and that Pauli ought to include something on that in his talk—this work made its way into [Fierz and Pauli (1939)], but the (unpublished) version for the Solvay Congress can be found in [von Meyenn (1993), pp. 897–901]. In the unpublished draft, Pauli utilizes the Fermi trick (supplementary conditions) to fix a gauge; he also names the quanta of the quantized field "gravitons". Moreover, Pauli explicitly refers to the limitations of the approach as a result of working in the linear approximation, which he relates to the divergence difficulties of field theory [in his words: "Es ist sicher eine Begrenzung der quantentheoretischen Seite dieser Betrachtung, daß man sich hier mit derjenigen Naherung begnügt, in welcher die allgemeinrelativistischen Feldgleichungen linear sind. Diese Begrenzung hängt mit den bekannten Divergenzschwierigkeiten der Feldtheorie aufs engste zusammen" (ibid., p. 901).

[30] A good discussion can be found in [Ortin (2004), p. 59].

that a contemporaneous theory of Wolfgang Pauli and Markus Fierz faced, both due to the result of mapping onto the weak field approximation of Einstein's theory only, thus neglecting the non-linear aspects that truly distinguish general relativity—we return to these flat space theories in Chapter 5.

2.3   Novelties in Quantum Gravity's History

Spencer Weart, former director of the American Institute of Physics [AIP], writes that "[o]lder physicists felt that younger students had a very imperfect feeling for how physics has been done, and how it should be done, and that history could be a great help here" ([Weart (1990)], p. 31). He goes on:

> [A]s we all know, much of our familiar lore is not history: it is myth...The public, our students, and some of our colleagues, perhaps even you and I, know things about the history of physics that simply are not accurate either to the historical facts or to the real spirit of doing science. And if we believe anything as scholars, it must be that an accurate story is better for everyone than a myth."
>
> ([Weart (1990)], p. 31)

Fortunately, the history of quantum gravity is not "big history" (as Weart has referred to it, [Weart (1990)], p. 39). Though the *content* is formidable, as a field of research it is rather well-contained and isolated. No big communities requiring big groups of historians. However, as we march through the twenty-first century, this luxury is fading away: quantum gravity research is finally making connections to experiments, and many of the experiments employed are going to be monsters. We had better get a grip on the story up to this point before the complexities cascade. Myths are already crystallizing in quantum gravity.

Despite a fairly rich century of lifespan, historical research on quantum gravity is virtually non-existent. Aside from a handful of small-scale studies [Stachel (1998), Gorelik (1992), Rovelli (2002)], historians have yet to venture into this territory. This is a little surprising since quantum gravity is as old as its ingredients, quantum field theory and general relativity, and they have (independently) received a good deal of attention. This can be excused to some extent, on the grounds that independently the theories have had a great deal of empirical success and form a greater part of the landscape of contemporary physics. However, fairly large and significant portions of the present ("successful") research landscape of physics are a result of ("unsuccessful") struggles with the problem of quantum gravity.[31] For this reason alone it is high time for a detailed study of the development of quantum gravity. In many ways, this period of quantum gravity's history can be seen as the emergence of a subject matter (a field of research) in its own right, peeling itself off the goings on of its ingredient theories (general relativity and quantum mechanics). Of course, the problem is still bound up with how we understand

---

[31] The unraveling of the structure of classical general relativity has already been mentioned, but we can also mention other more generally important advances, such as loop computations, ghost particles in gauge theories, the renormalizability of Yang–Mills theories, to name just a few.

these ingredient theories, but the field of research has earned some autonomy that was simply not there in the earliest phases of research.

In the case of quantum gravity, there is additional motivation in studying the history of the subject. Not least because the subject is so old: those responsible for the very earliest work are no longer with us, so no single living human has seen the entire development and cannot know what has and hasn't been done first hand. I wonder whether there is a precedent or parallel in the history of science, in which an unresolved puzzle such as this has origins that belong to past, dead masters only. In any case, this is all the more reason for detailed historical analyses, to unearth original ideas and motivations. But in addition to the benefits of history for quantum gravity, there are benefits flowing in the opposite direction too. I have already mentioned constraints in quantum gravity, as well as the apparent side-stepping of Whiggism,[32] and there are also interesting issues concerning the methodology of the field as it relates to this feature.

In one of the very few studies of methodological issues in quantum gravity, Jürgen Audretsch [Audretsch (1981)] relates this situation to Kuhn's analysis of the development of science as a punctuated cycle of normal science, with a single paradigm, followed by exceptional science (as a result of unacceptable, accumulating tensions). Certain elements of Kuhn's analysis of revolutions in science seem to provide a fitting model for quantum gravity. For instance, Kuhn writes:

> Confronted with anomaly or with crisis, scientists take a different attitude toward existing paradigms, and the nature of their research changes accordingly. The proliferation of competing articulations, the willingness to try anything, the expression of explicit discontent, the recourse to philosophy and the debate over fundamentals, all these are symptoms of a transition from normal to extraordinary research.
>
> ([Kuhn (1962), pp. 90–91])

However, Audretsch believes that quantum gravity does not in fact support Kuhn's approach. Audretsch describes the current situation in physics as "multi-paradigm" since there seem to be two paradigms each with an "all-claim" and "overlapping domains of application," yet he claims in this pair there are "neither anomalies nor is there a crisis" (ibid., p. 332). He then views the problem of quantum gravity as one of unifying these two paradigms *despite* the apparent lack of tension. He believes this shows Kuhn's scheme of the structure of science is incompatible with quantum gravity research.

I think this is a mistaken analysis. There are anomalies of a sort and there is a crisis of sorts. I do, however, agree that there exists something like a multi-paradigm situation, but it is propped up precisely by the *absence* of overlapping domains of application. It is precisely when we extend the two "paradigms" beyond their domain of applicability, and encroach on the other's, that the anomalies emerge, and we see an extremely serious

---

[32] Echoing Butterfield, Schweber puts it in terms better for our purposes: Whig history is "the writing of history with the final, culminating event or set of events in focus, with all prior events selected and polarized so as to lead to that climax" ([Schweber (1984)], p. 41). It is harder to sin in the context of quantum gravity: the big picture at present is a morass of parallel theories. We can perhaps use our present crop of preferred constraints to condemn or celebrate one or other of the past approaches, but we have no real agreement on these principles either.

crisis exists.[33] It is a crisis concerning the all-claims, and relates precisely to the esthetic unease evident in what appeal to be contradictory claims about the nature of reality. Now, of course, one can do as Rosenfeld did in his later life and suspend judgement about whether there is a clash or not, since there is no empirical data to settle the issue. But there is more to theories than their empirical consequences, as we have already witnessed, and if we look at the theoretical constructs themselves, then there is a *prima facie* conflict.

This conflict goes beyond mere "aversion to pluralism" then. Better is Audretsch's "logical motivation" for seeking unity, namely that, as a result of the overlap, there are scenarios the understanding of which would appear to require both paradigms working in tandem (e.g. in the physics of the very early universe). Here Audretsch seems to agree that the paradigms are *incompatible*, but not as a result of any experimental phenomenon. This is true, but given the spread of constraints we mentioned above, to restrict to experiment only is a gross over-simplification of scientific methodology.

2.4   The "Low Water Mark" and "The Renaissance"

A popular puzzle in recent history of general relativity is to explain how general relativity went from being a backwater of physics (what Jean Eisenstaedt has called a "low water mark" of the theory) to a thriving research topic (what Clifford Will calls "the renaissance"). Solutions to the problem range from the fairly sudden appearance of new high precision experiments, the discovery of quasars (leading to the emergence of relativistic astrophysics), to the refashioning of general relativity as a field theory along the lines of elementary particle physics—that is, disposing of the geometrical approach that, as Steven Weinberg puts it, drove a "wedge between general relativity and the theory of elementary particle physics" ([Weinberg (1977)], p. vii). We turn to this topic, and its relationship to the emergence of quantum gravity research, in Chapter 9 in which we present a slightly unorthodox (externalist) account of the renaissance (based on private philanthropy[34]), but let us end this chapter with a few brief words on the puzzle, since it provides useful background and motivation to subsequent chapters.

---

[33] In fact, Audretsch mentions two of the examples I have in mind here (ibid., p. 333), namely the UV divergences in quantum field theory and the singularities in general relativity, but argues that these are rendered unproblematic by renormalization/regularization and horizon effects respectively. So Audretsch believes that there are only the usual kinds of puzzle-solving problems in our two fundamental paradigms: there are neither technical problems nor empirical problems (no conflicts with the *facts* as we have them to hand) that warrant any claim of crisis. Yet Audretsch still notes that quantum gravity researchers are engaged in "extraordinary science," which for him becomes a curious puzzle. Further, while there is no experimental result like the perihelion of Mercury that renders general relativity problematic, there are observable factors. For example, the value that quantum field theory gives to the energy density of a vacuum (the cosmological constant) is very many orders of magnitude too large. Audretsch does mention the prediction of singularities by general relativity, but dismisses this as a problem since they are "hidden behind horizons" (p. 333), and therefore not empirically observable. It is certainly true that there are no "naked singularities" in general relativity, and therefore that one would not be measurable. However, the "cosmic censorship hypothesis" outlawing naked singularities might not be true in quantum gravity theories. Furthermore, it seems like a step back to positivism to bracket worrying about a direct theoretical prediction on the grounds that we couldn't measure it.

[34] Parts of this are heavily related to work done in collaboration with David Kaiser in [Kaiser and Rickles (2018)].

There was a general belief until the 1950s that general relativity was essentially "understood," with no mysteries left to unlock and explore. To pursue the study of gravitation in this period would have been professional suicide. For example, Christian Møller here describes how Bohr attempted to push him away from the study of general relativity for just these reasons:

> It was the famous paper on general relativity from 1916 I was studying, and he [Bohr] told me about the importance of Einstein's work, how it had changed our notions of space and time. But then he said, "Well, you see, all these things are now cleared up, and you should start to study quantum mechanics, because there are many things to be cleared up yet, and probably the transformation of our thinking, I mean the revolution in our epistemology, will be much deeper in quantum theory than in relativity theory." So I followed his advice, and started to study (quantum mechanics).[35]

No doubt many a young scholar eager to learn general relativity was put off in a similar way.[36] Einstein himself was well aware of the neglect of his theory, writing in the forward to Peter Bergmann's[37] 1942 textbook on relativity:

> I believe that more time and effort might well be devoted to the systematic teaching of the theory of relativity than is usual at present at most universities. It is true that the theory of relativity, particularly the general theory, has played a rather modest role in the correlation of empirical facts so far, and it has contributed little to atomic physics and our understanding of quantum phenomena.
>
> (Einstein, in [Bergmann (1942)], p. iii)

However, Einstein goes on to point out that general relativity may indeed have some hidden wonders that might be relevant to what were considered to be the more pressing issues in quantum theory:

> It is quite possible, however, that some of the results of the general theory of relativity, such as the general covariance of the laws of nature and their nonlinearity, may help overcome the difficulties encountered at present in the theory of atomic and nuclear processes.
>
> (ibid.)

---

[35] Interviewed by: Thomas S. Kuhn. Location: Copenhagen, Denmark. Interview date: Monday, July 29th, 1963: https://www.aip.org/history-programs/niels-bohr-library/oral-histories/4782.

[36] Charles Misner recalls John Wheeler, who we will meet in Chapter 8, making similar remarks: "This fits Misner's claim "that he always was interested in gravity only never put a lot of time into it until he taught a course in it, which was his way of learning it. In probably 1952, just the year before I came. But he said that during the thirties he wouldn't take a student to work on relativity because it would have been the end of his career — I mean the end of the student's career. There was just absolutely nothing in the universities, nobody interested in hiring somebody who had worked on relativity. And so it was definitely out, and I think it probably continued to be out at Princeton even as John was working on it, because other people thought it was less interesting than particle physics." (https://www.aip.org/history-programs/niels-bohr-library/oral-histories/33697)

[37] Of course, there is also the well known quip, told by Peter Bergmann to Abraham Pais, that "You only had to know what your six best friends were doing and you would know what was happening in general relativity" ([Pais (1982)], p. 268).

This idea—general relativity in the wider service of physics—was to become a major feature of the earliest work on quantum gravity, and provided one of the central motivations for studying it. Indeed, it provides the (internal) *raison d'etre* for quantum gravity in the period studied—we explore this issue in more depth in Chapter 7.

2.5   Conclusion

Though often looked at with wry amusement when mentioned in the same breath, "history" and "quantum gravity" fit remarkably well together. Not only is there more than enough in chronological terms, the episodes are closely intertwined with other important developments in the life histories of the ingredient theories, quantum theory and general relativity. Much of this book concerns this intertwining. However, there are also more sociologically interesting aspects having to do with the emergence of a community of quantum gravity scholars, itself piggybacking on the availability of funding sources. Finally, we note the special status of quantum gravity, historiographically speaking, as a rare case of a field of research with more than a century of history behind it, including within it various rejections and selections of results, but which has as yet no experiments of its own, and no final endpoint from which to interpret the past. This renders it an especially fruitful case study for historians of physics. Present day quantum gravitationalists would also benefit from a deeper understanding of the development of their chosen subject.

··

REFERENCES

Ashtekar, A. and A. Magnon (1975) Quantum Fields in Curved Space-times. *Proceedings of the Royal Society of London. Series A* **346**(1646): 375–94.
Audretsch, J. (1981) Quantum Gravity and the Structure of Scientific Revolutions. *Zeitschrift für Allgemeine Wissenschaftstheorie* **12**(2): 322–39.
Badino, M. and J. Navarro (2013) *Research and Pedagogy: A History of Quantum Physics through Its Textbooks*. Berlin: Edition Open Access.
Bensaude-Vincent, B. (2006) Textbooks on the Map of Science Studies. *Science and Education* **15**: 667–70.
Bergmann, P. G. (1962) Summary of the Colloque Internationale de Royaumont. In A. Lichnerowicz and M. A. Tonnelat (eds.), *Les Théories Relativistes de la Gravitation* (pp. 464–72). Paris: Centre national de la recherche scientifique.
Bergmann, P. G. (1956) Quantisierung Allgemeinen-Kovarianter Feldtheorien. In A. Mercier and M. Kevaire (eds.), *Fünfzig Jahre Relativitätstheorie Cinquantenaire de le Théorie de la Relativité* (pp. 79–97). Basel: Birkhäuser Verlag.
Bergmann, P. G. (1942) *Introduction to the Theory of Relativity*. New York: Dover Publications.
Birkhoff, G. D. (1944) Flat Space-Time and Gravitation *Proceedings of the National Academy of Sciences* **30**(10): 324–34.

Butterfield, H. (1965) *The Whig Interpretation of History.* New York: W. W. Norton & Company.

Carlip, S., D-W. Chiou, W-T. Ni, R. Woodard (2015) Quantum Gravity: A Brief History of Ideas and Some Prospects. *International Journal of Modern Physics D* 24(11): 1530028.

Curiel, E. (2001) Against the Excesses of Quantum Gravity: A Plea for Modesty. *Philosophy of Science* 68(3): S424–S441.

Darrigol, O. (2009) *Worlds of Flow: A History of Hydrodynamics from the Bernoullis to Prandtl.* New York: Oxford University Press.

DeWitt, B. S. (1975) Quantum Field Theory in Curved Space-time. *Physics Reports* 19(6): 295–357.

DeWitt, B. (1962) Quantization of Geometry. In L. Witten (ed.), *Gravitation: An Introduction to Current Research*, (pp. 266–381). New York: John Wiley and Sons.

Deser, S. and P. Van Nieuwenhuizen (1974) One-loop Divergences of Quantized Einstein-Maxwell Fields. *Physics Review D* 10(2): 400–10.

Deser, S. and B. Zumino (1976) Consistent Supergravity. *Physics Letters B* 62(3): 335–37.

Dimock, J. (1980) Algebras of Local Observables on a Manifold. *Communications in Mathematical Physics* 77: 219–28.

Duff, M. J. (1975) Covariant Quantization. In C. J. Isham, R. Penrose, and D. W. Sciama (eds.), *Quantum Gravity: An Oxford Symposium* (pp. 78–135). Clarendon Press: Oxford.

Einstein, A. (1918) Motiv des Forschens. In E. Warberg, M. von Laue, A. Sommerfeld, and A. Einstein (eds.), *Zu Max Plancks 60. Geburtstag: Ansprachen in der Deutsche Physikalische Gesellschaft* (pp. 29–32). Karlsruhe: Verlag der C. F. Müllerschen Hofbuchhandlung.

Fierz, M. (1939) Force-free Particles with any Spin. *Helvetica Physica Acta* 12: 3–37.

Fierz, M. and W. Pauli (1939) On Relativistic Wave Equations for Particles of Arbitrary Spin in an Electromagnetic Field. *Proceedings of the Royal Society of London. Series A, Mathematical and Physical Sciences* 173(953): 211–32.

Freedman, D. Z., P. van Nieuwenhuizen and S. Ferrara (1976) Progress Toward a Theory of Supergravity. *Physical Review D* 13: 3214–27.

Galison, P. (2012) Structure of Crystal, Bucket of Dust. In A. Doxiadis and B. Mazur (eds.), *Circles Disturbed: The Interplay of Mathematics and Narrative* (pp. 52–78). Princeton: Princeton University Press.

Galison, P. (1995) Theory Bound and Unbound: Superstrings and Experiments. In F. Weinert (ed.), *Laws of Nature: Essays on the Philosophic, Scientific, and Historical Dimensions* (pp. 369–408). Berlin and New York: Walter de Gruyter.

Galison, P. (1993) Feynman's War: Modelling Weapons, Modelling Nature. *Studies in History and Philosophy of Modern Physics* 29(3): 391–434.

Galison, P. (1987) *How Experiments End.* Chicago: University of Chicago Press.

Gorelik, G. (1992). First Steps of Quantum Gravity and the Planck Values. In J. Eisenstaedt and A. J. Kox. (eds.), *Studies in the History of General Relativity* (Einstein Studies, Vol. 3, pp. 364(379). Boston: Birkhäuser.

Goroff, M. H. and A. Sagnotti (1986) The Ultraviolet Behaviour of Einstein Gravity. *Nuclear Physics, B* **266**(3–4): 709–36.

Haag, R. (1986) Quantum Field Theory and Gravitation. In T. C. Dorlas, N. M. Hugenholtz, M. Winnink (eds.), *Statistical Mechanics and Field Theory: Mathematical Aspects* (pp. 258–68). Berlin and Heidelberg: Springer-Verlag.

Hacking, I. (1983) *Representing and Intervening: Introductory Topics in the Philosophy of Natural Science*. Cambridge: Cambridge University Press.

Hawking, S. W. (1975) Particle Creation by Black Holes. *Communications in Mathematical Physics* **43**: 199–220.

Isham, C. J. (1991) Conceptual and Geometrical Poblems in Quantum Gravity. In H. Mitter and H. Gausterer (eds), *Recent Aspects of Quantum Fields* (pp. 123–229). Berlin: Springer.

Isham, C. J. (1975) An Introduction to Quantum Gravity. In C. J. Isham, R. Penrose, and D. Sciama (eds.), *Quantum Gravity: An Oxford Symposium* (pp. 1–77). Oxford: Oxford University Press.

Isham, C. J., R. Penrose, and D. Sciama, eds. (1975) *Quantum Gravity: An Oxford Symposium*. Oxford: Oxford University Press.

Jacobsen, A. S. (2012) *Léon Rosenfeld: Physics, Philosophy, and Politics in the Twentieth Century*. Singapore: World Scientific.

Kaiser, D. (2013) Epilogue: Textbooks and the Emergence of a Conceptual Trajectory *Research and Pedagogy: A History of Quantum Physics Through Its Textbooks* (pp. 287–89). Max Planck Research Library: Edition Open Access.

Kaiser, D. (2005) *Drawing Theories Apart: The Dispersion of Feynman Diagrams in Postwar Physics*. Chicago: University of Chicago Press.

Kaiser, D. (1998) A ψ is just a ψ? Pedagogy, Practice, and the Reconstitution of General Relativity, 1942–1975. *Studies in History and Philosophy of Modern Physics* **29**: 321–38.

Kaiser, D. and D. Rickles (2018) The Price of Gravity: Private Patronage and the Transformation of Gravitational Physics after World War II. *Historical Studies in the Natural Sciences* **48**(3): 338–79.

Koertge, N. (1971) Inter-Theoretic Criticism and the Growth of Science. In R. C. Buck and R. S Cohen (eds), *PSA 1970* (pp. 160–73). Dordrecht: Springer.

Kuhn, T. S. (1977) The Essential Tension: Tradition and Innovation in Scientific Research. Reprinted in *The Essential Tension: Selected Studies in Scientific Tradition and Change* (pp. 225–39). Chicago: University of Chicago Press.

Kuhn, T. S. (1962) *The Structure of Scientific Revolutions*. Chicago: University of Chicago Press.

Lakatos, I. (1978) *The Methodology of Scientific Research Programmes*. Cambridge: Cambridge University Press.

Lalli, R. (2014) A New Scientific Journal takes the Scene: The Birth of Reviews of Modern Physics. *Annalen der Physik* **526**(9–10): A83–A87.

Misner, C. (1957) Feynman Quantization of General Relativity. *Reviews of Modern Physics* **29**: 497–509.

Moshinsky, M. (1950) On the Interactions of Birkhoff's Gravitational Field with the Electromagnetic and Pair Fields. *Physical Review* **80**(4) 514–19.

Ortin, T. (2004) *Gravity and Strings.* Cambridge: Cambridge University Press.

Pais, A. (1982) *Subtle is the Lord: The Science and the Life of Albert Einstein.* Oxford: Oxford University Press.

Peres, A. (1968) Canonical Quantization of Gravitational Field. *Physical Review* 171(5): 1335–44.

Rickles, D. (2014) *A Brief History of String Theory: From Dual Models to M-Theory.* Berlin and Heidelberg: Springer-Verlag.

Rovelli, C. (2002) Notes for a Brief History of Quantum Gravity. In R.T. Jantzen et al. (eds.), *Proceedings of the Ninth Marcel Grossmann Meeting on General Relativity* (pp. 742–68). Singapore: World Scientific.

Rüger, A. (1989) Complementarity Meets General Relativity: A Study in Ontological Commitments and Theory Unification. *Synthese* 79(3): 559–80.

Salecker, H. (1957/2011) Conceptual Clock Models. In C. DeWitt and D. Rickles (eds.), *The Role of Gravitation in Physics: Report from the 1957 Chapel Hill Conference* (pp. 171–85). Berlin: Edition Open Access.

Shaw, D. F. (1990) A Physicist Looks at the Resources of Research Libraries. In J. Roche (ed.), *Physicists Look Back* (pp. 3–28). Bristol: Adam Hilger.

Schweber, S. (2003) Quantum Field Theory: From QED to the Standard Model. In M. J. Nye (ed.), *The Cambridge History of Science Volume 5. The Modern Physical and Mathematical Sciences* (pp. 375–93). Cambridge University Press.

Schweber, S. (1984) Some Chapters for a History of Quantum Field Theory. In B. DeWitt and R. Stora (eds.), *Les Houches, 1983, Proceedings, Relativity, Groups and Topology, II* (pp. 37–220). Paris: Elsevier.

Smolin, L. (2007) *The Trouble with Physics: The Rise of String Theory, The Fall of a Science and What Comes Next.* New York: Mariner Books.

Stachel, J. (1998) The Early History of Quantum Gravity (1916–1940). In B. R. Lyer and B. Bhawal (eds.), *Black Holes, Gravitational Radiation and the Universe* (pp. 525–34). Dordrecht: Kluwer Academic Publishers.

Stachel, J. (1995) History of Relativity. In L. Brown, A. Pais, and B. Pippard (eds.), *Twentieth Century Physics* (pp. 249–356). New York: American Institute of Physics Press.

Stachel, J. (1994) Scientific Discoveries as Historical Artifacts. In K. Govroglu, J. Christianidis, and E. Nicolaidis (eds.), *Trends in the Historiography of Science* (pp. 139–148). Dordrecht: Kluwer Academic Publishers.

't Hooft, G. and M. Veltman (1974) One Loop Divergencies in the Theory of Gravitation. *Annales de l'Institut Henri Poincaré, A* 20: 69–94.

Trautman, A. (1982) Yang–Mills Theory and Gravitation: A Comparison. In R. Martini and E. M. de Jager (eds), *Geometric Techniques in Gauge Theories* (pp. 179–89). Berlin: Springer.

Unruh, W. (2014) Has Hawking Radiation Been Measured? *Foundations of Physics* 4: 532–45.

Vizgin, V.P. (1994) *Unified Field Theories in the First Third of the 20th Century.* Boston: Birkhäuser.

von Meyenn, K. (1993) *Wissenschaftlicher Briefwechsel mit Bohr, Einstein, Heisenberg u.a. Band III: 1940–1949 / Scientific Correspondence with Bohr, Einstein, Heisenberg, a.o. Volume III: 1940–1949*. Berlin: Springer.

von Meyenn, K. (1985) *Wissenschaftlicher Briefwechsel mit Bohr, Einstein, Heisenberg u.a. Band II: 1930–1939 / Scientific Correspondence with Bohr, Einstein, Heisenberg, a.o. Volume II: 1930–1939*. Berlin: Springer.

Wald, R. M. (1994) *Quantum Field Theory in Curved Spacetime and Black Hole Thermodynamics*. Chicago: University of Chicago Press.

Wallace, D. (2019) The Case for Black Hole Thermodynamics Part II: Statistical Mechanics. *Studies in History and Philosophy of Modern Physics* **66**: 103–17.

Weart, S. R. (1990) Preserving and Making Known the History of Physics: The American Institute of Physics Center for History of Physics. In J. Roche (ed.), *Physicists Look Back* (pp. 29–43). Bristol: Adam Hilger.

Weinberg, S. (1977) Search for Unity: Notes for a History of Quantum Field Theory *Daedalus* **9**(18): 17–35.

Weinberg, S. (1972) *Gravitation and Cosmology*. New York: J. Wiley and Sons.

Weinberg, S. (1964) Photons and Gravitons in S-Matrix Theory: Derivation of Charge Conservation and Equality of Gravitational and Inertial Mass. *Physical Review* **135**(4B): B1049–B1056.

Wüthrich (2019) Are Black Holes about Information? In R, Dardashti, R. Dawid, K. Thébault (eds.), *Why Trust a Theory?* (pp. 202–23). Cambridge: Cambridge University Press.

3

"Prehistoric" Quantum Gravity

In this chapter we consider what might best be called 'prehistoric quantum gravity.'[1] We show that, even before the concept of the *quantization* of the gravitational field in 1929, there was a fairly lively investigation of the relationships between gravity and quantum stretching as far back as 1916.

3.1 *Pourparlers* for Quantum Gravity

In a lecture delivered to the British Institute of Philosophy, November 15th, 1932, Sir Arthur Eddington wrote in surprisingly modern terms of the problem of merging quantum theory and the general theory of relativity:

> At present theoretical physics is divided into macroscopic theory and microscopic theory, the former dealing with systems on a scale comparable with our gross senses, and the latter dealing with the minute substructure underlying the gross phenomena. Broadly speaking, relativity theory covers macroscopic phenomena and quantum theory the microscopic phenomena. The two theories must ultimately be amalgamated, but at present we have not got much beyond the *pourparlers* for amalgamation. So the gap exists—not, however, as a gap in the external world, but as a gap in our understanding of it.
>
> [Eddington (1933), 30]

---

[1] The expression was suggested to me by Stanley Deser (private communication). Curiously, the expression "quantum gravity" (as opposed to "quantum theory of gravity," or other such extended expressions) would itself take a rather long time to take hold. It seems that the first usage of the term in the title of a talk was John Klauder's "Soluble Models of Quantum Gravity" [Klauder (1969)], presented at the "Relativity Conference in the Midwest" on June 2nd, 1969, while Klauder was at Bell Labs (the published version in the proceedings curiously bears the different title "Soluble Models of Quantum Gravitation"—Klauder wrote a related paper the following year, with his student Erik Aslaksen, entitled "Elementary Model for Quantum Gravity" [Klauder and Aslaksen (1970)]. Charles Misner also uses the expression, just once, in his paper on quantum cosmology, from the same Midwest conference [Misner (1969), p. 76]. However, the 'quantum gravity' label was only really solidified as the description of the field as a whole following the Oxford Symposium on Quantum Gravity, held in 1974 [Isham, Penrose, and Sciama (1975)].

Covered with Deep Mist: The Development of Quantum Gravity (1916–1956). Dean Rickles, Oxford University Press (2020). © Dean Rickles.
DOI: 10.1093/oso/9780199602957.001.0001

This suggests that even as far back as 1932, the problem of quantum gravity already possessed some historical pedigree (there *had been* "*pourparlers*," as Eddington puts it). Though in somewhat different terms, given the rapid and radical developments in physics at that time, Eddington himself had been thinking about the problem since at least 1918 (see §3.4.1, below). Yet, in the handful (if that...) of historical studies of quantum gravity that exist (e.g. [Stachel (1998), Rovelli (2002)]), it is claimed that quantum gravity research originated with Léon Rosenfeld's 'pioneering' pair of papers from 1930, forging, as they indeed did, both the canonical and covariant quantizations of the gravitational field (although even this is not quite accurate either, as we shall see in later chapters)—though lip service has at least been paid to Einstein's own early prophetic remarks about the potential conflict between general relativity and quantum theory, in papers of his from 1916 onwards[2], but this is still followed by the claim that nothing was really done about it until Rosenfeld tackled the problem (prompted by Pauli).

However, though the version of quantum theory involved was significantly different (and certainly less systematic and coherent—ditto general relativity) from the present framework (or frameworks)—stemming from the pivotal 1925–1928 developments that produced quantum mechanics and quantum field theory, more or less in their modern form—there was nonetheless a rich debate about the relationship between (old) quantum theory (and the atomic physics that preceded it[3]) and gravity. This interaction is to be expected:

1. since the two frameworks were constructed over much the same period of time, often by the same architects

2. since the right-hand side of the Einstein field equations must naturally include (in some way) contributions from the matter best described by the quantum theory.

It is true that this work rarely, if ever, involved consideration of *quantum properties of the gravitational field*, but this could equally be said of some recent approaches that we are nonetheless perfectly content to label 'quantum gravity'.[4] When I speak of quantum gravity in this chapter, I shall mean it very liberally to indicate some approach

---

[2] On which see §3.3 below. See, in addition, [Stachel (1998)], in which he also considers an earlier and slightly more diverse group of actors. Another excellent study—though with a focus more on unified field theories than quantum gravity—is [Goldstein and Ritter (2003)]. This traditional view that quantum gravity originates with Rosenfeld can be traced back at least as far as Bryce DeWitt's brief historical review in the first of his three *Physical Review* papers on quantum gravity in 1967: [DeWitt (1967)].

[3] Of course, 'atomic' certainly does not mean 'quantum'. However, many of the issues that were discussed in the context of merging atomic and gravitational physics are nonetheless relevant for later work on quantum gravity since they often involve overlapping concerns: implementing discreteness, singular behaviour, and so on, in the context of a theory like general relativity. For this reason I shall often discuss non-quantum, atomic physics. It will be clear from the context when this is the case. It is my contention that if we are looking for sources of the earliest examples of genuine quantum gravity research then we cannot ignore these non-quantum examples.

[4] An example of such quantum gravity without the *quantization* of gravity are the so-called "emergent gravity" proposals. See [Hu (2009)] for a nice review of these.

that involves dealing with the problem of the coexistence of quantum systems[5] and gravitation.

In some ways, this early work closely mirrors what would come later, and, indeed, many of the key notions of the later work—the importance of the Planck scale in demarcating the domain of applicability of classical general relativity, the experimental inaccessibility of quantum gravity, and the potentially radical revision of spacetime concepts—were discussed even at this early stage. This must surely have contributed to later work, at least in establishing a mindset for thinking about the amalgamation of quantum and gravity. This work deserves to be discussed under the banner of 'quantum gravity' just as much as any other that would come later, and any attempt to force quantum gravity into the mould of the *quantization of the gravitational field* (which is implicit in the Rosenfeld origin story) misses out on both an important source of later ideas and an important set of ideas in their own right.

In this chapter we discuss these early skirmishes into and around the problem of quantum gravity, from their prehistory (close to the creation of general relativity) to the emergence of wave-mechanics and just prior to the Dirac equation—at which point the course of quantum gravity research, quite naturally, radically shifts, and we have an appropriate context for the kind of work carried out by Rosenfeld (this is the topic for the next chapter). We see that these early approaches also offer a very useful probe for investigating several important agendas that were in operation at that time, including a desire to meld the cosmological and microscopic, and to unify both physics and our *knowledge* of physics via axiomatic foundations. A future task is to consider whether and how these agendas continued to play out in the later developments.

For reasons of space and convenience (there are a lot of papers that might be included, given the liberality of 'quantum gravity' employed here), I focus heavily on papers appearing in the journal *Nature* during this period. Though it has a potential to introduce an English bias, it nonetheless gives us a good snapshot of the general state of research, since it was common practice to supplement more technical publications (e.g. in *Zeitschrift für Physik* or the *Proceedings of the Royal Society*) with a brief note in *Nature*,[6] describing the key findings. Indeed, it perhaps offers a closer glimpse into the state of play since a feature of *Nature*, especially in these early issues, is the almost direct personal correspondence between individuals that can be found in its pages, via responses in notes and letters—something that could be easily carried out on account of the weekly publication frequency.

Though quantum gravity is today viewed as a slightly strange problem on the frontiers of physics (no doubt because of its highly theoretical and mathematical nature), in the early days surrounding their creation, there existed a fairly natural dialogue between

---

[5] Where, as I already indicated, this might sometimes be stretched to strictly non-quantum situations such as early atomic physics when this physics has some properties relevant to the full quantum context, such as discreteness of spectra and indeterminism.

[6] These rather non-technical accounts were often duplicated in German in *Naturwissenschaften*.

quantum theory and general relativity.[7] Again, I am here adopting an enormously *liberal* characterization of *the problem of quantum gravity* so as to offer as inclusive an account as possible, thus minimizing the risk of sidelining what may have been important sources of later ideas, Eddington's *pourparlers* for amalgamation. Whether any of the approaches I discuss amount to quantum gravity in themselves is of course highly debatable, but then we do not have *any* such theory yet.

3.2   The Torch of Unification: Mie, Hilbert, Weyl, and Haas

I don't wish to cover the already well-trodden ground of the genesis of quantum mechanics and general relativity.[8] However, many of the initial forays into quantum gravity were, in many ways, extensions of debates that were conducted *before* the creation of these theoretical frameworks—hence, we are concerned with the *pre*-pre-history here. One can also find alternative accounts of gravitation—in particular those based on an electromagnetic ether—persisting well beyond the creation of general relativity. One such approach that was clearly of importance (not least to Hilbert, Weyl, and Eddington) was that of Gustav Mie. We begin with a brief account of Mie's approach since it marks an approach to the amalgamation of the central theories of physics, as it was then, on the precipice of radical changes.

In 1912 and 1913 Gustav Mie sought to develop a (reductive) unified field theory in which both matter and force could be derived from the electromagnetic field (understood as an emergent property of the ether) alone. The core object was a 'world function' and the derivation of gravity and electromagnetism would proceed from this via the calculus of variations—he was not able to get a fully unified theory of both phenomena: Mie's theory was based on two scalar potentials. The general thrust of Mie's program stemmed from his belief in the significance of the new empirical facts about the behavior of atoms, that had recently emerged. In 1912 he wrote:

> The significance of the recently acquired empirical facts about the nature of the atoms ultimately amounts to something essentially only negative, namely that in the atoms' interior the laws of mechanics and Maxwell's equations cannot be valid. But regarding what should replace these equations in order to encompass from a single standpoint the profusion of remarkable facts associated with the notion of quantum of action, and in addition the laws of atomic spectra and so forth, the experimental evidence is silent. In fact, I believe that one must not expect anything like that from experiment alone. Experiment and theory must work hand in hand, and that is not possible as long as the theory has no foundation on which it can be based. Thus it seems to me absolutely necessary for further progress of our understanding to supply a new foundation for the theory of matter. With this work, I have tried in the following to make a start, but in

---

[7] Or, in somewhat less anachronistic terms, in the case of the very earliest such work, between the puzzling behavior of discrete matter then observed to behave in an increasingly curious way, and gravitation.

[8] We will have a little more to say in the next chapter, but for good comprehensive treatments of both, consult: [Galison, et al. (2001)] and [Renn (2007)] respectively.

view of the difficulty of the matter one should not right away expect results accessible to experiment. The immediate goals that I set myself are: to explain the existence of the indivisible electron and: to view the actuality of gravitation as in a necessary connection with the existence of matter. I believe one must start with this, for electric and gravitational effects are surely the most direct expression of those forces upon which rests the very existence of matter. It would be senseless to imagine matter whose... smallest parts did not possess electric charges, equally senseless however matter without gravitation. Only when the two goals I mentioned are reached will we be able to consider making the connection between the theory and the complex phenomena mentioned above.

[Mie (1912), 1554]

We see quite clearly here the extent to which, in Mie's mind, gravitation, electromagnetism, and matter are all manifestations of one and the same 'world-building' stuff—this worldview would spread to Weyl, Hilbert, Eddington, and many others. Out of this mixture, Mie expected the phenomenological facts of quantum theory to emerge—this theory might somewhat perversely be viewed as an approach to quantum gravity that *predates* both general relativity and quantum mechanics. At this time, in 1913, there were three other (serious) alternative theories of gravitation, those of Einstein and Grossman (the 'entwurf' theory), Abraham, and Nordström. Abrahams' theory was inadequate in several ways, chief amongst these being the incompatibility between the variable light speed adopted by Abraham and his usage of the Lorentz transformations. We will return to Nordström's theory later in this chapter, for now we quickly see how Mie's theory influenced Hilbert.

David Hilbert is not a name usually associated with quantum gravity research, but he figures centrally in several early episodes, some of which played a crucial role in later work (up to the present day, in fact).[9] Hilbert was, of course, directly involved in aspects of the birth of general relativity, using his beloved variational principles to derive the field equations. Hilbert was led to this approach by a rather indirect route involving a modification of Gustav Mie's electromagnetic theory of matter and force: Hilbert made direct use of Mie's theory in his derivation of general relativity. This imposed a severe restriction on the form of the theory, since it depended upon a specific matter-source. As Pauli put it in his encyclopedia article on general relativity:

Hilbert's presentation... was not quite comfortable for the physicists, because in the first place he axiomatically defined the variational principle, and, which is more important, his equations were expressed not for an arbitrary material system, but were based on Mie's theory of matter.

[Pauli (1921), 211]

Einstein[10] famously found Hilbert's approach "childish" since it didn't show a proper awareness of the "pitfalls of the real world"—the reason Einstein adopted 'principle

---

[9] I'm indebted in this section to the hard work of Leo Corry (see Corry (1997), Corry (1999)), David Rowe, Tilman Sauer, and others, in unpacking the complex relations between Mie, Hilbert, and Weyl.

[10] Letter to Weyl, dated November 23rd, 1916, cited in [Rowe (2003), p. 65]—see also: [Sauer (2002)].

theory' approaches was, of course, precisely to avoid such pitfalls (in this case "risky hypotheses about the structure of the electron"). The axiomatic method was therefore not a good basis for physical theorizing according to Einstein.

Hilbert's ideas about the foundations of mathematics (and his emerging ideas about the unity of scientific knowledge) were combined with this physical background in his celebrated work on general relativity in 1915. What is interesting about this work, for our project, is that, like Mie, he quite clearly believes that the physics of gravity will be able to unlock the secrets hidden in atomic processes:

> As one sees, the few simple assumptions expressed in Axioms I and II suffice by sensible interpretation for the development of the theory: through them not only are our conceptions of space, time, and motion fundamentally reformulated in the Einsteinian sense, but I am convinced that the most minute, till now hidden processes within the atom will become clarified through the fundamental equations herein exhibited and that it must be possible in general to refer all physical constants back to mathematical constants— just as this leads to the approaching possibility, that out of physics in principle a science similar to geometry will arise: truly, the most glorious fame of the axiomatic method, while here, as we see, the mighty instruments of analysis, namely the calculus of variations and invariant theory, are taken into service.
>
> [Hilbert (1915), 407]

Whether it was through interactions with Einstein or Weyl, or self-realization about the magnitude of the task he had set himself, Hilbert was a little more subdued a couple of years later. In 1917 Hilbert spoke on "Axiomatic Thought" to the Swiss Mathematical Society in Zürich. At the root of his talk was a belief in the unity of scientific knowledge, with mathematics the linchpin holding it all together:

> I believe that everything which can be the subject of scientific thought, as soon as it is ripe enough to constitute a theory, falls within the scope of the axiomatic method and thus directly to mathematics. By pursuing ever deeper-lying layers of axioms ... we gain ever deeper insights into the essence of scientific thought itself and we become ever more conscious of the unity of our knowledge. In the name of the axiomatic method, mathematics appears called upon to assume a leading role in all of science.
>
> [Hilbert (1918), 156]

Weyl was in attendance at this talk, and would adopt a formal approach to the problem of unification of gravitation and electromagnetism (though not quantum theory) that was methodologically similar to Hilbert's. Weyl generalized Riemannian geometry, adding a principle of calibration or gauge ("eich") invariance to account for the non-integrability of length (over non-infinitesimal distances). Einstein had similar gripes with this approach: in this case the theory had the absurd consequence that objects taken around different paths having identical origins and termini will, at the point of termination, be found to have different sizes and rates.[11]

---

[11] See [Scholz (2001)] for a fine discussion of this episode along with a translation of Weyl's text.

The torch of 'unity through axiomatization' was carried on, in a rather different way, by Arthur Haas. Haas was a strong advocate of Hilbertian axiomatization, and in his case it led him to early speculations about matters related to the problem of quantum gravity. For example, as early as 1919, Haas writes (on the basis of 'unification' ideals) that:

> Arguably, one of the most important future tasks of the axiomization of physics is the implementation of quantum theory in the system of the general theory of relativity.
>
> [Haas (1919), 749][12]

Though he doesn't explicitly name the individual constants associated to the ingredient theories (viz. $c, \hbar, G$), it is reasonable to surmise that this is what Haas had in mind in the following passage:

> The main task of the axiomatization of physics will be the problem concerning the integration of the universal constants of physics. Also the solution of this question may be expected to reveal deeper knowledge of the relations, only intimated by Hilbert, holding between gravity and electricity, and of a further integration of these relations with the quantum hypothesis.
>
> (ibid., 750)[13]

This interpretation is somewhat strengthened by the fact that Haas went on to consider the various possible combinations of other constants in other contexts, investigating the way they demarcate domains [Haas (1938)]. In many ways, this idea that we must consider the integration of the fundamental constants to solve the problem of the relationship between quantum theory and gravitation coincides with the modern understanding.[14]

3.3   Einstein on the Relationship Between Gravity and Quantum

> The intellect seeking after an integrated theory cannot rest content with the assumption that there exist two distinct fields totally independent of each other by their nature.
>
> Einstein's Nobel lecture (1923).

---

[12] "Eine der wichtigsten Zukunftsaufgaben, die in dieser hinsicht der physikalischen Axiomatik gestellt ist, ist wohl die Einfügung der Quantentheorie in das System der allgemeinen Relativitätstheorie."

[13] "Aufgabe der physikalischen Axiomatik sein wird; es ist das Problem des Zusammenhanges zwischen den universellen Konstanten der Physik. Auch die Lösung dieser Frage darf vielleicht erhofft werden von einer tieferen Erkenntnis der von *Hilbert* erst angedeuteten Beziehungen zwischen Gravitation und Elektrizitat und von einer Verknüpfung dieser Beziehungen mit der Quanten-hypothese."

[14] I might add that [Gorelik (1992)] assigns the discovery that these fundamental constants might point to the limits of present physical theory to Matvei Bronstein. While I agree that Bronstein was the first to produce an explicit account of the nature of this limitation in the 1930s, I'll show later that Eddington too made similar claims in 1918.

It is a little curious that so many great revolutionary episodes happened almost simultaneously at the beginning of the twentieth century. Perhaps one revolution made it easier for others to follow, via some kind of snowball effect? Whatever the reason, the revolution that resulted in general relativity and the revolution that resulted in quantum theory were close neighbours in time. Einstein was profoundly involved in the creation of both theoretical frameworks, though the former more so than the latter. At the time of the construction of the general theory of relativity he firmly believed in the existence of quanta of radiation. But this only involved a belief in the property of discreteness (with no real sense of ontological substrate beyond this), rather than belief in what would become quantum mechanics (or quantum field theory—though here too his contributions on emission and absorption of radiation proved crucial). Most physicists believe *another* revolution is required to bring quantum theory and general relativity together—*cf.* [Rovelli (2000)].

Since such quanta, with their discrete energies and other properties, would inevitably couple to the gravitational field (in however small a way), Einstein couldn't ignore the fact that *something* would need to be said about the nature of this interaction.[15] Even before General Relativity had been published in its final form, Einstein was in correspondence with Arnold Sommerfeld about its possible relationship with quantum theory.[16] Einstein heard about Sommerfeld's new theory of spectral lines first-hand while he was still working on general relativity. Sommerfeld thought that the general theory of relativity might be able to offer some help in resolving problems caused by the Stark effect (see [Sommerfeld (2000)], 438). Einstein was not positive, writing:

> General relativity is unlikely to be able to assist you, because it practically coincides with the more restricted theory of relativity for those problems.... [A]ny other theory that corresponds with relativity in the restricted sense can be taken over in the general theory of relativity through simple transformation, without the latter delivering any new criteria. Thus you see that I cannot help you in the least.
>
> [Einstein, letter to Sommerfeld, [Berlin] December 9th, 1915 ([Schulmann *et al.* (1998)], p. 159)]

It is quite likely that Sommerfeld's willingness to consider the relationship between what looked at this stage like disparate fields of inquiry was grounded in a similar belief system to that of Hilbert and Haas (and the Göttingen school), though with a far more empirical basis. Indeed, Schweber ([Schweber (2009)], pp. 269–278) has noted that a "doctrine of pre-established harmony" (between mathematics and physics and mathematics and nature) can be found running through much of Sommerfeld's earlier work. Given this,

---

[15] A little later it would also come to be understood that there is a 'formal interaction' between general relativity and quantum objects stemming from the peculiar nature of fermions: including objects with half-integer spins imposes a variety of constraints on the spacetime structure, and therefore on the gravitational field (resulting in a slightly modified theory of gravitation). This was a rather slow lesson.

[16] Though it appears that it was Sommerfeld who led this exchange, fired up, no doubt, by the success of his application of *specially* relativistic principles to quantum theory.

it is reasonable to expect some inner unity holding between so fundamental a pair of frameworks as relativity and atomic theory.[17]

Almost as soon as general relativity was completed, Einstein became aware of a possible conflict between it (or, more specifically, the existence of gravitational waves) and the principles of quantum theory,[18] and, therefore, the need to say something about the problem of quantum gravity. Thus, he writes that

> [A]s a result of the internal-atomic movement of electrons, atoms must radiate not only electromagnetic but also gravitational energy, if only in minuscule amounts. Since this cannot be the case in nature, then it appears that the quantum theory must modify not only Maxwellian electrodynamics but also the new theory of gravitation
>
> ([Einstein (1916)], 696).[19]

In this case Einstein is clearly aware[20] of the potential clash between the theoretically predicted gravitational radiation combined with the empirically observable stability of atoms: *any* moving mass (even the electrons in atoms) will radiate gravitational energy (given the right kind of motion, that is). In other words, something like Planck's law of radiation would have to be found for gravity in order to account for the stability. He repeated this claim again in 1918, stating that "an improved version of quantum theory would lead to changes in the gravitational theory" ([Einstein (1918)], 167).

---

[17] Norbert Wiener, who would spend much time in Göttingen also seems to have become caught up in the general need for a harmonious structure at the foundations of physics. He writes in the second volume of his autobiography that "By, 1925...the world was clamouring for a theory of quantum effects which would be a unified whole and not a patchwork" ([Wiener (1956)], 105). Wiener's interest in gravitation might have been piqued by a chance encounter with Einstein on the train between Frankfurt and Basel, July 24th, 1925. It seems Einstein had discussed his new reduction of gravitation and Maxwell's equations to a single minimization problem. He also notes that, during a conversation on the train journey they shared, Einstein said that "He does not expect relativity in its present form to last many decades, and hopes that further work will soon go beyond it"—though he also mentioned his dissatisfaction with quantum theory it seems (letter to Bertha, MIT archive). We return to Wiener's views in the next chapter.

[18] As Kragh has pointed out, the version of quantum theory that Einstein would have been thinking about at this early phase of general relativity's development was precisely the Bohr–Sommerfeld theory—see [Kragh (2000)], 965. Einstein would have been particularly impressed with the way the Sommerfeld theory integrated (special) relativity and quantum theory. Helmut Rechenberg claims that Sommerfeld published his results after Einstein informed him that, as one might expect, general relativity would not modify the results in any appreciable way ([Rechenberg (1995)], 160).

[19] "Gleichwohl müssten die Atome zufolge der inneratomischen Elektron-enbewegung night nur electro-magnetische, sondern auch Gravitations-energie ausstrahlen, wenn auch in winzigem Betrage. Da dies in Wahrheit in der Natur nicht zutreffen dürfte, so scheint es, dass die Quantentheorie nicht nur die Maxwellsche Elektrodynamik, sondern auch die neue Gravitationstheorie wird modifizieren müssen." The clarity with which this passage expresses, more or less, the modern problem of quantum gravity accounts for my beginning the history of quantum gravity proper in 1916 (as indeed have other authors, such as Gorelik, op. cit., and John Stachel [Stachel (1998)]). This idea of Einstein's implanted for a long time the belief that seeking an interplay of gravity and quantum was a *necessary* task.

[20] Though apparently not too troubled. In a letter dated July 19th, 1916 he writes breezily to his friend Heinrich Zangger ([Buchwald *et al.* (2006)], Vol. 8, 237a, 25) after just completing this work (and his contemporaneous work on the quantum theory of emission and absorption of radiation), showing no signs of real concern over the fate of general relativity—though it is also very possible that his mind was preoccupied with the breakup of his marriage at this point.

This looks like a potential *empirical* motivation for pursuing quantum gravity. However, as Gorelik correctly points out, whilst atomic radiation (computed along the lines of Maxwell's theory) leads to the collapse of the atom in (order of) 10^{-10} seconds (a fact inconsistent with observations), atomic *gravitational* radiation, computed using Einstein's formula, has a collapse time of the order of 10^{37} seconds. Therefore, there would in fact be no empirical inconsistency as a result of gravitational radiation and we should not be puzzled by the stability of atoms in this case.

[Gorelik (1992), p. 365] argues that an "analogy with electrodynamics" lay behind this comment of Einstein's. This analogy was a persistent feature of early research on quantum gravity, as we will see in Chapter 5. One must also bear in mind that the issue of absorption and emission of radiation must have occupied a central place in his thinking at the time of writing, for his paper on the emission and absorption of radiation in quantum theory appeared very shortly afterwards—replete with the statement that "it seems no longer doubtful that the basic idea of quantum theory must be maintained." What is remarkable, given what we know of the certainty he professed about general relativity, is that he openly considered the possibility that the quantum theory would demand some kind of 'modification' of general relativity![21]

Quantum theory was invoked several times (in discussions of general relativity, and unified field theories) to mark some kind of boundary of the *applicability* of a theory.[22] Einstein himself expressed just this view, in a lecture entitled "Ether and the Theory of Relativity" at the University of Leyden, in October 1920. This address is interesting for many reasons, historical and philosophical. For our purposes it is interesting because Einstein once again speculates on the possible restrictions that the quantum theory might place on general relativity:

> Further, in contemplating the immediate future of theoretical physics we ought not unconditionally to reject the possibility that the facts comprised in the quantum theory may set bounds to the field theory beyond which it cannot pass.

Indeed, we can find several examples of Einstein expressing this kind of sentiment. Inasmuch as his comments (here and in his 1916 paper) have been investigated by historians, it has tended to be in the context of the study of gravitational waves. It is true that gravitational waves are naturally involved here, but since Einstein is considering the possibility that the radiation of such waves is quantized, we ultimately have what can also be seen as heralding the beginning of research investigating the possible *quantization* of gravity.

---

[21] This openness of Einstein to the possibility of a quantum theoretical modification of general relativity would not last for long, of course, and was already beginning to sour at this stage. His taste for quantum theory soon soured to the extent that towards the end of his life he was searching for ways to reproduce quantum mechanical phenomena using a purely classical field theory. Suraj Gupta (who developed a special-relativistic theory of quantum gravity in the 1950s) has a different (inverted) interpretation of Einstein's underlying reasons for distrusting quantum mechanics: "Because his theory is different from other field theories, he tried to construct unified field theories and because he could not see how his theory in the curved space could possibly be quantized, he criticized quantum mechanics" ([Gupta (1962)], 253).

[22] For example, Goldstein and Ritter note how Weyl adopts this position in his *Raum, Zeit, Materie* ([Goldstein and Ritter (2003)], 104).

3.4   Quantum Meets Gravity in the Pages of *Nature*

The pages of *Nature*, in the period immediately following the construction of general relativity, were littered with a variety of suggestions involving some kind of connection between gravitation and quantum phenomena.[23] For example, the period following the well-publicized 1919 observation to test Einstein's predicted value for the amount of deflection of starlight by the Sun resulted in a steady flow of papers probing the possible relationship between quantum theory and general relativity. This is a fairly natural line of inquiry given the context since by this time light was, of course, understood in quantum theoretic terms and since gravity was having a direct effect on the propagation of light, it follows that there must be some relation between gravitation and quantum systems. The atomic theory of matter based on quantum theory was becoming established, which further deepened the need to consider the connection between gravity and matter in this form.

This episode is of wider historical interest since in many cases the articles were part of their authors' wider agenda, be it the unity of nature, a distaste for relativity, adherence to the axiomatization programme, or some other underlying motivation. Indeed, what is is striking about the issues of *Nature*, in and around our chosen period, is that there is seen to be no real division between the sciences, and certainly not between atomic physics and gravitational physics. We work in a largely chronological fashion, except where there are thematic links across years.

3.4.1   Eddington on Fundamental Length

Eddington discussed the relationship between gravitation and quantum from the period following the creation of general relativity to the end of his life.[24] Eddington very frequently refers to quantum theory in the context of gravitation and vice versa. In fact, he begun to consider the relationship between gravity and quantum at least a year *before* the deflection observation. It is highly likely that it was as a result of his (and Einstein's) work on gravitational waves that he was initially led to think about the problem for, as we have seen, Einstein had already contemplated the potential clash between quantum theory and general relativity as a result of his own work on gravitational radiation. Indeed,

---

[23] To reiterate what was said in the introduction, I am not solely focusing on proposals that aim to *unite* quantum theory and gravitation in a common framework. Rather, I am concerned with showing how quantum theory and quantum phenomena and general relativistic phenomena and general relativity did not occupy isolated conceptual schemes in the minds of physicists at this time but were very much intertwined. This often manifests itself in ways that have persisted into modern thinking on the problem of quantum gravity, such as the notion that merging quantum theory and general relativity might serve to resolve some internal problem with one or another ingredient theory. However, I also include less obviously interesting examples indicating merely that the problem of linking the two theories together was 'in the air' so that the later work of Rosenfeld, for example, is seen is a fairly natural problem to focus on.

[24] This quest has been discussed in some detail by several authors. Clive Kilmister (a student of a student of Eddington, namely George McVittie) focuses directly on Eddington's concern with the relation between gravity and quantum theory [Kilmister (1994)]. Ian Durham [Durham (2003)] focuses more on Eddington's desire to achieve an *objective* account of physics independently of human measures.

there are elements of Eddington's writing, in discussing the gravitational red shift, as far back as 1916 that suggest an emerging awareness of quantum theory's relevance: "The vibrations of an atom must be slower in an intense field, so that the lines of the solar spectrum should be displaced slightly to the red as compared with terrestrial spectra" [Eddington (1916), p. 330].

The second instalment of this article (appearing in 1918) shows an even greater appreciation of the relationship. Certainly, one of the more remarkable things that emerges from Eddington's early work on general relativity is his claim that a fundamental length can be formed from the three basic universal constants, and that this length will inevitably form a piece of the future theory blending quantum and gravity:

> From the combination of the fundamental constants, G, c, and h it is possible to form a new fundamental unit of length $L_{min} = 7 \times 10^{-28} cm$. It seems to be inevitable that this length must play some role in any complete interpretation of gravitation.... In recent years great progress has been made in knowledge of the excessively minute; but until we can appreciate details of structure down to the quadrillionth or quintillionth of a centimetre, the most sublime of all the forces of Nature remains outside the purview of the theories of physics.
>
> (March 14th, 1918, 36)

This is a remarkably prescient passage; though it appears somewhat clumsily disconnected from the rest of the article, appearing as the final paragraph. In it Eddington has clearly targeted what we now label 'the Planck length,' $\sqrt{hG/c^3}$.[25] This length is, of course, a fairly generic feature of all modern approaches to quantum gravity.[26] That Eddington believes this length to *inevitably* play a role in a future interpretation of gravitation plainly implies that he sees quantum theory as essentially bound up with the physics of gravitation.

---

[25] Note that the value he derives is some six orders of magnitude off from the value we have today (namely $6.6 \times 10^{-34} cm$). However, in his *Report on the Theory of Relativity* (also from 1918), he gets the correct order, 4×10^{-33} cm. Though there is no suggestion relating this fundamental length to such things as gravitational contraction and so on, it nonetheless stands as a limit on spatiotemporal events imposed by the three core theories of physics. Note also that Eddington does not mention Planck's name, so one might wonder whether he came to the result (of a system of unique scales from combining these three constants) independently of [Planck (1899)]. (Of course, the idea that there is a smallest length would later become a common feature of quantum gravitational physics—Helge Kragh [Kragh (1995)] provides a useful study of the earliest work (based on quantum considerations) on the notion of a minimal length (see also [Carazza and Kragh (1995)].) I might also add that Eddington is a missing figure from Gennady Gorelik's [Gorelik (1992)] otherwise superb recounting of the history of the role of the Planck units in the early history of quantum gravity research.

[26] Of course, these units derive initially from Planck's system of 'absolute units' [Planck (1899)]. But in that paper Planck does not link this to any synthesis of quantum theory and gravitation, nor did he suggest that the absolute unit of length imposed any lower limit on objects and processes. He was, rather, impressed with their *independence* from the usual conventional elements involved in 'terrestrial' units. That the units are just 'pure numbers' encoding the laws of physics must have impressed Eddington, for just this connection would form the basis of his later (near-numerological) work on deriving the laws of physics from such pure numbers, in his last book *Fundamental Theory*, for example. I suspect (though it is not the place to discuss it here) that this early realization about absolute units and their relationship with (objective descriptions of) physical laws might have played a greater role in Eddington's later work than has previously been realized.

In the paragraph immediately preceding the above quoted paragraph, Eddington states (again, rather presciently) that "we know that in consequence of the undulatory theory of light, a ray traversing a heterogeneous medium always takes the path of least time; and one can scarcely resist a vague impression that the course of a material particle may be the ray of an undulation in five dimensions" [Eddington (1918), p. 36]. Eddington clearly has in mind here a notion of the particle as a 'projection' of a wave phenomena down from five to four dimensions (as in the later more well-known Kaluza–Klein theories). One might immediately latch upon the work of Nordström [Nordström (1914)] as a precedent for such 5-dimensional speculations; though, of course, Nordström's theory was based on a scalar theory of gravitation. However, given Nordström's isolation it is highly likely that Eddington was not aware of his paper.[27] It is quite likely that Eddington's offhand remark might have sparked higher-dimensional thoughts in Kaluza and others.

Joseph Larmor [Larmor (1919a)], in discussing the possible application of quaternions to general relativity, suggests something strikingly similar (again independently, it seems, of Nordström and Kaluza[28]), and in a way that makes projective notions more explicit. He labels it a "hyperspacial version of the Einstein gravitational theory" (p. 357). Larmor initially develops a symbolic geometrical calculus (devised by W. J. Johnston) to talk about electromagnetism in *flat* spacetime (i.e. with $\delta s^2 = \delta x^2 + \delta y^2 + \delta z^2 + (ict)^2$). However, he is concerned with introducing gravitation into his scheme, and notes that this can be achieved by introducing a new dimension ('preferably of space'–p. 353) ξ, such that

$$\delta s^2 = \delta x^2 + \delta y^2 + \delta z^2 + \delta \xi^2 + (ict)^2 \tag{3.1}$$

Since this includes electromagnetism too, an additional component is received by the vector potential. The idea is to have the physics of flat four-dimensional spacetime "as a hypersurface within our auxiliary flat five-dimensional scheme, in which both the electrodynamic and the gravitational theory shall exist". He develops this idea as follows:

> Now any continuum of four dimensions, having a quadratic line-element, however complex, is expressible as a hypersurface in this homaloid continuum of five dimensions. If these considerations are correct, the Einstein generalization, made with a view to include gravitation within his four dimensions, must be interpretable as the geometry of some type of hypersurface constructed in this extended homaloid of five dimensions. For the previous homaloid theory of Minkowski which ignored gravitation, this hypersurface, existing in the five dimensions, in which the world-process is represented, is flat; or more conveniently in some connections it may be taken as a closed region (hypersphere) of

---

[27] *Cf.* [Halpern (2004)] for a discussion of the (lack of) impact of Nordström's proposal.

[28] Daniela Wuensch points out that Larmor's paper appeared before Kaluza's, but argues that because Larmor had used flat spacetime it didn't excite much interest ([Wuensch (2003)], 526). I don't think this can be the right explanation: Larmor's approach involves flat Minkowski space only as a *projective* feature of the world. I find it more likely that the cumbersome nature of the approach, the novelty, and the heavily abstract formulation were more likely to be the reasons that the approach didn't catch on.

assigned uniform extremely small curvature, instead of the unlimited hyperplane. The problem then is to include in the scheme the influence—actually very slight in realizable cases—of gravitation; and this is to be done by recognizing slight local deformations on this hypersphere in order to represent that effect. *Now in the four-dimensional Minkowski map of the historical world-process, the rays of radiation are the curves of minimum length on the locus for which the analytic element of length δσ vanishes; and the paths of particles when gravitation was neglected were the curves (then straight lines in the flat) for which the length between assigned terminal points is minimum.* If the hypersurface, which is very nearly uniform of very small curvature in the actual problem as presented in nature, can be so chosen that these two relations persist—namely, that the rays of light shall be geodesics on the locus determined by δσ vanishing, and the free orbits of particles with gravitation now introduced shall be the paths of minimum length on the hypersurface—then one way of absorbing the universal phenomena of gravitation, into the mixed space-time scheme which has arisen from and has transcended and obliterated the previous idea of relativity of positions and motions, will have been accomplished. ([Larmor (1919a), p. 354] [the section "On Gravitational Relativity," from which this set of quotations were taken, was added by Larmor on November 20th, 1919]—my emphasis).

Larmor associates this idea of generalizing dimensions (and dealing with the properties of one as projections in another) with Clifford. These several proposals for 'dimensional expansion' indicate that when Theodore Kaluza formulated his five-dimensional formulation of gravitation and electromagnetism, he did so in a period when such speculations were not entirely rare. Of course, this idea of increasing the number of spacetime dimensions is a central feature of string theoretic approaches to quantum gravity.

3.4.2 Larmor's Paradox

In the Christmas day edition of *Nature*, 1919, Joseph Larmor drew attention to a potential conflict (a paradox, in fact)[29] between the quantum theory of light and the manner in which light is treated in general relativity in order to raise doubts about the veracity of the latter. As Larmor sees it ([Larmor (1919b)], 516), Einstein's general theory demands on the one hand (given an undulatory description of light) that the velocity of light will be diminished as it nears the Sun, but that "the scale of time" must undergo a compensatory expansion so that, overall, there is no change in wave-length. Larmor refers to such a notion of time as "heterogeneous time" and argues that given this notion (and given that space is almost flat) the path of a ray of light will be determined "fundamentally by minimum number of wave, and not by minimum time". In this case, claims Larmor, there ought (on kinematical grounds) to be no such deflection of light passing the Sun.

   However, Larmor then considers a dynamical explanation for the deflection test, drawing in Einstein's work on the quantum theory of light. According to this description, the velocity of light ought to increase, and (according to Einstein's theory, in just the amount observed):

---

[29] J. Sanchez-Ron [Sanchez-Ron (1995)] provides an investigation of Larmor's gripes with general relativity.

Dr. Einstein requires in another connection that light should consist of discrete bundles or *quanta* of energy. Let it also be granted that inertia and gravitation are attributes of energy. It seems to follow that each of these bundles of energy will swing round the sun in a hyperbolic orbit, and that its velocity will be *increased* when near the sun. It is well known that this would account for half the observed deflection. But, again, physical optics could not exist without the idea of transverse waves and their phases, which must be grafted on somehow to the bundles of energy.

[Larmor (1919b), p. 412]

Larmor can be seen to be clearly grappling with the puzzling 'wave–particle' nature of light, and seizes upon the opportunity of applying this puzzle to general relativity to render it less certain. Given this conflict, he argues that the recent deflection test conducted by Eddington should be looked upon as a "guide rather than a verification" ([Larmor (1919b)], 412). Of course, Einstein's own path would involve an engagement with just such issues. His approach was to consider the quantum, particulate aspects as merely an emergent phenomenon (as special solutions) from an underlying classical field theory. He had, moreover, already (by this stage) considered the possibility of a 'quantum correction' to general relativity.

3.4.3   The Cavendish Lab's Intervention: Radioactivity and Gravitation

An experimental venture into the interaction of the gravitational field with what were slowly becoming viewed as quantum properties of particles was conducted by Ernest Rutherford and Arthur Compton at the Cavendish Laboratory ([Rutherford and Compton (1919)]). It appeared in the same Christmas Day issue as Larmor's article discussed above. Their paper constituted a response to an article by 'Prof. Donnan' from the previous week's issue over the behaviour of radioactive substances in strong gravitational fields.[30] They note that pretty much the same question was put to them by 'Dr. Schuster' some years earlier. The problem considered was whether the intensity of gravitational field strength could modify the rate of transformation of various radioactive substances. Before they had a chance to put Arthur Schuster's question to the test,[31] the First World War intervened.

Rutherford and Compton (1919) bypassed the need to use a large heterogeneous mass to generate a suitable gravitational field by using the equivalence between gravitational acceleration and centrifugal acceleration. They therefore performed the experiment by placing radioactive substances at the edge of a rapidly spinning disc (generating

---

[30] F. G. Donnan deduced a relation between "the variation of mass in a physical change of state or chemical reaction and the rate of variation with gravitational potential of the corresponding change of total internal energy" ([Donnan (1919)], 392). Note, there was a general concern around this time with the source of stellar energy. It is thus fairly natural to consider the possible dependence of the rate of emission on the star's gravitational field.

[31] They got as far as a method of testing radioactive decay rates (over a period of up to 100 days), and were planning on sending their various radioactive substances to parts of the world with significant differences in their gravitational field strength.

20 000 times the strength of the Earth's gravitational field) and measured γ-ray rate responses (using a balance method), looking for (significant) discrepancies. However, as he noted, on the basis of Donnan's calculations, no change in rate was to be expected since, if it existed, the effect would be "very much smaller than can be detected by measurements of this character" [Rutherford and Compton (1919), p. 412].[32]

Note that their method followed Einstein's own suggestion, in his "popular account" of relativity, from 1916 [Einstein (1920)]. He considers a setup in which a clock is situated on a spinning disc, a certain distance γ from the centre. The clock's velocity, relative to a frame K at rest with respect to the moving clock, is $v = \omega \gamma$ (where ω is the angular velocity). Where v_0 represents the "number of ticks of the clock per unit time...relative to K when the clock is at rest," the rate of the clock when it is moving relative to K with velocity v (but at rest relative to the disc) is:

$$v = v_0 \sqrt{1 - \frac{v^2}{c^2}} = v_0 \left(1 - \frac{1}{c^2} \frac{\omega^2 \gamma^2}{2} \right) \tag{3.2}$$

He then considers the difference of potential of the centrifugal force between the clocks position and the disc's centre, written ϕ:

$$\phi = -\frac{\omega^2 \gamma^2}{2} \tag{3.3}$$

Which, on substitution into Eq. 3.2 gives:

$$v = v_0 \left(1 + \frac{\phi}{2} \right) \tag{3.4}$$

From which one derives the time dilation as a result of the centrifugal acceleration. The equivalence principle leads one to the result that an observer rotating with the disc will find themselves in a gravitational field with potential ϕ. This is then applied to an atom that is emitting spectral lines, which can be viewed by analogy with the clock. The expectation will then be that:

> An atom absorbs or emits light of a frequency which is dependent on the potential of the gravitational field in which it is situated.
>
> [Einstein (1920), p. 389]

But after considering a centrifugal (acceleration) example, Einstein then switches (for obvious practical reasons) to consideration of an atom on the surface of a heavenly body, noting that its frequency will be a little less that the frequency of the same element on a

---

[32] Sánchez-Ron [Sanchez-Ron (1992)] claims that Rutherford and Compton "did not make any effort to see whether or not their experimental results agreed with the predictions of general relativity" (p. 68). However, the previous remark clearly states that, for processes of such microscopic nature, it would be a practical impossibility to compare the experimental results with the theory beyond the very broad fact that no result is expected to be seen on the basis of general relativity.

smaller, less massive body. One could test this with spectral lines originating on the Sun and the Earth respectively. Though there is no Planck's constant in this example, and so this is not by any means a quantum gravitational phenomenon, at root we are talking about something (spectral lines) that was central in discussions of the old quantum theory. In intervening in atomic frequencies, the gravitational field was surely intervening in quantum processes.

There were, then, some early experimental suggestions concerning the influence of gravity on elementary processes, but these quickly died out. It is possible that this was due to Eddington's theoretical calculations and these null results from the Cavendish Lab.[33] However, it was, of course, already known that the gravitational effects on single atoms would be minuscule simply by inspecting the size of the gravitational constant. In this sense, the Cavendish lab's results merely confirmed what was already believed.[34]

Before we leave this section, we note that it seems that the debate discussed here was in many ways a direct continuation of an earlier one over the relationship between gravitation and temperature following experiment work by E. Shaw [Shaw (1916)]. This work stretches back to a period before general relativity was established and that remained largely independent of general relativity even when it did become better established. Shaw had conducted experiments in 1915, with a Cavendish torsion balance. These pointed to a positive temperature coefficient for the gravitational constant. Shaw's theoretical position was roundly criticized, not least by Oliver Lodge [Lodge (1916), Lodge (1917)] who pointed to several problems with momentum non-conservation and potential empirical inadequacies. However, Shaw's experimental work was positively received and it was this that filtered through into the later work on the relationship between radioactivity and gravitation.

3.4.4   Einstein on the Development of Relativity

Following an outline of how he arrived at the form of general relativity, Einstein concluded his account with a list of "important questions which are awaiting solution" [Einstein (1921), p. 784]:[35]

> Are electrical and gravitational fields really so different in character that there is no formal unit to which they can be reduced? Do gravitational fields play a part in the constitution

---

[33] However, the same question was tackled in 1942, with the benefit of new theoretical knowledge and improved experimental techniques (then able to produce centrifugal fields of 1 000 000g), by Freed, Jaffey, and Schultz [Freed *et al.* (1942)]—they explicitly cite the Cavendish results in their work. Even at these centrifugal field strengths, no effect was seen that could be distinguished from experimental error. For a discussion on how this early experimental research developed into the modern era of quantum gravity research, see [Gillies and Unnikrishnan (2002a,b)] (especially 127 of the latter).

[34] Curiously, J. Joly [Joly (1920)] returned to the general issue of the connection between rates of radioactive decay and the principle of relativity posing the question of whether radioactive clocks might offer an invariant way to measure absolute time, or whether "radio-activity [is] also 'in the conspiracy'" (p. 468). Clearly the centrifugal experiments were not sufficient to determine the answer one way or the other. Of course, it was only fairly recently that the question was answered and the effect of gravitational time dilation on atomic clocks was confirmed [Hafele and Keating (1972)].

[35] Interestingly, as we see in Chapter 8, John Wheeler's 'Geon' project would engage with all three of these questions.

of matter, and is the continuum within the atomic nucleus to be regarded as appreciably non-Euclidean?

As I understand it, here Einstein is, firstly, hinting at a unified field theory, through which both gravitational and electrical forces are described. On the basis of this, the question is begged as to whether the gravitational force plays any role in holding atoms together.[36] Finally, and most interesting from our point of view, it is natural to consider what kind of gravitational field would exist in the interior of an atom—though Einstein thinks directly in terms of what spacetime would look like inside atoms. This has a bearing on the other questions since unless the spacetime is appreciably curved, there will be no work for gravitation to do in the structure of matter. What can be reasonably inferred from this is that Einstein was considering the possibility that general relativity might have something to say about the constitution of matter, and *a fortiori* the nature of quantum theory. This is backed up by remarks that Einstein is reported to have said following a lecture at King's College London in 1921:

> After the public lecture Prof. Einstein was the guest of the Principal... In responding to his health, Prof. Einstein made an interesting revelation of his attitude to the quantum theory. This theory was, he said, presenting a difficult problem to physics, but the very nature of the difficulty served to bring into relief the attractiveness and satisfaction of the principle of relativity. That principle had served to give a simple and complete explanation of experimental facts which under any other aspect were discordant. In the quantum theory as it stood at present we were faced with discordant experimental facts, and were searching for the principle on which to interpret them.
>
> (Reported in *Nature* 107(2694), 504)

One inference to make here is that Einstein expected that the general theory of relativity itself might be able to supply such a missing principle on which to found a satisfactory theory of matter, quantum or not. However, Tilman Sauer (private communication) interprets the passage as more likely indicating that Einstein had in mind, not that the principle of relativity would itself serve as a *guide* for quantum theory, but rather that something *analogous* to the principle of relativity, though of a different sort, more relevant for quantum theory (such as the correspondence principle or the adiabatic principle), might be required to interpret the experimental data then available. However, given that, at this time, Einstein considered the *direct* role of gravitation in the constitution of matter as one of his most important questions in need of resolution, it is perfectly possible that he intended the former. Note also that the issue of the possible role of gravitation in the structure of elementary particles (especially as a regulator, taming infinities and so on), would become something of an *ideé fixe* in work from the late 1940s until the late 1950s.

---

[36] Recall that at this time, electromagnetism alone was thought to be responsible for the structure of matter.

3.4.5   James Jeans on Indeterminism in GR and QM

In 1926, James Jeans presented a curious argument suggesting that the "unpalatable determinism" brought forth by relativity (in the form of the 'block universe') might be somehow *cured* by developments in quantum theory. To modern ears, this might sound the wrong way around, since it is the indeterminism of quantum theory that is unpalatable to many.

It is clear that Jeans has in mind *fatalism* here, since he writes that "Einstein's work on relativity changed the universe from a drama into a picture drama" ([Jeans (1926)], 311). Clearly he supposed that the random nature of atomic processes could inject some much needed randomness in all processes:

> [R]elativity is not the whole of natural science; it is not even the whole of Einstein's work. His contributions to science fall into two columns which, unhappily, are parallel and show no signs of meeting. The first column contains his contributions to the theory of relativity,... the second column contains his contributions to the theory of quanta... It is not yet altogether clear which of these columns will figure most prominently in the history of present-day science when this is finally written in its proper perspective. But it already seems possible that the second column of Einstein's work may contain the needed antidote to the determinism and automatism to which the first column, if it stood by itself, would seem to condemn us [Jeans (1926), p. 311].

Jeans is clearly well aware of the *problem* of bringing together quantum and gravity—that is, of the task of saying something about the "meeting" between the "two columns"—however, the probabilistic nature of quantum theory was still not fully appreciated by the wider scientific community,[37] nor was the dynamics of general relativity (*qua* theory describing the evolution of geometrical data) understood. Indeed, Jeans apparently viewed the four-dimensional nature of Einstein's theory as its core distinguishing feature. Or, as he puts it, "the dynamical explanation of a gravitational force crumbled in the hands of Einstein" (ibid., 310).

However dated Jeans' specific comments may strike us now, his comments mark a very clear expression of the problem of quantum gravity as a potential *conceptual* (rather than 'merely technical') clash. It is also an early example of a proposal to *utilize* one of the ingredient theories of quantum gravity in order to resolve some supposed problem with the other.[38] In this case it was the conceptual problem of the block-like nature of the universe according to general relativity and Jeans proposed that the theory of quanta might offer some assistance in changing the worldview "back into a drama."

---

[37] Jeans' examples are based on radiation, involving the disintegration of uranium and Einstein's own work on the emission and absorption of of light quanta (described by Jeans as "the statistics of their jumping about"—[Jeans (1926)], 311). With such phenomena, says Jeans, "we seem to be beyond the domain of...natural laws" (ibid.).

[38] Something that will occupy us in Chapter 7.

3.4.6 Klein on Five-Dimensional Quantum Relativity

Oskar Klein[39] came up with the idea of a five-dimensional approach while visiting Ann Arbor as instructor in theoretical physics at the University of Michigan. Klein began working on the approach in 1924, though returned to this initial foray when he returned to Copenhagen in 1925. He published a paper in *Nature* ([Klein (1926a)]) a little after the more well-known paper from *Zeit. Phys.* ([Klein (1926b)]), though both appeared in 1926. It was Pauli who, early in 1926, informed Klein that Kaluza had already published on a similar idea (*cf.* [Pais (2000)], 131). However, there's genuine novelty in Klein's approach in that Planck's constant emerges as a consequence of topological structure. Or, as Klein puts it, his result "suggests that the origin of Planck's quantum may be sought just in this periodicity in the fifth dimension" ([Klein (1926a), p. 516]).

This was a new development of the much older idea that geometry and topology could be used as a 'resource' in world-building. In fact, in his own later recollections of how he came to the five-dimensional idea Klein describes an approach broadly similar to that envisaged by Larmor and, earlier, Eddington (as described above). He notes how he was searching for "a wave background to the quantization rules" and had been playing with "the idea that waves representing the motion of a free particle had to be propagated with a constant velocity, in analogy to light waves—but in a space of four dimensions—so that the motion we observe is a projection on our ordinary three-dimensional space of what is really taking place in four-dimensional space" [Klein (1991), pp. 108–9].[40]

Klein was in discussions with Ehrenfest and Uhlenbeck during the spring of 1926, during a visit to Leiden at Ehrenfest's invitation, and the discussions were enough to lead to a paper on five-dimensional relativity by Ehrenfest and Uhlenbeck [Ehrenfest and Uhlenbeck (1926)], appearing around the same time as Klein's own note in *Nature* (*cf.* [Ford (2009)], pp. 9–10).[41] Ehrenfest had a long fascination with the concept of dimensionality long before Klein began thinking about his dimensional expansion. Ehrenfest [Ehrenfest (1917)] had written on the possible reasons for why space is three-dimensional, showing how various processes and the stability of orbits depend on it. This might well have been behind Klein's own suggestion that given that physical quantities would be periodic functions of the compact dimension and observables would be given as averages over the small circumference, ordinary space must be three-dimensional (see [Klein (1991)], 110).

---

[39] For a biographical sketch of Klein, see [Pais (2000), pp. 122–147].

[40] Klein claims that Bohr had earlier made similar suggestions [Klein (1991), p. 109]. However, he recalls his later discussions with Bohr and Heisenberg as being received with "kind skepticism" (ibid., 112).

[41] Klein notes ([Klein (1991)], 112) that Ehrenfest had asked Lorentz to invite Klein (on a Lorentz Fellowship) after having read a copy of Klein's paper that was given to him by L. H. Thomas (himself passing through Leiden on his way from Copenhagen to Cambridge). Uhlenbeck refers to an unpublished paper between himself, Ehrenfest, and Klein (see Interview of G. Uhlenbeck by T.S. Kuhn on December 9th, 1963, Niels Bohr Library & Archives, American Institute of Physics, College Park, MD USA, www.aip.org/history/ohilist/4922_5.html). Uhlenbeck recalls that at the time he believed Klein had something like a theory of everything: "it seemed then that one was very close to a world formula—one equation containing everything, you see. I remember that I had the feeling that 'Golly, we now perhaps know everything"' (ibid.)—though he notes that the same was not true of Ehrenfest.

As with several other approaches mentioned in this chapter, Klein's work on five-dimensional relativity might be seen to fall somewhat outside of the category of quantum gravity. However, also as with many of the other such approaches discussed, the influence of the work on later quantum gravity research cannot be underestimated. Further, it shows how, in some sense, the shape of space (a feature dynamically determined within general relativity) can determine what would be otherwise inexplicable features of the world (in this case the existence of a quantum of action). Of course, it was already known following general relativity that geometry offers a potentially exceptional explanatory resource, but Klein's work showed that this resource was more widely applicable than previously supposed.[42]

As mentioned above, Klein's approach was closely related to the earlier efforts of Theodore Kaluza [Kaluza (1921)], and Klein begins by outlining Kaluza's approach. Kaluza himself was inspired by Weyl's earlier modification of the metric of general relativity so as to have a total metric that could account for (what was then) "all physical phenomena." The introduction of a spacetime fifth dimension was necessary in such a theory since in four dimensions the only Christoffel symbols that are available are those of the gravitational field. Kaluza imposed a 'cylinder condition,' effectively eliminating variations with respect to the fifth dimension by regarding x^0 derivatives as having very small or zero magnitude.

Though Kaluza's approach was purely classical, he does conclude his 1921 article with a consideration of its *microscopic* significance:

> [M]atter, in its fundamental constituents at least, is not weakly charged; in the words of H. Weyl its "macroscopic placidity" stands in sharp contrast to its "microscopic turbulence," and this is true in particular for the new coordinate x^0: for the electron or H-nucleus the quantity $\frac{\rho_0}{\mu_0}$ [the ratio of charge density to rest-mass density, or the "specific charge" of matter] and with it the "velocity"-component is anything but small! In the form demanded by Approximation II [very small specific charge] the theory can describe most macroscopic phenomena and the key question is whether it can be used for the above elementary particles.
>
> If one tries to describe the motion of electrons by geodesics in R_5 one encounters immediately a difficulty that threatens to destroy the whole structure. The problem is that, if one takes the earlier assumptions literally, the fact that $\frac{e}{m} = 1.77 \times 10^7$ (in lightseconds) means that the quantity u^0 is so large that the last term in $[\bar{v}^\lambda = \frac{dv^\lambda}{d\sigma} = \Gamma^\lambda_{\rho\sigma} v^\rho v^\sigma + 2\alpha F^\lambda_\kappa u^0 v^\kappa - \mathfrak{h}_{,\lambda}(u^0)^2 {}^{43}]$, instead of disappearing, takes a value much greater than is observed experimentally and becomes the leading term.... [I]t would seem to be impossible to proceed in the old manner without some new hypotheses.
>
> [Kaluza (1921), p. 57].

---

[42] This was taken to extremes in the 'topological tricks' of Wheeler and his group (see Chapter 8, in which an attempt was made to extract all properties of quantum particles from the geometry and topology of space).

[43] The $\mathfrak{h}$ term here is Kaluza's expression for the g_{00}th component (i.e. the 'corner potential') of the metric tensor; $d\sigma$ is a five-dimensional Riemannian line-element (given by $d\sigma = \sqrt{\sum \gamma_{ik} dx^i dx^k}$); $\alpha = \sqrt{\frac{\kappa}{2}}$, with κ the gravitational constant.

Kaluza's solution to the problem was to throw out the gravitational constant "so that gravitation would appear as a sort of difference-effect" (ibid.). This, Kaluza argues, would have as an "attractive feature" the fact that a "statistical role" could be attributed to the gravitational constant. He finishes by remarking that "for the moment the consequences of this hypothesis cannot be foreseen; and of course there are other possibilities to consider. And threatening all universal hypotheses is the Sphinx of modern physics, the quantum theory" [Kaluza (1921), p. 58].

Klein focuses directly on Kaluza's 'Sphinx,' on the microscopic description. He also diverges from Kaluza in assigning a definite scale to his x^0 and treating it realistically. The approach involves the establishment of a link between Kaluza's unified theory and the, then, new work on quantum mechanics of de Broglie and Schrödinger—we briefly return to this in the next chapter. He characterizes Kaluza's approach as a unified theory in which the unification is achieved via the coefficients γ_{ik} of the five-dimensional Riemannian line-element mentioned in footnote 43. Klein shows how one can view the equations of motion for charged particles propagating in an electromagnetic field (constructed by Kaluza) as radiation equations (that is, according to which matter is a wave-phenomenon, following de Broglie's idea). When this viewpoint is adopted, a generalization of the Schrödinger wave-equation follows. Restricting to a class of solutions in which the fifth dimension has a period related to Planck's constant, then, Klein argues, one finds that the standard quantum mechanical laws drop out. Hence, one has a unified theory of electromagnetism and gravitation, *and* one has an elementary notion of quantum theory that appears *as a consequence* of the theory. In a sense we find in Klein's approach geometry being used as a *resource* in the construction, deduction, or explanation of other puzzling phenomena.

3.5   On the Way to Quantum Geometry

The relationship of gravity (and indeed general relativity) to the phenomena revealed by quantum theory was used strategically by those who opposed relativity around the 1920s. Oliver Lodge, for example, thought that "if posterity is forced to accept and employ devices...for dispensing with the ether I fear that a damaging blow will have been dealt at physics" [Lodge (1919), p. 62].

However, Lodge signalled an early warning for those who might wish to link up the discontinuity of matter with spacetime:

> May I parenthetically urge philosophers to be on their guard against any system which introduces discontinuity into space or time, or even energy? Matter is discontinuous, electricity is discontinuous, I venture to say that real number is discontinuous; but space and time and ether are continuous. Energy may acquire a discontinuous aspect in its relation with matter, and the quantum is an important metrical fact, but it is explicable in terms of the atom or electron, and is not a feature in energy itself. Time is absolutely continuous, however it be measured and expressed numerically.
>
> [Lodge (1919), p. 62].

Developments from as early as the 1930s—that have persisted to the present day, packaged in the concept of "quantum geometry" (some of which we will discuss in Chapter 7)—would follow just the path attacked by Lodge, himself too strongly committed to classical physics and the ether theory to budge.

An even earlier statement of potential short-distance gravitational distortions was given in an editorial of 1919:

> If the distortion of space were very great, the customary methods of dynamics might lose their significance; and the question arises: Will, on Einstein's theory, the space inside an atom be so far from Euclidean that ordinary dynamical methods are unjustifiable? The answer to this question is "No." There are two lengths which have special significance in connection with the atom; one of these is what we call the radius of the atom, and is of the order 10^{-8}cm; the other we call the radius of the electron, and is about 10^{-13}cm. Even at the smaller of these distances the gravitational potential due to the mass of the atom, and therefore the distortion from Euclidean space, would be exceedingly small compared with the corresponding quantities due to earth at its surface, so that there is no special distortion inside the atom, except at distances from the centre which are infinitesimal even when compared with the radius of an electron.
>
> (*Nature*, December 4th, 1919, p. 362)

Not unrelated is the suggestion made by Norman Campbell in 1921, arguing that better sense could be made of the interior of atoms if the distinction between stationary and moving electrons were abolished by arguing that time ceased to make sense in the interior of atoms[44]:

> The suggestion that I made is that, by means of a generalized principle of correspondence, the distinction between moving and fixed electrons might be abolished and the conceptions that have proved so fruitful in explaining spectra made available immediately for explaining also such things (if there are such things) as are only explicable by fixed electrons. Thus the distinction would be abolished if "time" had no meaning inside the atom. For the difference between electrons following an orbit and electrons fixed at points on that orbit can only be expressed in terms of temporal conceptions; if all such conceptions are totally invalid in dealing with problems of atomic structure the distinction vanishes.
>
> [Campbell (1921), p. 170].

One wonders whether such suggestions could have been conceptualized in this way were it not for the parallel debates in the context of general relativity.

The gravitational field of the electron (though classically conceived) received an interesting early, speculative treatment by G. B. Jeffery [Jeffrey (1921)]. As with many other papers we have already discussed, Oliver Lodge [Lodge (1921)] thought fit to comment on this paper of Jeffrey's in *Nature*. Jeffrey's paper concerned the gravitational

---

[44] Hence, this is far more radical than Einstein intimated in his question over whether the interior of atoms is non-Euclidean.

field's possible influence on the structure of the electron. George Barker Jeffery (May 9th, 1891—April 27th, 1957) was one of the few physicists to directly acknowledge Einstein's remarks about the probable modifications that quantum theory would bring about in the theory of gravitational radiation.[45] The main result of the paper is not entirely novel, though he appears to have discovered his results independently: Jeffery himself notes that he had been unaware until publication that Nordström had only just derived similar results to his. The central outcome is an extension of Schwarzschild's line element (of space surrounding a point) to the combined electromagnetic and gravitational case (considering particles with charge and inertial mass), done by modifying the γ-term as:

$$\gamma = 1 - \frac{2\kappa\,m}{c^2 r} + \frac{\kappa\,\epsilon^2}{4\pi\,c^4 r^2} \tag{3.5}$$

In this case the point is a singularity of both fields (i.e. there are two singularities), rather than the gravitation field alone:

$$ds^2 = -\gamma^{-1}dr^2 - r^2 d\theta^2 - r^2 sin^2\theta\,d\phi^2 + \gamma\,c^2 dt^2. \tag{3.6}$$

Oliver Lodge again translates this, as in other cases, as an attempt to "ascertain something about the state of the aether close to an electron" ([Lodge (1921), p. 392]), which indicates something about the transitionary state of physics at the time. But he also rightly points out that the paper involves thinking about the status of fields near to points, and the question of whether the elementary particles really have structure, which itself leads to the additional question of radiation's interaction with elementary particles in such cases (especially as regards absorption of radiation by elementary particles). As he puts it: "a study of what happens to radiation when it impinges on, or penetrates between the ultimate elements of matter—in fact, a study of the whole behaviour of a stream of radiation at its two ends, the source and the sink—is obviously of great importance" (ibid.). Such reasoning is hardly shot down by the act of bringing those 'ultimate elements of matter' in line with quantum principles.

   As mentioned, the ultimate aim of Jeffery's paper was to show how the gravitational field might be involved in the structure of the electron, with the conclusion that the electrical and mass potentials would offer some kind of stabilizing effect by opposing each other.[46] This brings him to Einstein's earlier rather pessimistic remarks about the

---

[45] Like Eddington, Jeffery was a Quaker, and had spent time in prison in 1916 as a conscientious objector—Einstein was aware of this, writing wittily in his letter to Jeffery that it "is a highly welcome fact that a considerable portion of England's learned world upholds the pacifist ideal" ([Kormos-Buchwald et al. (2009), p. 85]).

[46] This, he thought, might be the result of an analysis of the two-body problem in his scheme: a subject for future work. Of course, this problem (exact solutions for the gravitational two-body problem for two point singularities) would ultimately have to wait for new methods in the initial value problem and numerical techniques—though the Einstein, Hoffmann, Infeld paper ([Einstein, Infeld, and Hoffmann (1938)]) is precisely along these lines and, as we will see in Chapter 7, Peter Bergmann used this method precisely for its ability to avoid interaction divergences. Curiously, Jeffery argues that without including electric charge, the point singularity blows up so that one has "not a solution with a single point singularity, but a solution with a point singularity surrounded by a spherical surface of singularity" (p. 131). Adding charge eliminates all zeros

fate of general relativity at the hands of quantum theory, regarding the instability of the atom in the face of continuous classical gravitational radiation. Jeffery believed that the problem could be evaded. Einstein was not convinced by Jeffery's idea:

> I unfortunately cannot share your optimism regarding the solution to the quantum problem. I believe that the theory of relativity does not bring us a step closer, at least in its current form. I am convinced that the two-body problem will not lead to a discrete manifold of paths but to a continuous one. (Einstein, letter to Jeffery [Berlin], March 18th, 1921: ([Kormos-Buchwald et al. (2009), p. 85]))

But, as we have seen, Einstein had already started down a path that followed the spirit if not the letter of Jeffery's approach (namely, using GR, or some modification of it, to recover quantum behaviour). In other words, Einstein's earlier assumption about quantum restrictions of gravitation was replaced by the view that quantum phenomena are to be derived from general relativity, and so not fundamental. However, here Einstein does not indicate this aspect of his thinking to Jeffery, and indeed the quote above looks largely negative as far as the entire project of getting anything quantum from relativity goes.

Though Einstein was followed by several others along the 'unified field theory' path, the majority view was that his earlier pessimism should and could be responded to without rejecting quantum theory, as we see in the subsequent chapters. In the next chapter, the two ingredient theories are worked out in more detail along exactly the lines of figuring out *how* one can have generally relativistic principles and quantum principles co-existing, at least in a formal-structural sense. These are essentially extensions of the kind of examination Jeffrey provides, though with quantum matter instead of the classical electron.

Uhlenbeck and Goudsmit's [Uhlenbeck and Goudsmit (1926)] introduction of the hypothesis of quantized angular momentum of electrons (to explain several puzzling results in quantum theory and experiment) radically altered the landscape in dramatic ways, both for the quantum theory and for the kinds of models needed in general relativity. Eddington [Eddington (1926)] discusses a potential conflict between the spinning electron hypothesis and relativity theory. He notes that some have been perplexed by what seems like a straightforward conflict between relativity's prohibition of superluminal velocities and the fact that the electron's periphery apparently moves at just such velocities. Eddington dissolves the perplexity in two ways: firstly, the prohibition applies to the propagation of *signals*, but clearly no such signalling is possible by utilizing the electron's angular velocity. Secondly, the spin is a quantum number: it represents, as Eddington says, "a state of the world" (ibid., 652). Finally, he notes that the idea that

---

and infinities other than that occurring at $r = 0$. Even more curiously, Jeffery speculates that given the light deflection features of general relativity "it would seem to be not impossible that a ray which passed sufficiently close to an attracting particle might be so strongly deflected that it would be permanently entrapped by the particle" (ibid.). Again, the introduction of charge is invoked to prevent such light sinks (it seems clear that something like a black hole is being suggested here).

the electron has a spacelike (superluminal) $\mathcal{J}^\mu$ vector was already postulated by Weyl in connection with his investigation into the relationship between gravitational and electrical fields, and was deduced purely from his action principle.

Again, this clearly points to the fact that the domains of the large and small, gravitational and atomic, were not seen to be disconnected in any fundamental way, but rather were part of a unity. The view that the world of the quantum and of gravity might well be a split-brained one came with later (failed) attempts to directly *quantize* general relativity.[47] We turn to the impact of 'the new quantum mechanics,' including wave mechanics, spin, and (especially significant in terms of establishing structural coherence between general relativity and quantum theory) Dirac's equation, in the next chapter.

3.6   Conclusion

In this chapter we have examined the very earliest work on the problem of quantum gravity (understood very liberally). We have seen that there was a very lively debate even in this early stage, and no suggestion that such a theory would not be forthcoming. Indeed, there are, rather, many suggestions explicitly advocating that an integration of quantum theory and general relativity (or gravitation, at least) is *essential* for future physics, in order to construct a satisfactory foundation. We have also seen how this belief was guided by a diverse family of underlying agendas and constraints, often of a highly philosophical nature. However, it must be borne in mind that the quantum theory in operation in the period < 1925 was 'the old quantum theory,' at this stage not yet in possession of a rigorous, meaningful framework—this, of course, is very much in line with the polysemicity issue of Chapter 1. The next chapter considers the impact of advances in the understanding and formulation of quantum theory (and, to a lesser extent, general relativity and cosmology) in the period ≥ 1925. These shifts highlight the enormously fragile nature of the problem of quantum gravity itself in its earliest phases.

....

REFERENCES

Ashtekar, A. (2005) The Winding Road to Quantum Gravity. *Current Science* **89**: 2064–75.
Buchwald, D., T. Sauer, Z. Rosenkrantz, J. Illy, and V. I. Holmes (2006) *The Collected Papers of Albert Einstein. Volume 10. The Berlin Years.* Princeton University Press.
Campbell, N. (1921) Atomic Structure. *Nature* **107**(2684): 170.
Carazza, B. and H. Kragh (1995) Heisenberg's Lattice World: The 1930 Theory Sketch. *American Journal of Physics* **63**: 595–605.

---

[47] See, e.g., [Ashtekar (2005)].

Corry, L. (1999) From Mie's Electromagnetic Theory of Matter to Hilbert's Unified Foundations of Physics. *Studies in History and Philosophy of Modern Physics* **30**(2): 159–83.

Corry, L. (1997) David Hilbert and the Axiomatization of Physics (1894–1905). *Archive for History of Exact Sciences* **51**: 83–198.

DeWitt, B. S. (1967) Quantum Theory of Gravity. I. The Canonical Theory. *Physical Review* **160**(5): 1113–48.

Donnan, F G. (1919) Heat of Reaction and Gravitational Field. *Nature* **104**(2616): 392–3.

Durham, I. T. (2003) Eddington and Uncertainty. *Physics in Perspective* **5**: 398–418.

Eddington, A. S. (1933) Physics and Philosophy. *Philosophy* **8**: 30–43.

Eddington, A. S. (1926) Spinning Electrons. *Nature* **117**(2949): 652.

Eddington, A. S. (1918) Gravitation and the Principle of Relativity II. *Nature* **101**(2524): 34–6.

Eddington, A. S. (1916) Gravitation and the Principle of Relativity I. *Nature* **98**(2461): 328–30.

Ehrenfest, P. (1917) In What Way Does It Become Manifest in the Fundamental Laws of Physics that Space Has Three Dimensions? *Proc. Amsterdam Acad.* **20**: 200–9.

Ehrenfest, P. and G. Uhlenbeck (1926) Graphische Veranschaulichung der De Broglieschen Phasenwellen in den Fünfdimensionalen Welt von O. Klein. *Zeitschrift für Physik* **37**: 895–906.

Einstein, A. (1921) A Brief Outline of the Development of the Theory of Relativity. *Nature* **106**(2677): 782–4.

Einstein, A. (1920) On the Special and General Theory of Relativity (A Popular Account). In *The Collected Papers of Albert Einstein. Volume 6. The Berlin Years: Writings, 1914–1917* [Doc. 42: authorized translation by R. W. Lawson, 1961] (pp. 247–420). Princeton: Princeton University Press.

Einstein, A. (1918) Über Gravitationswellen. *Sitzungsberichte der Königlich Preußischen Akademie der Wissenschaften Berlin*: 154–67.

Einstein, A. (1916) Näherungsweise Integration der Feldgleichungen der Gravitation. *Sitzungsberichte der Königlich Preußischen Akademie der Wissenschaften Berlin* **XXXII**: 688–96.

Einstein, A., L. Infeld, and B. Hoffmann (1938) The Gravitational Equations and the Problem of Motion. *Annals of Mathematics* **39** (1): 65–100.

Ford, G. (2009) *George Eugene Uhlenbeck, 1900–1988: A Biographical Memoir.* Washington, DC: National Academy of Sciences.

Freed, S., A. H. Jaffey, and M. L. Schultz (1942) High Centrifugal Field and Radioacive Decay. *Physical Review* **63**: 12–17.

Galison, P., M. Gordin, and D. Kaiser (eds.) (2001) *Quantum Mechanics: History of Modern Physical Science, Volume Four.* New York: Routledge.

Gillies, G. T. and C. S. Unnikrishnan (2002a) Experimental Perspectives on the Interplay of Quantum and Gravity Physics. In G. Bergmann and V. de Sabbata (eds.), *Advances in the Interplay Between Quantum and Gravity Physics* (pp. 103–22). Dordrecht: Springer.

Gillies, G. T. and C. S. Unnikrishnan (2002b) Quantum Physics-motivated Measurements and Interpretation of Newtonian Gravitational Constant. In G. Bergmann and V. de Sabbata (eds.), *Advances in the Interplay Between Quantum and Gravity Physics* (pp. 123–32). Dordrecht: Springer.

Goldstein, C. and J. Ritter (2003) The Varieties of Unity: Sounding Unified Theories 1920–1930. In A. Ashtekar et al. (eds.), *Revisiting the Foundations of Relativistic Physics* (93–149.). Dordrecht: Kluwer.

Gorelik, G. (1992) The First Steps of Quantum Gravity and the Planck Values. In J. Eisensaedt and A. J. Kox (eds.), *Studies in the History of General Relativity* (pp. 364–79). Boston: Birkhaüser.

Gupta, S. (1962) Quantum Theory of Gravitation. In P. Bergmann (ed.), *Recent Developments in General Relativity* (pp. 251–8). New York: Pergamon Press.

Haas, A. (1938) The Dimensionless Constants of Physics. *Proceedings of the National Academy of Sciences of the United States of America* **24**(7): 274–6.

Haas, A. (1919) Die Axiomatik der Modernen Physik. *Naturwissenschaften* **7**(41): 744–50.

Hafele, J. C. and R. E. Keating (1972) Around-the-World Atomic Clocks: Predicted Relativistic Time Gains. *Science* **177**(4044): 166–8.

Halpern, P. (2004) Nordström, Ehrenfest, and the Role of Dimensionality in Physics. *Physics in Perspective* **6**: 390–400.

Hilbert, D. (1918) Axiomatisches Denken. *Mathematische Annalen* **79**: 405–15.

Hilbert, D. (1915) The Foundations of Physics (First Communication). In R. S. Cohen et al. (eds), *The Genesis of General Relativity* (pp. 1925–38). Dordrecht: Springer, 2007.

Hu, B.-L. (2009) Emergent/Quantum Gravity: Macro/micro Structures of Spacetime. *Journal of Physics: Conference Series* **174**: 1–16.

Isham, C. J., R. Penrose, and D. W. Sciama (eds.) (1975) *Quantum Gravity: An Oxford Symposium*. Oxford: Clarendon Press.

Jeans, J. (1926) Space, Time, and Universe. *Nature* **2939**(117): 308–11.

Jeffrey, G. B. (1921) The Field of an Electron on Einstein's Theory of Gravitation. *Proceedings of the Royal Society of London. Series A* **99**(697): 123–34.

Joly, J. (1920) Relativity and Radio-activity. *Nature* **104**(2619): 468.

Kaluza, T. (1921) Zum Unitätsproblem der Physik. *Preussische Akademie der Wissenschaften (Berlin)Sitzungsberichte*: 966–72. Reprinted in L. O'Raifeartaigh (ed.), *The Dawning of Gauge Theory* (pp. 53–8). Princeton: Princeton University Press, 1997.

Kilmister, C. (1994) *Eddington's Search for a Fundamental Theory*. Cambridge: Cambridge University Press.

Klauder, J. (1969) Soluble Models of Quantum Gravitation. In M. Carmeli, S. Fickler, and L. Witten (eds.), *Relativity: Proceedings of the Relativity Conference in the Midwest, held at Cincinnati, Ohio, June 2–6, 1969* (pp. 1–18). New York: Plenum Press.

Klauder, J. and E. Aslaksen (1970) Elementary Model for Quantum Gravity. *Physical Review D* **2**: 272–6.

Klein, O. (1991) From My Life of Physics. In G. Ekspong (ed.), *The Oskar Klein Memorial Lectures* (pp. 103–17). Singapore: World Scientific.

Klein, O. (1926a) The Atomicity of Electricity as a Quantum Theory Law. *Nature* **118**: 516.

Klein, O. (1926b) Quantentheorie und Fünfdimensionale Relativitätstheorie. *Zeitschrift für Physik A* **37**(12): 895–906.

Kormos-Buchwald, D., Z. Rosenkranz, T. Sauer, J. Illy, and V. I. Holmes (2009) *The Collected Papers of Albert Einstein, Vol. 12*. Princeton: Princeton University Press.

Kragh, H. (2000) Relativity and Quantum Theory from Sommerfeld to Dirac. *Ann. Phys. (Leipzig)* **9**(11–12): 961–74.

Kragh, H. (1995) Arthur March, Werner Heisenberg, and the Search for a Smallest Length. *Revue d'histoire des sciences* **48**(4): 401–34.

Larmor, J. (1919a) On Generalized Relativity in Connection with Mr. W. J. Johnston's Symbolic Calculus. *Proceedings of the Royal Society of London. Series A* **96**(678): 334–63.

Larmor, J. (1919b) Gravitation and Light. *Nature* **104**(2617): 412.

Lodge, O. (1921) The Gravitational Field of an Electron. *Nature* **107**(2691): 392.

Lodge, O. (1919) Time, Space, and Material, II. *Symposium: Time, Space, and Material: Are They, and If so in What Sense, the Ultimate Data of Science? Proceedings of the Aristotelian Society, Supplementary Volumes, Vol. 2, Problems of Science and Philosophy*: 58–66. London/Oxford: Aristotelian Society/Oxford University Press.

Lodge, O. (1917) Gravitation and Thermodynamics. *Nature* **99**: 104.

Lodge, O. (1916) Gravitation and Temperature. *Nature* **97**(2433): 321.

Mie, G. (1912) Foundations of a Theory of Matter (Excerpts). In M. Janssen, J. D. Norton, J. Renn, T. Sauer and J. Stachel (eds.), *The Genesis of General Relativity, Vol. 4* (pp. 1554–619). Boston Studies in the Philosophy of Science, Volume 250, Part 8. Dordrecht: Springer, 2007.

Misner, C. (1969) Classical and Quantum Dynamics of a Closed Universe. In M. Carmeli, S. Fickler, and L. Witten (eds.), *Relativity: Proceedings of the Relativity Conference in the Midwest, held at Cincinnati, Ohio, June 2–6, 1969* (p. 55–80). New York: Plenum Press.

Nordström, G. (1914) Über die Möglichkeit, das Elektromagnetische Feld und das Gravitationsfeld zu Vereinigen. *Physikalische Zeitschrift* **15**: 504–6.

Pais, A. (2000) *The Genius of Science*. Oxford: Oxford University Press.

Pauli, W. (1921) *Relativitätstheorie*. Enzyklopädie der Mathematishen Wissenschaften. B. G. Teubner, Berlin, Vol. V. (English translation: *Theory of Relativity*. New York: Pergamon, 1958.)

Planck, M. (1899) Über Irreversible Strahlungsvorgänge. *Sitzungsberichte der Preussische Akademie der Wissenschaften Berlin* **5**: 440–80.

Rechenberg, H. (1995) Quanta and Quantum Mechanics. In L. M. Brown, A. Pais, and B. Pippard (eds.), *Twentieth Century Physics, Volume I* (pp. 143–248). Bristol: Institute of Physics Publishing.

Renn, J. (ed.) (2007) *The Genesis of General Relativity (4 Volumes)*. Dordrecht: Springer.

Rowe, D. E. (2003) Hermann Weyl, the Reluctant Revolutionary. *Mathematical Intelligencer* **25**(1): 61–70.

Rovelli, C. (2002) Notes for a Brief History of Quantum Gravity. In V. Gurzadyan, R. T. Jantzen, and R. Ruffini (eds), *The Ninth Marcel Grossmann Meeting On Recent Developments in Theoretical and Experimental General Relativity, Gravitation and Relativistic Field Theories. Proceedings of the MGIXMM Meeting. Held 2–8 July 2000 in The University of Rome "La Sapienza", Italy (pp. 742–768)*. Singapore: World Scientific.

Rovelli, C. (2000) The Century of the Incomplete Revolution: Searching for general relativistic quantum field theory. *Journal of Mathematical Physics* 4(6): 3776–801.

Rutherford, E. and A. Compton (1919) Radio-activity and Gravitation. *Nature* 104(2617): 412.

Sanchez-Ron, J. M. (1995) Larmor versus General Relativity. In H. Goenner, J. Renn, J. Ritter, and T. Sauer (eds.), *The Expanding Worlds of General Relativity* (pp. 405–430). Basel: Birkhaüser.

Sanchez-Ron, J. M. (1992) General Relativity Among the British. In J. Eisenstaedt and A. J. Kox (eds.), *Studies in the History of General Relativity* (pp. 57–88). Basel: Birkhaüser.

Sauer, T. (2002) Hope and Disappointments in Hilbert's Axiomatic 'Foundations of Physics'. In M. Heidelberger and F. Stadler (eds.), *History of Philosophy of Science* (pp. 225–237). Dordrecht: Kluwer.

Scholz, E. (2001) *Hermann Weyl's Raum-Zeit-Materie and a General Introduction to his Scientific Work*. Basel: Birkhaüser.

Schulmann, R, A. J. Kox, M. Janssen, and J. Illy (1998) *The Collected Papers of Albert Einstein, Volume 8*. Princeton: Princeton University Press.

Schweber, S. S. (2009) Weimar Physics: Sommerfeld's Seminar and the Causality Principle. *Physics in Perspective* 11: 261–301.

Shaw, P. E. (1916) The Newtonian Constant of Gravitation as Affected by Temperature. *Philosophical Transactions of the Royal Society of London, Series A* 216: 349–92.

Sommerfeld, A. (2000) *Wissenschaftlicher Briefwechsel. Band 1. 1892–1918*. M. Eckert and K. Märker (eds.). Deutsches Museum, Verlag für Geschichte der Naturwissenschaften und der Technik.

Stachel, J. (1998) The Early History of Quantum Gravity (1916–1940). In B. R. Iver and B. Bhawal (eds.), *Black Holes, Gravitational Radiation and the Universe* (pp. 525–534). Dordrecht: Kluwer Academic Publishers.

Uhlenbeck, G. and S. Goudsmit (1926) Spinning Electrons and the Structure of Spectra. *Nature* 117(2946): 264–5.

Wiener, N. (1956) *I am a Mathematician: The Later Life of a Prodigy*. Cambridge, MA: MIT Press.

Wuensch, D. (2003) The Fifth Dimension: Theodor Kaluza's Ground-Breaking Idea. *Ann. Phys. (Leipzig)* 12(9): 519–42.

4

The Shock of the New

New discoveries in quantum mechanics in 1925–1928 radically altered the nature of the problem of quantum gravity. Before this, as we have seen, there was a vague sense that gravitation might be modified by the discreteness of energy spectra and the transitions between states, and also that indeterminacy might have some role to play in gravitation. There was also some consideration of potential effects of gravity on properties of quantum systems. But there was no firm quantum framework established to enable consideration of the more formal issue of whether the respective structures could coexist in harmony, or whether there was instead some basic conflict. In other words, the *problem* of quantum gravity was manifest without yet having a means of formulating it in any concrete way, let alone pursuing genuine solutions. This is a fundamental compatibility question: special and general relativity were considered solid enough to demand—Einstein's own earlier pessimistic remarks notwithstanding—that any satisfactory theory of quantum matter should at the very least allow such matter to exist in a relativistic world.

There were two directions pursued along these lines: (1) studying the compatibility of the respective structures of quantum theory and general relativity (e.g., focusing on whether matter waves, spin and other quantum properties could coexist with general relativity, or at least curved spaces [including properties on a cosmological scale], and seeing if one could 'general relativize' wave mechanics or the Dirac equation[1]); and (2), coming later, investigating whether one could directly *quantize* the gravitational field itself. Another more conceptual question, related to (1), concerned whether there was a conflict between general relativity and the Copenhagen interpretation of

---

[1] As G. C. McVittie put it, in his review and extension of this line of attack, this work "enables us to take into account, in the Quantum Theory, the effect of gravitational fields on the properties of the electron" ([McVittie (1932)], p. 868)—indeed, McVittie is able to deduce features of gravitation (the red-shift behavior of spectral lines of light emitted by atoms in strong gravitational fields) from quantum mechanics; though he notes in conclusion that there are observable effects of the new work that the weakness of gravity would preclude experimentally verifying. Later work would, quite naturally, generalize this to particles other than electrons and, indeed, to a wide variety of spacetime structures—a fairly prolific Japanese school pursued this into the 1950s, considering, e.g., mesons in a variety of cosmologies, including de Sitter, by a generalization of the earlier wave mechanical ideas: [Goto (1951)], [Goto (1952)], [Goto (1954)] [Raje (1951)], [Ikeda (1953)], [Watanabe (1949)].

Covered with Deep Mist: The Development of Quantum Gravity (1916–1956). Dean Rickles, Oxford University Press (2020). © Dean Rickles.
DOI: 10.1093/oso/9780199602957.001.0001

quantum mechanics, as explored in the famous 1927 Solvay debate between Einstein and Bohr.[2]

Recall that Bohr's early quantum theory simply involved the classical theory of the electron (for a single electron in a hydrogen atom) with quantum conditions tagged on. Developments, starting with the matrix and wave mechanics of Heisenberg (and Born and Jordan) and Schrödinger in 1925 and 1926 respectively,[3] and then later, in 1928, with Dirac's equation for the electron, supplied a new physical framework, essentially by providing a new mathematical structure couched in non-commutativity (initially represented within the matrix language of Heisenberg, Born, and Jordan). There were discovered relatively simple rules for converting the dynamical variables of a classical theory into operators (obeying commutation rules), resulting in a physical picture associated with particle–wave duality (i.e. complementarity): particulate systems could be rendered, through complementarity, into a wavelike system and vice versa (i.e. so that one could view a classical field as particles, through the method of quantization).

Heisenberg and Pauli, in 1929, would then produce the first *field* quantization (for the complete electromagnetic field), supplying the tools and formal and conceptual framework for first considering an actual quantization of the gravitational field.[4] Hence, until 1929 the problem of quantum gravity was not (and *could not* be) understood at all in terms of the quantization of a classical field system, but in terms of the behavior of quantum objects (whose properties were then still being refined in radical ways, such as electrons) in a gravitational field (along with the two-way traffic between these). It is no surprise that these new radical evolutions in the conception of matter, energy, field, and particle would be tested against the then more firmly established—at least in the sense of a *completed* theoretical structure—theory of gravitation, not least given the universal coupling of gravity. Moreover, demonstrating such compatibility *a fortiori* demonstrates the invariance of the quantum theory or property of such. The present chapter will discuss quantum gravity's development prior to field quantization, focusing on the impacts of relevant major stages of the evolution of quantum theory, and also cosmology, during this time.

4.1   Enter Wave Mechanics

Heisenberg published his initial paper on what would become matrix mechanics, "Über Quantentheoretische Umdeutung Kinematischer und Mechanischer Beziehungen," on

---

[2] These approaches are largely concerned with whether one *could* harmonize gravity and quantum theory. Still later work would consider whether, in light of evidence to the contrary, there was reason to suppose that we *must* quantize the gravitational field at all. This point would become a major bone of contention at the Chapel Hill Conference on the Role of Gravitation in Physics (see http://www.edition-open-sources.org/sources/5/, especially Session 8).

[3] Heisenberg (1925), Born and Jordan (1925), Schrödinger (1926), Dirac (1926).

[4] Of course, this stage can hardly be considered as the *completion* of quantum theory, since there is still no notion of such basic modern principles as, e.g., renormalization that would once again transform the problem of quantum gravity—the latter also providing a much needed 'principle' to guide theory construction, as we saw in Chapter 2.

29th July, 1925. This, coming ten years after the creation of general relativity, was the essential first step in what we tend to think of as 'modern' quantum mechanics (or "the new" quantum mechanics in the terminology of Bohr), usurping the 'old quantum theory' of Bohr and Sommerfeld—though all the while giving firm quantitative expression of Bohr's correspondence principle linking classical and quantum. It was notoriously difficult to physically interpret, e.g. dealing in *transitions between states* rather than the states themselves (which are simply not represented in the formalism). The observables, rather than the states were the stars of the show.

Matrix mechanics wasn't around in isolation long enough to be pulled into the problem of quantum gravity,[5] followed shortly after its creation by Schrödinger's more intuitive wave mechanical approach, based on Louis de Broglie's 'ondes de phase' (phase waves), and published in March the year following Heisenberg's paper.[6]

Einstein's influence, especially his own views about the similarities between radiation and gases, were pivotal in the development of Schrödinger's wave equation—and indeed, it was Einstein who put Schrödinger onto de Broglie's work. However, though initially receptive, Einstein soon turned against wave mechanics. Central to this debate, and Einstein's interest, is the nature of physical space. Certainly, Einstein's eventual distaste for Schrödinger's wave mechanics stemmed from the fact that it involved waves in (multi-dimensional) phase space and so to him "does not smell like something

---

[5] It was hard enough to apply matrix mechanics to problems in atomic physics (*cf.* [Mehra and Rechenberg (1982)], p. 4), let alone gravitational systems. Though Jordan did almost immediately apply the method of matrix mechanics to the (free) electromagnetic field in 1925. (We might find an exception to the lack of matrix mechanics considerations of quantum gravity issues, very loosely understood, via Dirac's first 1926 paper on the Compton effect, in which he treats the time variable on an equal footing with other dynamical variables (leading him to the time-energy uncertainty relations) *on relativity considerations.* As Blum and Salisbury point out, this innovation bears a striking resemblance to his later usage of the parameter formalism in the field-theoretical context, which was directly relevant to quantum gravity research [Blum and Salisbury (2018), p. 459]—we return to this in the next chapter. This work of Dirac's, together with matrix mechanics, might also be viewed as joined up to the earliest work on non-commutative spacetime. In this approach, initiated by Hartland Snyder in 1947, spacetime coordinates are represented by matrices instead of numbers—we discuss this in Chapter 7.) Note that Heisenberg himself suggested a link between a non-commutative space scheme (with non-commuting coordinates) and the uncertainty relations—without a postulated universal length—in an early letter to Rudolph Peierls (13th June, 1930: [von Meyenn (1985), p. 16]).

[6] The two formulations were, of course, proved equivalent by Schrödinger shortly after his first paper outlining wave mechanics. Recall that de Broglie's approach, outlined in his 1924 PhD thesis, *Recherches sur la Théorie des Quanta*, involved associating a wave with each particle, with wavelength $\lambda = \frac{h}{p}$ (for a particle of momentum p). Schrödinger's wave mechanics then essentially provided an equation for such waves for which he attempted to provide a physical interpretation associating the eigen-oscillations of the wave system with stable energy states of an atom—Feynman's approach would then later simplify by dealing with the *solutions* to the wave equation given by path-integrals over wave histories. Schrödinger in fact had originally derived a special relativistic version of this equation (preempting the so-called 'Klein–Gordon equation'), but found empirical flaws when attempting to recover the fine structure of the hydrogen atom, understood later to be related to electron spin—indeed, Schrödinger was well aware that what was missing was "certainly Uhlenbeck and Goudsmit's idea" (letter to Lorentz, 6th June, 1926: [Przibram (1934), p. 73]). Recall that in 1925 Uhlenbeck and Goudsmit had already postulated the existence of a quantum version of angular momentum, later termed 'spin' by Pauli [Pauli (1927)], to explain the structure of spectral lines and their anomalous Zeeman effect, and this had been incorporated into matrix mechanics, though there were complications in making calculations in using this method.

real".[7] Schrödinger expressed similar sentiments with respect to the lack of visualizability of matrix mechanics (see, e.g., Moore (1989), p. 205).[8]

Born's statistical interpretation (suggesting a fundamental statistical randomness of outcomes, represented by the wavefunction)[9] was also to become a key part of the new quantum mechanics, and provided a serious deterrent for Einstein with respect to the new quantum mechanics. There is a step against a direct representation of physical reality in Born's manoeuvre (eradicating the physical pictures that appealed to Schrödinger), since now the wavefunction is related (as a probability amplitude) to the *probability* density $|\Psi|^2$ of finding outcomes (such as particles at specific positions). A further deterrent came from Bohr's analysis of the physical and epistemological interpretation of this new mechanics, as we will discuss in §4.3. 1927 would bring Heisenberg's uncertainty relations paper, as well as the transformation theories of Jordan and Dirac, piecing together the interpretation of Born's statistical and Heisenberg's uncertainty ideas in the contexts of wave mechanics and matrix mechanics respectively—though both noticed that the results of Born and Heisenberg were independent of wave or matrix specifics (see [Duncan and Janssen (2013)] for a good discussion of this episode). This work culminated, later that same year, in John von Neumann's Hilbert space formulation of quantum mechanics.

Gravity entered in two key ways during this phase: (1) in Einstein's debates with Bohr over the nature of objective reality; (2) in assessments of the structural coherence of general relativity (with its curved spaces) and quantum theory. Unlike Heisenberg's matrix mechanics, Schrödinger's view, *prima facie* at least, provided a fairly natural context in which to consider the gravitational field, viewing as it did particles as "a kind of 'wave crest' on a background of waves".[10] Schrödinger's view was based on an extension of William Hamilton's optical-mechanical analogy (cf. [Joas and Lehner (2009)]), linking particle motion with light rays. Since De Broglie had suggested that

---

[7] (EA 10–138) and (June 18, 1926, to Ehrenfest, EA 16–607) respectively (quoted in [Howard (1990)], p. 83). Note that Lorentz had expressed this point (and a preference for something like the wave equation in four-dimensional space) already in a letter to Schrödinger dated May 27th, 1926. He writes that: "If I had to choose now between your wave mechanics and the matrix mechanics, I would give the preference to the former, because of its greater intuitive clarity, so long as one has to deal with the three coordinates, x, y, z. If, however, there are more degrees of freedom, then I cannot interpret the waves and vibrations physically, and I must therefore decide in favour of matrix mechanics" ([Przibram (1934)], p. 48). (In his response, Schrödinger points out that he was well aware of the difficulty and thinks he has resolved it by assigning the physical meaning to a quadratic function of the wavefunction instead, interpreting the result as the electric charge density in real space: see [Przibram (1934)], p. 62).

[8] However, though there was an absence of visualizable elements, matrix mechanics shared some lineage with Einstein on a more conceptual and methodological level, based as it was on a focus on observables, much along the lines of Einstein's own reasoning about 'point-coincidences' forming the ontology of general relativity (cf. [Kragh (2000)], p. 967).

[9] This view was in opposition to Schrödinger's own preferred, more direct way of interpreting his wavefunction as electron density, as mentioned above.

[10] Quoted in Martin J. Klein, "Einstein and the Wave–Particle Duality, *The Natural Philosopher,* 1964, 3: 3–49 (p. 43). We might also note that this marks a clear difference between De Broglie's view and Schrödinger's; whereas Schrödinger desired a reductive, monistic theory, treating particles as waves, De Broglie sought a dualistic theory of matter waves and particles (cf. [Kragh (1984)], p. 1025).

matter could have wavelike properties too, Schrödinger reasoned that the optical analogy could be applied to these matter waves too, eventually coming up with a wave equation for the hydrogen atom:[11]

$$\nabla\Psi + \frac{2m}{K^2}\left(E + \frac{e^2}{r}\right)\Psi = 0 \tag{4.1}$$

This approach had the surprising implication that the integral values (the discreteness) of atomic systems, inserted by hand in matrix mechanics, falls out as a result of the wave representation, much as integral nodes emerge from vibrating strings.

As mentioned, Schrödinger had begun his work on wave mechanics by first writing down a relativistic version of the wave equation, which he had in his notebooks by New Year's Day, 1925 (see [Kragh (1981)], p. 33, footnote 7).[12] He didn't publish since he realized that it could not reproduce the correct experimentally observed energy levels (fine structure) of the hydrogen atom.[13] Perhaps this failure served as a deterrent from also pursuing a generally relativistic version of his wave mechanics—though, as mentioned, he did eventually turn to the generally relativistic Dirac wave equation in 1932, and indeed persisted with it from then on, even eventually discussing his wave mechanics in general relativity in the context of Eddington's ideas about the connections between the large scale structure of the universe (including its expansion: [Schrödinger (1939)]) and the wave equation. It is also possible that the problems with giving a direct spacetime interpretation, suggested by both Born's probabilities and Heisenberg's

[11] Here, Ψ is the wavefunction; m is the electron mass; e is the electron charge; r is distance between electron and nucleus in the hydrogen atom (where $r^2 = x^2 + y^2 + z^2$, and $K = h/2\pi$ (the reduced Planck's constant, chosen for agreement with empirical data: the Balmer series)). E is the total energy of the system which initially appears in Schrödinger's analysis as the solution of the Hamilton–Jacobi equation: $H\left(q, \frac{\partial S}{\partial q}\right) = E$.

[12] Schrödinger later attempted to provide an account of Dirac waves in a general, Riemannian space [Schrödinger (1932)]. According to McVittie, what Schrödinger achieved was to *define* "what is to be meant by an electron and an atom in a gravitational field" ([McVittie (1932)], p. 869). McVittie took this theory to be a crucial departure from the other unified field theories of gravitational and electromagnetism in that it generated genuinely novel results. There are features that can be deduced from it, that cannot be otherwise deduced from quantum mechanics—there might be some grounds to quibble with this statement, especially as concerns the interesting results such as charge quantization from Klein's five-dimensional theory. However, the point is well taken that in Schrödinger's approach there is the possibility for convergent deductions of known results in general relativity from quantum theory. McVittie also refers to the cosmological implications, such as the invariance of the frequency of emitted spectral lines from light as a result of the expansion or contraction of the universe. Linda Wessels has argued Schrödinger increasingly distanced himself from quantum mechanics (and focused on more cosmological matters) after Heisenberg's 1927 uncertainty relations revealed that a continuous picture of the kind he had in mind simply couldn't work [Wessels (1983), p. 265]. [Rüger (1988)] disagrees with Wessels, arguing instead that wave mechanics infiltrated his later work as part of a grand unification of atomic physics and cosmology (we side with Rüger—Schrödinger continued his researches of wave mechanics well beyond 1927. We return to this in the next and also the final sections of this chapter).

[13] Of course, this was the so-called "duplexity phenomenon" [Dirac (1928a), p. 610] resolved by Goudsmit and Uhlenbeck with the concept of spin angular momentum of a half, and then incorporated into quantum mechanics by Pauli [Pauli (1927)] with his matrices. However, it was difficult initially to come up with models that accounted for this phenomenon, rather than simply fitting it in an *ad hoc* manner. Dirac's equation did precisely this by combining relativity with his transformation theory.

uncertainty relations, of the waves pointed, in Schrödinger's mind, to too great a conflict with the physically robust and visualizable space of general relativity.

Others, however, were not deterred, and constructed similar schemes (for single spinless particles, and therefore suffering from the same defect as Schrödinger's initial attempt). The earliest attempts to construct a specially relativistic version of Schrödinger's equation by others (now usually called the Klein–Gordon equation) were in fact often linked with broadly general relativistic considerations.[14] Walter Gordon [Gordon (1926)], however, was led to the relativistic version of Schrödinger's equation by simply quantizing the relativistic version of the Hamiltonian for a classical point electron by replacing the appropriate terms by their operator versions—Gordon's target was the Compton Effect. Thus starting from:[15]

$$F \equiv \left(\frac{W}{c} + \frac{e}{c} A_0 \right)^2 + \left(\mathbf{p} + \frac{e}{c} \mathbf{A} \right)^2 + m^2 c^2 \tag{4.2}$$

one can simply make the substitutions:

$$W = ih \frac{\partial}{\partial t} \tag{4.3}$$

$$p_r = -ih \frac{\partial}{\partial x_r} \quad (r = 1, 2, 3) \tag{4.4}$$

In order to generate:

$$F\Psi \equiv \left[\left(ih \frac{\partial}{c \partial t} + \frac{e}{c} A_0 \right)^2 + \Sigma_r \left(-ih \frac{\partial}{\partial x_r} \mathbf{A} \right)^2 + m^2 c^2 \right] \Psi = 0 \tag{4.5}$$

One of the main problems with this approach is that it is not linear (in W), so that the wavefunction at some initial time will not determine it at any later time (cf. [Dirac (1928a)]). The other problem was that causing Schrödinger to reject his equivalent version: it neglects spin.

Oskar Klein (Klein (1926a)) developed his relativistic version of the equation in the context of the five-dimensional theory that we already met in the previous chapter.[16]

---

[14] Helge Kragh has studied the history of the Klein–Gordon equation in [Kragh (1984)]—Pauli referred to this equation as "the equation with many fathers," since not only Schrödinger, Klein, and Gordon, but also Théóphile de Donder and Frans Henry van den Dungen, Vladimir Fock, and Johan Kudar also discovered it—note that Dirac refers to the equation as the 'Gordon–Klein' equation in his 1928 paper on the electron, which also explicitly names Gordon and Klein as independent discoverers.

[15] Here, A_0 and $\mathbf{A}$ are the scalar and vector potentials of an electromagnetic field, W is the work function, and $\mathbf{p}$ is the momentum vector.

[16] Léon Rosenfeld, who would later write the first papers on the quantization of the gravitational field was also working on a relativistic five-dimensional wavefunction—this was carried out in Paris, simultaneously but independently of Klein's work. He eventually worked with DeDonder on this problem at the end of 1927, and indeed DeDonder introduced Rosenfeld and his five-dimensional work at the 5th Solvay Congress, which was enough to lead to an invite to work with Born in Gottingen (see [Jacobsen (2012)], p. 18).

I think we should take the links with general relativity as not quite as significant as has sometimes been suggested in the case of Klein's five-dimensional approach (and other kindred approaches from the same time). The link is somewhat indirect and amounts to a strategy for explaining away the problematic four-dimensional wave picture through the five-dimensional scheme. He derives the Schrödinger equation from a wave equation of a five-dimensional space with no Planck constant h inserted *a priori*, associating it instead with the periodicity in x^0, the fifth dimension—the small size of the length of the period he finds, with x^0's compactness explaining its empirical absence via a kind of averaging effect over it. This period was determined by a combination of Planck's constant and the gravitational constant, coming out close to the Planck length, 10^{-30} cm.

Klein's main aim was to gain a better understanding of quantum phenomena using a classical theory. This came on the back of Klein's previous work on ways of finding discrete, whole numbers in physics, using interference ideas, to match the new quantum phenomena, which Klein traces back to 1920.[17] He mixed this very early on with the idea of the waves propagating in a higher dimension (now with definite velocities), so that the four-dimensional world, with its indefinite velocities, is a projection: the paths of the particles would be geodesics of the five-dimensional space.[18] He also was close to Schrödinger's track, associating the quantum conditions with eigen-vibrations. He had the link with the Hamilton–Jacobi equation through Edmund Whittaker's book *A Treatise on the Analytical Dynamics of Particles and Rigid Bodies*. Following some dealings with relativistic particles moving in electromagnetic fields, he noticed an analogy between the Hamilton–Jacobi equation for this kind of motion and the Hamilton–Jacobi equation for ordinary particle motion in a space of a higher dimension. This pushed Klein more along the five-dimensional research program.[19] Hence, Klein was on a parallel track when Schrödinger's work came in 1925. However, Klein was fixed on a non-linear equation, and was fixed on the idea that one could not get a sensible theory in four dimensions. The solutions to his five-dimensional wave equation, however, enabled an interpretation in terms of matter waves moving in gravitational and electromagnetic fields in four dimensions which could then be viewed as particles.

Klein's 1927 paper (published in 1928) was rather more visionary in outlook. While he had given up hopes of using the five-dimensional theory to explain quantum theory, he pursued the idea that he might get a picture of quantum field theory out of the approach. Here Klein explicitly points to the conflicts apparent between quantum theory and general relativity: the wave–particle duality poses problems in giving meaning to the older spacetime concepts, grounded in terms of spacetime coincidences [Klein (1928),

---

[17] Interview of Oskar Klein by J. L. Heilbron and L. Rosenfeld on February 20, 1963, Niels Bohr Library & Archives, American Institute of Physics, College Park, MD USA: www.aip.org/history-programs/niels-bohr-library/oral-histories/4709–2.

[18] He credits Bohr with influencing his belief that a four-dimensional theory was impossible: Interview of Oskar Klein by T. S. Kuhn and J. L. Heilbron on July 16, 1963, Niels Bohr Library & Archives, American Institute of Physics, College Park, MD USA: www.aip.org/history-programs/niels-bohr-library/oral-histories/4709–6.

[19] Indeed, Klein claims to have learned relativity theory through this exercise, having known relatively little beforehand (ibid.).

p. 188]. It is quantum theory, then, rather than general relativity that is responsible for the "dissolution" of spacetime, and it is therefore the general theory of relativity that must adapt. Again, Klein invokes the five-dimensional formulation, now as a "natural starting-point for a general quantum field theory" (ibid., p. 190). Klein would return to this problem of a general quantum field theory in 1955 in his contribution to the Bohr Festschrift, though there is no mention of five dimensions, instead focusing on the incorporation of general invariance into quantum theory.

Dirac's equation dealt the death blow to Klein's work on the five-dimensional theory in 1928. As he put it:[20]

> I gave it up when Dirac's theory came. Before that I had tried to explain the spin with some kind of (tensor-vector) equation, but then, when Dirac's very nice solution came, I gave up these things for quite a number of years. I remember that Pauli and I drank some wine on the death of the fifth dimension in '28.

Klein points out (ibid.) that he also realized that his five-dimensional theory failed, like Schrödinger's radiation theory, to recover Planck's radiation formula (and only Rayleigh's)—i.e. it could not account for the black-body radiation formula at the very root of quantum theory. However, it seems that Dirac's equation closed this path off for Klein and, at the same time, terminated his grounding of wave equations in the context of general relativity (ibid.).[21]

Vladimir Fock (completed, July, 24th 1926: [Fock (1926)]) also employed a five-dimensional formulation, though he viewed the fifth coordinate as "superfluous," a useful parameter for setting up a relativistically invariant wave equation. Again, he produces a relativistic (generally covariant) version of the Schrödinger equation equivalent to the Klein–Gordon equation. Again, as with Klein, the equations of motion are given by geodesic motion in the five-dimensional space.[22] A key difference to Klein's paper lies in its identification of a wider group of transformation, namely what Fock labeled "gradient transformations" (which we would now call gauge transformations):

$$A_\mu \to A_\mu + \partial_\mu f(x) \tag{4.6}$$

$$\Psi(x) \to \exp\left(\frac{ie}{\hbar}f(x)\right)\Psi(x) \tag{4.7}$$

---

[20] Interview of Oskar Klein by J. L. Heilbron and L. Rosenfeld on February 25, 1963, Niels Bohr Library & Archives, American Institute of Physics, College Park, MD USA, www.aip.org/history-programs/niels-bohr-library/oral-histories/4709-3

[21] This is not quite true. Following the discovery of the pi meson (pion) in cosmic ray products, in 1947, Klein [Klein (1947)] returned to his five-dimensional formulation in a bid to incorporate meson theory in the scheme.

[22] Fock's article was submitted to the *Zeitschrift für Physik* just two weeks following the publication of Klein's paper (predating Gordon's version), in the same journal (Klein (1926b)). Curiously, in the edition that would feature Fock's article, both Heinrich Mandel [Mandel (1926)] and Ehrenfest and Uhlenbeck [Ehrenfest and Uhlenbeck (1926)] had papers on five-dimensional theories, the latter including an assessment of Klein's theory—Uhlenbeck was Ehrenfest's new assistant in Leiden, and Klein had been to visit them just before their publication [Ford (2009), pp. 9–10].

In other words, the classical freedom of choosing electromagnetic potentials (adding gradients) can be extended to quantum mechanics, so long as the appropriate change of wavefunction is carried out.

De Donder expresses the thrust of his own work in the discussion period after Schrödinger's report about wave mechanics to the audience of the fifth Solvay Congress in 1927. There he notes that one can recover the Schrödinger equation as a special case of a theory in which the gravitational field describes the interactions of n charged particles [Bacciagaluppi and Valentini (2009), pp. 426–7]. This insistence on structural correspondences can be found more strongly in the papers in which he (together with van den Dungen) develops a relativistic wave equation more or less identical to standard Klein–Gordon equations. However, de Donder is led to its formulation by a mathematical resemblance between the gravitational interaction modeled by a Fredholm equation which allows solutions with periodic behavior and the integral values of Schrödinger's equation. He views this as a logical deduction of quantization, in a similar way to Klein with his topological deduction from the compactification of his x_0. In a second paper de Donder considers an electron moving in a gravitational field that reduced to Schrödinger's when gravity was ignored. This work was explicitly presented as an extension of Schrödinger's wave mechanics to relativistic systems. De Donder and van den Dungen claim to get quantization *thanks to* general relativity. At the root of the procedure was a variational principle, related to Einstein's theory, that allowed for the derivation of a wave equation.[23]

The mathematician Dirk Struik was invited to MIT by Norbert Wiener. Struik had experience in the mathematics of general relativity, and had assisted in the development of parts of the differential geometric and group theoretic aspects. Wiener had met Struik on a visit to Göttingen—*cf.* [Rowe (1989), p. 23]. Together they worked on a unified theory of general relativity, electromagnetism, and quantum theory. The methodology was as above: to subsume quantum theory (in this case, specifically Schrödinger's wave mechanics) within general relativity. Using Schrödinger's wave theory, Wiener and Struik claimed to have found a result that "indicates an inner unity between the quantum theory and gravitational relativity" ([Wiener and Struik (1927)], p. 853).[24] They believed that they could show how the quantization of the relativistic Schrödinger equation—which they associate with de Donder—follows from the Einstein equations, though they have a different strategy to get an equation in the same form as de Donder's, the basis of which is Émile Cotton's theory of the differential invariants of linear partial differential equations of the second order.

---

[23] For a detailed analysis of de Donder's rather terse work, see [Peruzzi and Rocci (2018)].

[24] Wiener turned his attention to the "harmonization" of general relativity and quantum theory over the next three years, switching to consideration of the Dirac equation once that emerged, as we see in the next section. Wiener and Struik produced a more detailed version of their results in [Wiener and Struik (1927)]. It is possible that Wiener's interest in unified theories was triggered by a chance encounter he had with Einstein on a train journey in July 1925, in which Einstein explained his new ideas for reducing electromagnetism and gravity to a single theory (as described in a letter to his sister Bertha: https://libraries.mit.edu/archives/exhibits/wiener-letter/ [MIT, Norbert Wiener Collection: 14N-118].).

They see the fundamental problem as one of reconciling the fact that the matter content of general relativity is given in terms of a classical kinematics of particles (the invariants: rods, clocks, and yardsticks) with the new wave character of matter presented by Schrödinger. The fundamental physical phenomena of the two frameworks thus diverge in a radical way. Wiener and Struik argue that one should study the Schrödinger equation from the point of view of the invariance principle of general relativity, though couched not in the language of relativity (i.e. the quadratic form) but in the language of differential equations of the Schrödinger equation itself.[25]

Wiener returned to the issue of wave mechanics and relativity in the context of a highly speculative reassessment of Leibniz's philosophy [Wiener (1932)] (cf. [Galison (1994)], p. 255). He sought "historical support" for resolving the problems of wave-particle duality and the many-body problem, which he saw mirrored in the debates over the corpuscularian philosophy of the 17th century:

> As I have said, some of the Leibnizian monads mirror the world more clearly, some less clearly. This lack of clearness in mirroring is responsible for our impression that there is chance and indetermination in the world. Now, in the modern quantum theory, the indetermination which is an essential feature of the world, as represented in the ordinary found dimensions of time and space, is resolved, according to Heisenberg, if a sufficient number of additional, unperceived dimensions are superadded. This is the meaning of the five-dimensional theories of Fock and Klein, and is even more clearly brought out by the study of the problem of many bodies. This problem, which at present possesses no complete solution either from the standpoint of quantum theory or from that of relativity, even in the simple case of two bodies, can only be treated on the supposition that each electron carries with it its own three dimensions of spatiality, or what is more likely, its own complete space-time world. Thus, each electron possesses its own world of dimensions, which mirrors the many-dimensional universe of perfect cause and effect in an imperfect, four-dimensional, non-causal image. It is surely not fanciful to see in this a parallel to the Leibnizian monads, which live out their existence in a self-contained existence in a pre-established harmony with the other monads, yet mirror the entire universe.
>
> [Wiener (1932), p. 202]

In other words, somewhat reminiscent of Klein's notion of projection, Wiener sees Leibnizian monadology at work in the relationship between the connected, deterministic five-dimensional world, and the indeterministic "non-causal four-dimensional image": the electron is a monad that mirrors the five-dimensional world.[26] However, from here

---

[25] Manuel Vallarta (another Wiener collaborator) wrote to Wiener shortly after the paper was published (and, in fact, while he was working with Schrödinger on the derivation of the fine structure of the hydrogen atom, noting that spin seemed to be the crucial element), pointing out that he had shown it to Einstein who was impressed but thought it not a proper field theory ("keine eigentliche Feldtheorie ist"). The problem was that Wiener and Struik had inserted a constant *m* that referred to a *specific* given particle, which for Einstein was not allowed since a field theory ought to apply to all such particles [November 16, 1927: MIT, Norbert Wiener Collection: Folder 26].

[26] Such broadly philosophical ideas stand in close proximity to the closing puzzle of his long paper with Struik: "One matter of a considerable amount of importance is that of forming some sort of a well-defined four-dimensional space-time from the multi-dimensional world of the problem of several bodies. Perhaps the solution of this problem is to be found in the notion of the exploring particle" [Wiener and Struik (1928)].

Wiener was led[27] via pre-established harmony and then Leibniz's theory of automata, to his theory of cybernetics (cf. [Gale (1997)]); his work with Struik does not seem to have led anywhere.

We can view many of these approaches, showing how quantum waves and particles can exist in general spacetimes, as a kind of early flowering of 'hybrid' theories, mixing quantum and classical. Interestingly, though credit is often assigned to Schrödinger,[28] in his work on quantum cosmology (quantum fields in a closed universe), the study of quantum fields on curved spacetime began with Théophile de Donder in his investigation to determine whether Schrödinger's wave mechanical formulation of quantum mechanics was compatible with (the more established) general relativity.[29]

Arthur Eddington had a curious view of the relationship between wave mechanics and general relativity. He viewed any attempt at combination of wave mechanics and curved space as a kind of category mistake [Eddington (1943), p. 12, fn. a]. He sees the task as purely mathematical, having nothing whatsoever to do with general relativity. He charitably interprets such attempts as seeking an extension in generalized coordinates in flat space (by putting things into tensor form, as he puts it). For him, wave mechanics and general relativity were simply alternative ways of representing the distribution of energy (see [Eddington (1943), p. 46])—indeed, Eddington even claimed the Pauli exclusion was simply a wave mechanical analog of gravitation (ibid., p. 82).[30]

Though coming much later, related to this general line of studying the formal or structural relations between quantum mechanics and general relativity, was Eugene Wigner's [Wigner (1955)] later approach of looking at the concept of *particle* as linked to the symmetry group of the spacetime it occupies (or, rather, of the world in which the particle's equations hold)—Wigner first laid this out at the Einstein Jubilee conference in Berne, in 1955. For example, a particle in the context of special relativistic quantum field theory would be defined in terms of the irreducible representations of the Poincaré group, while in orthodox classical physics the Galilean group is the appropriate source of symmetries. Hence, symmetry becomes the fundamental theoretical concept in world-building, so one should look to the symmetry group of general relativity for clues as to the status of particles in such a context. He considers amongst his spacetimes de Sitter space as a probe (an "intermediate point for a future transition to the general theory of relativity," p. 211) for investigating the fate of the particle concept as one moves from Galilean and Lorentzian symmetries.

---

[27] Though not before turning his attentions to the Dirac equation as a structural link between gravitation and quantum, as we see in the next section.

[28] See e.g. [Birrell and Davies (1984), p. 7] and [Rüger (1988), p. 378]—though they have in mind particle creation effects in a background gravitational field.

[29] An Argentinian, Guillermo Knie, wrote a treatise on Schrödinger's work, entitled *Wave mechanics in a Curved Space* [Mecánicá ondulatoria en el espacio curvo] (Union Mathmatica Argentina. Memorias y Monografias (2) 1, (1942). 1–15. The book attempts to provide an expository and detailed account of some of the then recent work of Schrödinger which attempts a "fusion" of the relativity and quantum theories. The primary focus was Schrödinger's paper on quantum cosmology [Schrödinger (1940a)], which appeared amongst a group of papers in the *Proceedings of the Royal Irish Academy* in 1940. The principal topic dealt with is that of the transmission of electromagnetic and matter waves in a gravitational field.

[30] This is to be understood *functionally*: both the exclusion principle and the gravitational theory arrive at the same value for the (standard) mass (see Chapter 7 of [Kilmister and Tupper (1962)] for the full argument).

In the context of general relativity, Wigner extracts two key principles: the principle of point-coincidences (i.e. that only coincidence quantities should be involved in the theory's measurement analysis), and also the equivalence principle. This means, according to Wigner, that we should restrict our focus to *events* only, so that coordinates are not primitive but depend on degrees of freedom of the theory: coordinates have no direct physical meaning. Since these events will be quantum mechanical, there will be unavoidable fluctuations in coordinates. This pushes Wigner to a relational theory dispensing with coordinates entirely[31], making do with correlations between events only. Yet Wigner does not get very far into extending his scheme into general relativity at all.[32] He notes that the kind of theory one would end up with, in combining principles in this rather simple manner, would not deal with coordinate derivatives of quantities, but of probability amplitudes with respect to one another. Hence, his tentative solution to the problem of harmonizing quantum mechanics and general relativity is to eliminate coordinates:

> Now that we realize that coordinates are no more than parameters that must be eliminated in determining relations between observables, it becomes natural to ask whether we are using the most advantageous parameters, or even whether any such parameters are necessary.
>
> (ibid.)

His approach was to formulate a version of quantum mechanics without coordinates, using just field components.[33] The procedure is essentially what would nowadays be called the method of 'intrinsic coordinates,' in which the coincidence of a set of reference fields is used to invariantly localize other degrees of freedom (thus removing the coordinates).[34]

Max Born presented the core conceptual problem in very simple, stark terms, rooting it in Wigner's (itself, we might add, close to Wiener and Struik's earlier remarks) focus on the distinct conceptions of observability in the two theories:

---

[31] This is in contradistinction to Arthur Eddington who saw the uncertainty in reference frames as a central task.

[32] Wigner returned to the issue of relativistic invariance of quantum mechanics in his retiring presidential address to the American Physical Society in January 1957 [Wigner (1957)], where he bemoaned the lack of common ground between general relativity and quantum mechanics (by contrast with special relativity).

[33] Fred Hoyle was in agreement with Wigner, calling the use of coordinates "a psychological survival from the Newtonian era" ([Wigner (1955)], p. 224).

[34] In the question period following Wigner's talk, David van Dantzig mentions that he suggested essentially the same idea in his earlier talk, namely that one can *define* spacetime points by events. However, van Dantzig ends up defending a discrete model (what he calls a "flash" model) consisting only of a discrete set of possibly observable events, rather than a continuum of events as Wigner believes. Wigner adds, in a note included in the proofs of the proceedings, that not only van Dantzig, but also John Synge and Jules Géhéniau had also presented similar ideas.

> Professor Wigner has put his finger exactly on the spot where the difficulty of reconciling general relativity and quantum mechanics lies, mainly that relativity is based on the concepts of coincidences as the only observable things. Atomic physics has to do with collisions of particles; but quantum mechanics treats a collision as a spatially extended process of which only the asymptotic limits are observed.
>
> (Born's comments in [Wigner (1955), p. 225])

However, while agreeing on the diagnosis of the difficulties, Born disagrees on the remedy. Rather he argues that there is really no structural harmonization to be had here, because of the *kinds* of theory quantum mechanics and general relativity are. General relativity deals only with the macroscopic superstructure enabling the interpretation of microscopic goings on, along the lines of Bohr's ideas about the necessity of classical realm for making sense of the quantum. Wigner eventually came to defend this same view as Born, that, as he puts it, "the essentially nonmicroscopic nature of the general relativistic concepts seems . . . inescapable" [Wigner (1957), p. 68] (we return to this again briefly in §4.3). Such conclusions of disharmony did not come quickly, however, and the general consensus in the wake of the new quantum mechanics was that the respective structures were in fact compatible, and indeed often mutually supporting, as we saw with the case of de Donder, Klein, and others.

Dirac sought a (special) relativistically invariant theory of an electron (i.e. a relativistic wave equation) during 1927. The widespread belief was that this problem had already been resolved by Klein (in the Klein–Gordon equation of 1926). What is crucial, from the point of view of its impact on the problem of quantum gravity, is the half-integral spin that it attributes to electrons. Dirac's equation in 1928 would radically alter the challenge of integrating general relativity and quantum mechanics, forcing the issue of compatibility into physicists' minds as a result of (1) the linear nature of the new equation (in favour of compatibility), but (2) the apparent structural conflict between the tensorial relativity theory and the spinorial quantum theory. Non-classical two-valuedness also posed an intriguing problem for general relativists: was it possible to have such properties in a curved space, as described by general relativity? Are the respective mathematical structures even compatible?

4.2　Dirac's Challenge and its Solutions

> Obviously it is unlikely that one will ever come to a satisfactory unification of general relativity and microphysics unless it is possible to incorporate spin into general relativity theory.
>
> [Bergmann (1956), p. 93]

By 1955, Peter Bergmann—in his review of quantum gravity at the Bern conference for the 50th anniversary of special relativity—viewed the incorporation of spin, and the Dirac equation, as a mostly mathematical exercise. However, Rosenfeld describes the reception of Dirac's equation as "an absolute wonder."[35] For many,[36] the Dirac equation was a shock. As Clive Kilmister has noted [Kilmister (1966), p. 69], some of this shock could be traced back to the fact that the Dirac equation was invariant without being a tensor equation. And yet, it was commonly believed up to that point that tensor calculus was "all embracing" precisely because it was the tool for discussing the all-embracing general theory of relativity. Yet the Dirac equation was Lorentz invariant but appeared not to be generally invariant: it required the incorporation of spin, as well as translation.

General covariance (the principle of general relativity) was considered firmly established enough to function as a constraint on theory building at the end of the 1920s, so that testing for compatibility with it bestowed a higher degree of confidence on any new theory.[37] Moreover, general relativistic formulations often yielded manifestly covariant theories, as opposed to other more cumbersome procedures involving infinitesimal transformations of data and so on. Dirac's spin theory of the electron was quickly put through the generally relativistic mill by a great many physicists in short order. This can be seen as a natural continuation of the study of quantum theory in curved (general) spaces begun in the wave mechanical context, with many of the same people involved.[38]

First on the scene seems to have been Hugo Tetrode [Tetrode (1928a)], inspired by Einstein's work on 'teleparallelism' (or 'distant parallelism'), which sought to extend general relativity by employing a tetrad formalism in which the metric is represented by an arbitrary basis of local orthonormal vector fields.[39] Tetrode began by simply replacing the Minkowski metric η by the general metric of Einstein's general relativity, so that

---

[35] Interview of Leon Rosenfeld by Thomas S. Kuhn and John L. Heilbron on July 1, 1963, Niels Bohr Library & Archives, American Institute of Physics, College Park, MD USA, www.aip.org/history-programs/niels-bohr-library/oral-histories/4847–1. Dirac's equations (of which there is a family of four linear partial differential equations) describe the behavior of what was then a new object: the *spinor* (named by Paul Ehrenfest: see [Veblen (1934)] for, what is in my opinion, the friendliest introduction to the spinor concept). Spinors are a quantum affair. In the context of classical physics it suffices to give a coordinate system representation of systems (whether they are given by a scalar, vector, or tensor). Initially, the focus on spinors was from the point of view of the small scale structure of the world. Later work, primarily initiated by Roger Penrose, would view spinors as being of more general importance in gravitational physics, across the board—see, e.g., [Penrose (1962)] (this more general spinor approach appears to have been suggested by Louis Witten: equivalence is established between the Riemann curvature tensor and two spinors [Witten (1959)]. As mentioned earlier, a rather different set of related concerns led to Wheeler's 'geon project,' in which one of the pressing problems was whether a gravitational structure (space) could have half-integer spin—we discuss this further in Chapter 8.

[36] Not least, Eddington, as we return to later.

[37] We find a similar process in action in early work on direct quantizations now most naturally associated with the canonical approach to quantum gravity, as we will see in Chapter 6.

[38] Interestingly, the Dirac equation in general relativity would be put to use again as a kind of toy model for studying the neutrino geon by John Wheeler's group (see Charles Misner's 1956 lecture notes from Wheeler's course "Physics 581: Advanced Quantum Mechanics": John Archibald Wheeler Papers, Box 155; also: [Brill and Wheeler (1957)]. Given that this study was conducted in the overarching framework on Wheeler's geometrodynamics (in which charge is a property of the global topology: the multiple connectedness), Misner transfigures the historical discussions of the Dirac equation in general relativity into a different notation capable of treating global problems.

[39] This is a fairly obvious technical move given that spinors can only be defined within a local orthogonal frame. However, this common wisdom might not be strictly true. In order to understand gravitational interactions of fermions in the context of Gupta's theory, which they view as difficult from a tetrad perspective, Ogievetskiĭ and Polubarinov sketched a (non-tetradic) "group-theoretic approach" to spinors in a gravitational

the Dirac matrices were defined by these instead, generating coordinate-independent Gamma matrices.

$$\gamma_i \gamma j + \gamma_k \gamma_i = 2g_{ik} \tag{4.8}$$

Tetrode's speedy reaction was likely due to his pre-existing program involving the investigation of the invariance properties of the Dirac equation (see, e.g., [Tetrode (1928b)]. Hence, he wasn't concerned with the relationship between general relativity and quantum theory, so much as he was with the structure of the Dirac theory.

Indeed, teleparallelism seems to have ignited a small burst of research activity in the study of the relationship between Dirac's theory and general relativity—they were both published the same year (cf. [Goenner (2004), §7.2]). After Tetrode, Eugene Wigner [Wigner (1929)] also noticed a link between teleparallelism and Dirac's theory of electron spin, and extended the theory to include the setting of the relative orientation of the various local coordinate systems to each other by the electromagnetic field. Weyl [Weyl (1929)] added to this discussion, also employing a tetrad (vierbein) formalism in his bid to harmonize the Dirac theory with general relativity. However, since he was opposed to the notion of teleparallelism,[40] he sought a different approach to Tetrode and Wigner, basing it instead on the standard gravitational theory of Einstein, based on the metric tensor. Weyl thought the challenge of the generally relativistic Dirac equation went beyond a mere mathematical one. Since the Dirac equation contained a mass term, and mass is a gravitational effect, it is necessitated to consider general relativity. Not to do so would be akin, in Weyl's mind, to introducing charge without involving the electromagnetic field [Weyl (1929), p. 323].[41]

Weyl was followed by Wiener, this time with his student Manuel Vallarta. Wiener carries through the same theme of harmonization as with wave mechanics, though this time he believed that the vierbeine furnished the perfect tool for this program, allowing for a seamless integration, essentially by allowing coexisting geometries for the macro- and micro-worlds:[42]

field [Ogievetskiĭ and Polubarinov (1965)]—cf. [Pitts (2012)] for a historico-philosophical discussion of this point.

[40] His reasons: "my mathematical intuition objects to accepting such an artificial geometry; I find it difficult to understand the force that would keep the local tetrads at different points and in rotated positions in a rigid relationship. There are, I believe, two important physical reasons as well. The loosening of the rigid relationship between the tetrads at different points converts the gauge-factor $e^{i\lambda}$, which remains arbitrary with respect to Ψ, from a constant to an arbitrary function of space-time. In other words, only through the loosening of the rigidity does the established gauge-invariance become understandable" (Weyl, cited in [O'Raifeartaigh and Straumann (2000), p. 6]). Weyl was, of course, against any notion of action-at-a-distance, of which this clearly smacked.

[41] Jan Arnoldus Schouten attempted to generalize Weyl's idea, on getting a mass term, to five-dimensional spacetime, by then a standard strategy [Schouten (1931a)] [Schouten (1931b)]. Eddington too would later view the gravitational field as an essential part of the story of quantum theory precisely because it provides the inertial field that accounts for particle masses (cf., e.g., [Eddington (1943), p. 14]).

[42] Note that Henry Flint also referred to the link between $\gamma_{\mu\nu}$ and the metric (which he calls "Tetrode's relation") as one holding between micro- and macro-scales [Flint (1960)].

A much more pressing need of general relativity theory [than Einstein's unified field theory of electricity and gravitation—DR] is a harmonization with quantum theory, particularly with Dirac's theory of the spinning electron. On the basis of Levi-Čivita's parallelism the task seems hopeless, inasmuch as we have no adequate means of comparing spins at different points. On the other hand, the notion of a parallelism valid for the whole of space and of Einstein's *n*-uples enables us to carry over the Dirac theory into general relativity almost without alteration. All that we need do is to interpret Dirac's P_0, P_1, P_2, P_3, not as differentiation with respect to four variables x, y, z, t defined throughout space-time, but as differentiation along the lines of the quadruple (Einstein's '4-Bein'). That is, the quadruples need not be integrable so as to furnish us with a co-ordinate system throughout space, for such a co-ordinate system is absolutely inessential in the proof of the invariance of Dirac's equations under a Lorentz transformation. In other words, the quantities $^s h_\lambda$ of Einstein seem to have one foot in the macro-mechanical world formally described by Einstein's gravitational potentials and characterized by the index λ, and the other foot in a Minkowskian world of micro-mechanics characterized by the index s. That the micro-mechanical world of the electron is Minkowskian is shown by the theory of Dirac, in which the electron spin appears as a consequence of the fact that the world of the electron is not Euclidean, but Minkowskian. This seems to us the most important aspect of Einstein's recent. work, and by far the most hopeful portent for a unification of the divergent theories of quanta and gravitational relativity.

[Wiener and Vallarta (1929), p. 12]

In a slightly different way, Fock also attempted to merge the Dirac equation with the geometry of general relativity, also writing the equation in curvilinear coordinates, thus refuting Wiener and Vallarta's claim. His strategy (parallel transport of what Fock called 'semi-vectors,' also discovered independently by Weyl) was to modify the structure of the manifold so as to allow for spin—by adding (local) spinor structures. Fock desired (and thought he'd achieved) a "geometrization of Dirac's theory of the electron and its subsumption within general relativity" [Fock (1929a), p. 275]. Together with Dmitri Iwanenko [Fock and Iwanenko (1929)], they labeled their theory "quantum linear geometry." The basic idea was to modify the geometry in such a way as to include the properties of the Dirac matrices. They suggested introducing a linear differential form $ds = \Sigma \gamma_\nu dx_\nu$ that when squared would deliver the standard Riemann interval ds^2.

Fock and Iwanenko believed that this approach could lead to solutions of the most pressing problems in quantum field theory at that time (negative-energy solutions and the ubiquitous divergences, albeit by possibly bypassing them through the creation of a new kind of theory). Though he devised a near-identical theory to Fock, Weyl would later distance himself from the geometrization program—see [Scholz (2005)] for a nice account of this episode.

Fock viewed the problem of general relativizing the Dirac equation as involving the imposition of four conditions on the (Dirac) wave equation ([Fock (1929b)], p. 122):

1. be invariant with respect to an arbitrary coordinate change;
2. be invariant with respect to an arbitrary rotation of the *n*-hedron;

3. coincide with the adjoint equation and also be such that a time-like vector and vector of zero divergency can be defined in order to be interpreted as a current vector;

4. be reduced to the ordinary Dirac equation in the case of the Minkowski space.

The expression they end up with is for a Dirac operator, $\mathcal{F}$, in curvilinear coordinates (ibid., p. 125):

$$\mathcal{F}\Psi = \sum_j e_j \alpha_j \frac{1}{H_j}\left[\frac{h}{2\pi i}\frac{\partial \Psi}{\partial x_j} - \frac{e}{c}\psi_j + \frac{h}{4\pi i}\frac{\partial}{\partial x_j}\left(\log\frac{\sqrt{g}}{H_j}\right)\Psi\right] - mc\alpha_4 \qquad (4.9)$$

They note that Tetrode's earlier expression fails to satisfy all of their conditions, and so cannot be considered to constitute a genuinely generally relativistic generalization (ibid., p. 125, fn. 7). What they believed they demonstrated was "that the Dirac equation is perfectly compatible with the notion of a Riemann space" (ibid., p. 131), so that one can apply macroscopic geometrization methods at the micro-scale. Fock believed this might provide a basis for a quantization of gravity.

As Alex Blum notes [Blum (2018), p. 52], there are two motivations at work here: on the one hand, we have attempts, such as Fock's, grounded in quantum mechanics, attempting to extend and generalize the formalism (here, Dirac's equation) to include curved spaces; on the other hand are those approaches that seek to utilize gravitational considerations in the service of quantum mechanics—we might add another, namely those coming from the geometrical programs such as the unified field theory approaches looking for harmonization of the macro- and micro-domains.

There were some that did not believe a harmonization of the Dirac equation and general relativity was achievable. For example, George Temple argued for a modification of the Dirac equation (of the electron) on pain of "abandoning the theory of relativity" ([Temple (1929)], p. 352). Temple's approach was to construct a system of wave equations which possessed "all the advantages as Dirac's equations and which shall be tensorial in form in accordance with the general theory of relativity" (ibid.). Edmund Whittaker likewise established that a correspondence between a self-dual six vector has properties identical to a spinor, with which he formulated a tensorized version of Dirac's equation in a general spacetime: "the calculus of spinors is included in the calculus of tensors" [Whittaker (1937), p. 41]. This allowed conversions to occur in both directions, allowing simplifications of difficult problems (not least in the transition from special to general relativistic problems).[43] All of this work that appeared to drift away from

---

[43] Whittaker's approach was based on the two-component spinor calculus of [van der Waerden (1929)], who laid down the formal foundations of spinors for the consumption of physicists—John Synge, in his review of Whittaker's work on relativity, called this a "tensorial model of a spinor" [Synge (1958), p. 54]. Ruse [Ruse (1938)] presented an account of the two-spinor theory in the context of general relativity in 1938, just after Whittaker's paper. Whittaker's approach was shown to be equivalent to Fock's and Ruse's in [Martin (1942)]. Abraham Taub [Taub (1939)] carried out a similar tensor analysis, for Dirac's equations for spin 1 and spin-1/2 particles.

the original intent of the introduction of spinors (to describe the electron), was in fact almost entirely in the service of learning more about the nature of spinors, and so the overlapping with, what we could now reasonably consider, quantum gravity issues, had beneficial spinoffs for central areas of physics.

Schrödinger too eventually joined this bandwagon [Schrödinger (1932)], coming rather late in 1932, with an attempted simplification of the work then available, and also in the mode of Fock's attempts at a more significant merging of gravitation and quantum mechanics (and electromagnetism) along the lines of a unified theory. He attacked the problem of the generalization of Dirac's equation to a Riemannian manifold, and considered the wider implications of such a generalization. In so doing he effectively ushered in the field of quantum cosmology. In his book *Expanding Universes* he considers waves in a general Riemannian spacetime, and considers the effect of expansion on waves in such a space. Schrödinger himself was not overly impressed with his efforts, which he viewed merely as a "smart exercise" that "came to nothing".[44] However, his initial waning interest was reignited by Eddington's ideas for linking the global structure of space with the nature of particles. We turn to this in §4.4 in the context of early quantum cosmology.

Eddington was rather more strongly influenced by the Dirac equation than most; it guided his research program for the remainder of his life. In the Introduction to his *Relativity Theory of Protons and Electrons* he wrote that: "To say that Dirac's wave equation was the first connecting link gives only a partial idea of its importance. *It was a challenge to those, who specialised in relativity theory*" [Eddington (1936), p. 1]. In his later Dublin lectures, Eddington remarked:

> In 1928 Dirac opened up the spin avenue, and we went down it in a rush. My path bifurcated from the others soon after the start. The others took shortcuts to the things that most interested them.
>
> [Eddington (1943), p. 1]

What so impressed Eddington was the fact that quantum theory could yield new insights (such as Dirac's deduction of spin from his wave equation) through the introduction of relativity. We see later how Eddington sought to extend this mindset to general relativity. In fact, Charles Darwin's almost parenthetical remarks about the relationship between Dirac's theory of the electron and general relativity, and the fact that Dirac's theory was not in tensor form ("unsymmetrical" as Eddington puts it, [Eddington (1928), p. 524]), appear to have been responsible for igniting Eddington's obsession with spin, viz.:

---

[44] Letter to Frederik Belinfante, March 16th, 1951: Schrödinger Nachlass, University of Vienna. His self-assessment was perhaps a little too harsh, since he did demonstrate, in this paper, that teleparallelism was not required for general Dirac equations, ending the tide of approaches that started with Tetrode, and leading to the alternative approaches of Schouten, Ruse, and also Veblen [Veblen (1933)].

It is of some interest to consider this invariance [with respect to Lorentz transformations–DR] a little further. The whole theory of general relativity is based on the idea of invariance of form, and here we have a system invariant in fact but not in form. Should it not be possible to give it formal invariance as well, and would not that be the right way to express our equations? It is so possible, but it is not hard to show that it requires no less than 16 quantities to do it, viz., two scalars, two four-vectors and one six-vector, and even so each will have a real and imaginary part, so that we may say that 32 quantities are required! It seems quite preposterous to think that a single electron should require 32 equations to express its behaviour, and, moreover, these 32 will involve a large number of arbitrary inter-relations of no influence on the four quantities which are actually sufficient to describe it. Now the relativity theory is based on nothing but the idea of invariance, and develops from it the conception of tensors as a matter of necessity; and it is rather disconcerting to find that apparently something has slipped through the net, so that physical quantities exist which it would be, to say the least, very artificial and inconvenient to express as tensors.

[Darwin (1928), p. 657]

Darwin goes on to suggest, in a footnote, that perhaps gravitation (not included in his four wavefunctions) is "the hole in net"—indeed, Eddington quotes this passage approvingly (ibid.). We turn to Eddington in the final section of this chapter; for now we simply note that Eddington's route with the Dirac equation was orthogonal to the others since he ended up viewing both quantum theory and general relativity as approximate theories, suited to their domains of natural applicability, but ultimately complementary. Eddington's route to their union then involved finding an appropriate problem where *neither* can be neglected.

There were several relevant forays into quantum gravity (or less contentiously, 'quantum relativity') of a more mathematical nature during the 1930s and 40s. These were, however, curiously isolated from the concerns of the physics community, and speak to the notion that such research was more suited to mathematicians at the time, though would later reintegrate. For example, Taub, Veblen, and von Neumann [Taub, Veblen, and von Neumann (1934)] extended the consideration of the Dirac equation into the realm of projective relativity, with the aim of achieving invariance under coordinate, spin, and gauge transformations. They showed that their theory is physically equivalent to Einstein's general theory and Maxwell's electromagnetic field. The Dirac equation was also used as a case study of sorts by Abraham Pais [Pais (1941)], who in studying projective relativity theory, and specifically the projective energy momentum tensor for a general Lagrange function, computes the energy momentum tensor for Dirac's equation from the framework. These forays would also serve to elucidate the properties of the ingredient theories, highlighting their limits, etc.[45] Dirac himself attempted to allow

---

[45] One of the many byproducts from this research topic—strictly speaking, stemming from the more general investigation of the invariance properties of the Dirac equation—was the discovery by Abraham Taub [Taub (1938)] (presented to the American Mathematical Society on September 9th, 1937) and, with further developments, by Pauli [Pauli (1940)], of the result that the generally relativistic Dirac equation is also conformally invariant (i.e. invariant under transformations of the form $g'_{ij} = f^2(x)g_{ij}$) for vanishing mass

for a tensor approach by switching to a conformal space with the aim of revealing the expanded symmetries of quantum mechanics [Dirac (1936)]—he also produced a version based on de Sitter space [Dirac (1935)].[46]

The later development of this more formal research area would basically involve the generalization of the results to other particle spins. Indeed, the debate made a direct cameo appearance during the ascendance of meson physics in the early 1950s, in which the Duffin–Kemmer–Petiau [DKP] algebra was studied by Brulin and Hjalmars who attempted a similar analysis of the Duffin–Kemmer equations.[47] In this case we can find a direct line between the modern approaches to quantum fields in curved space and the older work stemming from the Dirac (and earlier wave) equation.[48] We follow this development into cosmology in §4.4, in which consideration of the Dirac equation in closed universes is discussed.

4.3   Measurability Matters

The 'shock of the new' encountered in the new quantum mechanics did not stop at the level of the nature of purely formal aspects: new problems of interpretation of the formal representation also emerged, mainly concerning measurement and uncertainty. Perhaps the most famous example of this kind, concerning the interplay between gravity and quantum, prior to 1930 was Bohr's usage of general relativity to argue against Einstein's 'photon in a box' critique of the uncertainty relations (in a bid to uphold his criterion of reality), at the 1927 Solvay Congress. As Oskar Klein explains:

> We know from BOHR's account how ingeniously EINSTEIN defended his standpoint—
> the essential incompleteness of the quantal description of nature—and how BOHR
> refuted every one of his arguments with more than ingenuity. What impressed us younger
> people most was, I think, the "Einstein box," where BOHR successfully turned general

---

fermions. Hence, it is revealed *why* there is no nonsingular analog of the Dirac equation (a fact already known to Dirac) which is conformally invariant: it is the mass that breaks it. (Note that in 1935, Taub had received his PhD, under Howard Percy Robertson, on a kind of quantum cosmology involving the Dirac equation (published as [Taub (1937)])—this work was triggered by Schrödinger's visit to Princeton University, during the spring of 1934, in which he discussed the cosmological Dirac equation in a series of lectures.)

[46] This was carried out while Dirac was visiting Princeton and in discussions with Veblen (who wrote a paper on the same topic at the same time, [Veblen (1935)], and whose work on conformal theory Dirac directly employed) and other mathematicians working on mathematical aspects of the Dirac equation. As mentioned in the previous footnote, the Princeton mathematicians appear to have been influenced by a 1934 visit of Schrödinger who, just before Dirac's visit, had given lectures on the Dirac equation in general spacetimes (cf. [Kragh (1990), p. 168]).

[47] We will encounter Gérard Petiau again the the next chapter. Petiau had been a student of de Broglie's, whose work was focused on the analysis of wavefunctions of particles of various spins. His fellow student, Marie-Antoinette Tonnelat, coming from the de Broglian school (with a devotion to the light–matter analogy), had applied this approach to the study of the photon (understood as a composite of two fermions, following de Broglie) in a curved space in 1941: [Tonnelat (1941)].

[48] See, e.g., [Dowker and Dowker (1966)], in which they mark these developmental stages quite explicitly, and also link the story to the search for a unification of general relativity and quantum mechanics.

relativity theory against EINSTEIN.... And still EINSTEIN, who accepted all defeats with the utmost fairness but without changing his basic view, may have felt that on the side of the quantum physicists the importance of the general relativity claim in the search for the laws of the microworld was usually underestimated.

([Klein (1955)], p. 117)

Einstein used quantum theory and special relativity (specifically, the famous relation between mass and energy) to try to circumvent the 'Energy–Time' uncertainty relations. Bohr used a combination of quantum theory and *general* relativity in order to eliminate the inconsistency that Einstein derived. As Christian Møller recalls:

> Well I remember of course the excitement when Bohr was able to beat Einstein with his own weapon. That was at a Solvay meeting; Einstein invented a way of showing that quantum mechanics was not consistent. He proposed to determine the energy of the photon which had come out of the box by weighing the box before and afterwards. Then Bohr could show that if one takes Einstein's formula for the rate of a clock in a gravitational field then it comes exactly to making the thing consistent again. And Gamow even made a model of this box with a spring and clock and shutter, which opened at a certain time and closed again at a certain time.[49]

Bohr returned to this theme in his H. H. Willis memorial lecture, given at the University of Bristol in 1931 (October 5th).[50] He then returned to related matters in what is effectively measurement analysis in a famous paper with Rosenfeld in 1933, on the measurement of the quantized electromagnetic field, which heavily influenced measurement analysis in the quantum gravitational context. Let us now see how the earlier argument works, before turning to the quantum gravitational context.

Einstein constructed a thought experiment designed to show that the energy transition of an object could be measured exactly (by measuring the change in mass), as could the time at which such a change occurred. He asked the Solvay participants to imagine a box in which there is a clock, with the set up designed so as to allow the opening and closing of an aperture for a very short time Δt which is just long enough to let one single photon pass through. One can make Δt arbitrarily small, or, in other words, our knowledge of the photon's escape time arbitrarily accurate. Einstein then proposed to involve the gravitational field to weigh the box before and after the escape. There is no limitation on our ability to weigh the box, so Einstein concludes we can know both energy and time with as much accuracy as we please, in violation of the energy–time uncertainty relations.

Bohr, against this, did the physicist's equivalent of a judo move, using Einstein's theory of general relativity against him by "taking into account the principles of general relativity to a fuller extent, and then by going into greater detail about the actual process of

---

[49] Interview of Christian Møller by Thomas S. Kuhn on July 29, 1963, Niels Bohr Library & Archives, American Institute of Physics, College Park, MD USA: https://www.aip.org/history-programs/niels-bohr-library/oral-histories/4782.

[50] Reproduced in [Bohr (1931)].

weighing" [Bohr (1931), p. 369]. This, argues Bohr, would involve some kind of pointer whose reading is calibrated to the weight. Bohr suggests hanging the box from a spring with a gauge. He then points out that infinite accuracy with respect to this pointer reading is impossible, since it would involve complete knowledge of the momentum of the setup. Thus, a less than infinitely perfect determination of the box's mass is the best that can be hoped for. Bohr then imagines an experiment as follows: a reading is taken, over some time T. The aforementioned uncertainty in the mass of the box ΔM becomes entangled with the gravitational potential $\partial \phi / \partial x$ and the uncertainty in momentum ΔP via:

$$\Delta M \cdot \frac{\partial \phi}{\partial x} T = \Delta P_x = \frac{h}{\Delta x} \tag{4.10}$$

Multiplying out the differentials Δx gives us:

$$\Delta M \cdot \Delta \phi \cdot T = h \tag{4.11}$$

Hence, the initial uncertainty about the pointer reading propagates into an uncertainty about the gravitational potential at the box's location. Then Bohr uses a theorem of general relativity which connects the uncertainty in gravitational potential with uncertainty in the timescale presented on the clock ΔT:

$$\Delta T = \Delta \phi \cdot \frac{T}{c^2} \tag{4.12}$$

Taking (4.11) into account gives:

$$\Delta T = \frac{h}{c^2 \Delta M} \tag{4.13}$$

Which, making use of $E = mc^2$, gives:

$$\Delta T = \frac{h}{\Delta E} \tag{4.14}$$

Multiplying through by ΔE gives back the original energy–time uncertainty relations associated with non-gravitational situations. There is, in other words, no escape.

What is curious about this argument is that it was never utilized directly to motivate the study of quantum gravity despite the fact that one of the results of Bohr's argument is precisely that the uncertainty would infect the gravitational potentials. But perhaps the idea that gravity must be quantized was more or less universally accepted by most physicists at the time, so that one more demonstration made no real difference. What is also interesting about this account is the remark with which he ends the discussion, which might supply an answer as to why the study of uncertainty in the gravitational potentials was not further pursued:

The foregoing analysis of the paradox brought forward by Einstein shows that the phenomenon of gravitation which, up till now, in its mathematical treatment stands so far apart from the rest of modern physics does so by no means with respect to the primary physical concepts evolved by Einstein in his theory of general relativity in order to remove the obstacles which until then seemed to prevent the reconciliation of these two groups of physical phenomena.

(ibid., p. 369)

In other words, there is simply seen to be no formal clash between the two theories: gravitation easily succumbs to the quantum paradigm. While de Donder repeatedly attempted to steer the discussion into general relativity, at the Solvay Congress, it seems to have stirred up no enthusiasm. However, such questions of measurability would repeatedly arise just before and after the Second World War.[51]

Matvei Bronstein seems to have been the first to explicitly discuss such issues in a gravitational context, in his PhD thesis, which was largely reproduced as [Bronstein (1936)].[52] The central theme of Bronstein's work is limitations of measurability of the gravitational field. This was first considered for the electromagnetic field by Landau and Peierls in 1931 [Landau and Peierls (1931)], extending Heisenberg's uncertainty relations into relativistic regimes. They concluded that field strengths were in fact *not* measurable (so that there appeared uncertainties greater than the quantum mechanical uncertainty principle dictates). But Bohr and Rosenfeld [Bohr and Rosenfeld (1933)] reanalyzed the question for the linear electromagnetic theory, focusing on the effect of imposing quantum conditions on the measurement instruments, and showed that there was no such problem once one gave a more accurate and realistic picture of the measurement process itself—in particular questioning Landau and Peierls' idealization to point particles.[53] One needs to use test bodies of large numbers of particles (giving a large charge) to achieve precise measurements of the electromagnetic field: it would be impossible to achieve accurate measurements with, e.g., a single electron. Moreover, one can contrive compensatory devices that tame any problems caused by the field generated by the large charge.[54]

Bronstein was concerned with elucidating some of the physical characteristics of a quantum gravitational field, which he begins by presenting along the lines of Fermi's approach to quantum electrodynamics (quantized in Lorentz gauge). Using the weak field approximation (for which the spacetime's deviation from flatness is very small) he directly employs the Bohr–Rosenfeld analysis as a kind of conceptual probe. Einstein's (weak limit) field equations, in the classical theory for the three-index Christoffel

---

[51] For a recent discussion linking these earlier debates to the question of whether gravity *must* be quantized, see [Albers, Kiefer, and Reginatto (2008)].

[52] Bronstein was one of the many intellectual casualties of Stalin's regime, murdered by firing squad on February 18th, 1938, after his death warrant was signed off by Stalin, Voroshilov, Molotov, and Kaganovich. A detailed account of Bronstein's life and work is in [Gorelik and Frenkel (1994)].

[53] Wigner, together with Salecker, repeated a Bohr–Rosenfeld type analysis for the gravitational field, involving distance measurements, in 1958 [Wigner and Salecker (1958)].

[54] See [Brown (1993)] for a good discussion of the history of these measurement-theoretic explorations.

symbol $[00, 1]$ (functioning as the gravitational analog of the electric and magnetic fields used in Bohr and Rosenfeld's analysis), are:

$$\frac{d^2x}{dt^2} = \frac{\partial h_{01}}{\partial t} - 1/2\frac{\partial h_{00}}{\partial x} = [00, 1] \tag{4.15}$$

From this basis, Bronstein selects as an example the measurement of the spacetime average of the Christoffel symbol in a volume V and throughout a time interval T. For a test body of volume V and mass ρV, he notes that the Einstein equations allow one to make a measurement of the momentum throughout an interval, where the average is:

$$\frac{(p_x)_{t+T} - (p_x)_t}{\rho VT} \tag{4.16}$$

In which case the $[00, 1]$-measurement has an uncertainty (with Δp_x the momentum uncertainty):

$$\Delta[00, 1] \approx \Delta p_x / \rho VT \tag{4.17}$$

The duration of the momentum measurement is Δt, with Δx the uncertainty of the corresponding position measurement (incurred when measuring the momentum). He then notes that the uncertainty in h_{01} (from the uncertainty in "recoil speed": $\Delta x/\Delta t$) must be of order $\rho(\Delta x/\Delta t) \cdot \Delta t^2$ (from the Einstein equation: $\Box h_{01} = \rho v_x$). Therefore, $\Delta[00, 1]$ must be of order $\rho \Delta x$, implying that momentum measurements pick up an additional uncertainty in the gravitational field, $\rho \Delta x \cdot \rho V \Delta t$. The momentum uncertainty is:

$$\Delta p_x \approx \frac{h}{\Delta x} + G\rho^2 V \Delta x \Delta t \tag{4.18}$$

Bronstein notes that the second summand can be made arbitrarily small as compared with the first, though, he says, to get the best possible measurement of $[00, 1]$, one should make both of the same order. Hence one should choose Δx to be:

$$\Delta x \approx \frac{1}{\rho}\left(\frac{h}{GV\Delta t}\right)^{1/2} \tag{4.19}$$

Which then yields:

$$\Delta[00, 1] \gtrsim \frac{h^{1/2}G^{1/2}\Delta t^{1/2}}{V^{1/2}T} \tag{4.20}$$

As he puts it, this shows that "an absolutely precise measurement of the gravitational field would be possible only if an arbitrarily rapid measurement of momentum were possible" [Bronstein (1936), p. 276]. This is not possible since $\Delta x << V^{1/3}$ should hold, by construction. But this implies:

$$\Delta t \approx \frac{h}{\rho^2 G V^{5/3}} \tag{4.21}$$

Moreover, special relativity implies that Δx can never exceed $c\Delta t$, further implying:

$$\Delta t \gtrsim \frac{h^{1/3}}{c^{2/3} \rho^{2/3} G^{1/3} V^{1/3}} \tag{4.22}$$

All of which shows that $\Delta[00,1]$ must exceed:

$$\frac{h}{\rho T V^{4/3}} \text{ or } \frac{h^{2/3} G^{1/3}}{c^{1/3} \rho^{1/3} V^{2/3} T} \tag{4.23}$$

As he points out, the second inequality applies to heavy test bodies, while the first inequality holds only for very light test bodies of order $\rho V \lesssim h^{1/2} c^{1/2} G^{1/2}$ (about 0.01 mg).[55] Hence, he suggests that the uncertainty $\Delta[00,1]$ must respect the second inequality to get the most precise measurement. In the weak field ($h_{\mu\nu} << 1$) case, there is no limitation to the accuracy of gravitational field measurements. Here, as he points out, we have linearized field equations which means that the linear superposition principle holds, meaning one can generate solutions of arbitrarily large mass-density ρ. This Bronstein views as clear demonstration of the possibility of a consistent quantum theory of weak-field gravity (i.e. a special-relativistic theory of quantum gravity, as would be carried out by several people—discussed in the next chapter). But, this does not carry over to the non-linear case, where the spacetime's deviation from flatness is not small. In particular, he notes, the "gravitational radius" of the test particle used to make the measurement of the field, given by $(G\rho V/c^2)$, must be less than its linear dimensions, $V^{1/3}$, so that an upper limit is placed on the possible density of the particle: $\rho \lesssim c^2/G V^{2/3}$. In other words, the Schwarzschild radius of the test body used to make

---

[55] This combination of the three fundamental constants, hcG we would today associate with the 'Planck mass.' Note that James Anderson would independently derive what are essentially these same results in 1954, where he also suggests that they might be taken to indicate that revisions of one or the other theory, quantum mechanics or general relativity, might be needed since it appears that "the most obvious way of measuring the gravitational field classically, is inconsistent with the requirements of quantum mechanics" [Anderson (1954), p. 8]. Maury Osborne, then at the Naval Research Laboratory, wrote a similar study, earlier in 1949, though this time focusing on the determination of curvature by triangulation as the field measurement, though with much the same outcome: curvature makes sense only in the large. He writes that in any "attempts to unite quantum theory with the general theory of relativity, the curvature of space should arise as a statistical concept valid only for very large numbers of particles" [Osborne (1949), p. 1582]. Kilmister and Stephenson likewise took the point of these studies of quantum restrictions of the gravitational field to be that gravitation is necessarily a theory for large masses and spatial volumes, so that we should be "looking only for a unified field theory in the macroscopic sense" [Kilmister and Stephenson (1954), p. 93]. Misner would later draw attention to this feature, remarking that one possible motivation for studying quantum gravity is the "desire to understand the quantum limitations on measurements of the gravitational field" [Misner (1957), p. 500].

measurements must be less than its linear dimensions in order for the measurements to be possible.

Hence, we find a restriction on the accuracy of measurement of the Christoffel symbol (the gravitational field), as a result of special features of gravitation: there cannot exist arbitrarily large masses for some given volume.[56] Given this feature, Bronstein was led to his claim that serious revisions[57] would be required in the foundations of physics in order to extend quantization to the full non-linear (non-weak field theory), though the nature of the revisions is not in any way made clear.

Jacques Solomon was one of the very, very few to cite Bronstein's work early on. He took up Bronstein's revisionary ideas, consolidating the view that the superposition principle of quantum mechanics appeared inconsistent with full general relativity, so that the available quantization methods were simply not adapted to general relativity [Solomon (1938), p. 484].[58] Solomon's concern over whether the *status quo* of thinking quantum mechanics and gravitational physics as 'non-overlapping magisteria,' that don't interfere with one another on account of their "very distant domains" and the weakness of gravitation was valid.[59] Solomon notes that the results we saw in the previous section appear to rule out any real formal conflict too, since quantum mechanics could be established on any Riemannian manifold. Indeed, for this reason, Solomon notes, quantum gravity did not look "susceptible to important progress" (p. 479). Curiously, Solomon points to two recent reasons for increased interest: (1) the potential link between Pauli's neutrino theory and gravitational waves (which we consider in the next chapter), and (2) Eddington and Milne's cosmologically inspired investigations of quantum and gravity.

Hence, Solomon wished to probe afresh the relationship between quantum mechanics and gravitation. In doing so he draws attention to practical limitations relating to

---

[56] In terms of the Bohr–Rosenfeld analysis, the universal coupling of general relativity means that there are no possibilities for introducing 'compensatory' devices (springs and such like) to keep the large fields (resulting from the large test charges used to make accurate measurements) under control. Clearly too, as we increase the masses of the measurement equipment, there will be inevitable distortions which, thanks to the non-linearity, will be highly unpredictable.

[57] Kozhevnikov suggests that Bronstein was keen on revolution in physics, since it suggested a crisis of bourgeois science, which itself pointed to a broader revolutionary situation in the West [Kozhevnikov (2004), p. 87/fn. 16].

[58] Solomon was another of Pauli's students to tackle quantum gravity. Indeed, Solomon had been a student with Rosenfeld in Paris in 1926, before the latter had produced his quantum gravity work. Though Solomon appears to have already had interests in general relativity (see [Rosenfeld (1959)], pp. 297–8), it seems Rosenfeld passed his quantum gravity concerns onto Solomon, while Solomon passed on his Marxist concerns to Rosenfeld (cf. [Jacobsen (2012)], p. 16)—Solomon, a Marxist activist and member of the French resistance, was also killed, on May 23rd, 1942, by the special brigades of the German occupation (see [Colin (2010)] for a biographical study).

[59] Though Dirac believed quite firmly that general relativity and quantum theory would have something to say to each other (and indeed did important work on the subject, as we see in Chapter 6), he nonetheless accepted this 'non-overlapping magisteria' argument and gave particularly explicit defences of it, e.g.: "Since the time when Einstein's general theory of relativity first appeared, various more general spaces have been proposed. Each of these would necessitate some modifications in the scheme of equations of atomic physics. The effects of these modifications on the laws of atomic physics would be much too small to be of any practical interest, and would therefore be, at most, of mathematical interest" [Dirac (1935), p. 657].

measuring quantum gravitational forces between particles, remarking on the 10^{36} order of magnitude difference between the Coulomb repulsion and the gravitational attraction. He notes that measurements would require extremely long timescales, comparable with the age of the universe, though not inconsistent from the perspective of the theory itself. Most of his discussion is couched in Newtonian gravity. He briefly discusses Bronstein's results, noting the relative ease of dealing with the weak field case, wherein one can use "methods analogous to those of the quantization of the electromagnetic field" (p. 483). One can, as in the Bohr–Rosenfeld case, measure the field quantities with arbitrary precision simply by choosing a test body of sufficiently high density, to reduce the uncertainty.

He goes on to discuss Bronstein's remarks about the non-linear theory, noting that since the superposition principle is inapplicable in such a case, Bronstein's main analysis, leading to the second inequality in (4.23), is ambiguous. He finds a further argument, pointing to what he views as a genuine inconsistency between general relativity and quantum field theory, in a result of Nathan Rosen's [Rosen (1937)] purporting to show that the full theory does not admit solutions representing polarized planar waves of finite amplitude which are at the same time free of singularities in the metric. Without waves of this sort, the standard methods of field quantization can't go through. Solomon takes such considerations to "seriously call into question the possibility of reconciling present formalism of field quantization with the nonlinear theory of gravitation" (ibid., p. 484).[60]

However, the visionary calls to action of Solomon and Bronstein were not answered for well over a decade, as more and more evidence pointed away from peaceful coexistence between quantum theory and general relativity. The focus on measurement issues, however, remained part of the problem of quantum gravity. A major injection of new ideas came from advances in cosmology. These were quickly linked to previous debates about the relationship between gravity and quantum. In particular, the question of *coherence* reigned supreme: how do the new advances square with what we know about quantum systems (Schrödinger and Dirac wavefunctions and so on)? This question would become implicated in the methodology of cosmological theories, since a better meshing with quantum reality upped the credence of some such theory.

4.4 Cosmology and "Invertebrate Cosmythology"

In addition to new developments in quantum theory, there were also influences to the problem of quantum gravity coming from the application of general relativity (and the more general study of gravitation) at the largest scales. These also were brought to bear

---

[60] Hermann Bondi, Felix Pirani, and Ivor Robinson argued against Rosen's result after the war (after new mathematical machinery was developed), revealing the singularities to be coordinate artefacts rather than physical singularities ([Bondi, Pirani, and Robinson (1959)], cf. [Kennefick (2007)], p. 97). But perhaps Solomon was not alone in taking Rosen's result seriously since, as Bondi, Pirani, and Robinson put it, "plane gravitational waves do not exhibit their planeness in so clear a way as plane electromagnetic waves do" [Bondi, Pirani, and Robinson (1959), p. 520].

on quantum theory, again in a bid to establish a coherent worldview. Schrödinger, in a review of Eddington's book *Relativity Theory of Protons and Electrons,* referred to the problem of "how to reconcile Relativity Theory and Quantum Theory" as a question of "theoretical conscience" [Schrödinger (1937a), p. 742]. He went on to present the problem in remarkably modern terms, invoking a clash of basic principles:

> Increasing uneasiness grew from the fact that in the course of their development, both theories approached what many regarded as final states of perfection, but without reaching true Anschluss[61] to one another, or even reconciling their mutual discrepancies. Having, both of them, acquired the rank of inalienable knowledge, they seemed incapable of undergoing serious modifications, yet in urgent need of such, in view of the mutual inconsistency of their respective fundamentals.
>
> (ibid.)

Of course, the work on the wave equations in generally relativistic universes was precisely concerned with demonstrating such an 'Anschluss,' but as we see in the next chapters, the later methods of field quantization tended to spoil what looked like a peaceful coexistence, revealing new incompatibilities, beyond those broad clashes suggested here by Schrödinger and recognized by only a very few at this time. Cosmological considerations were seen by a handful of physicists to offer some way of bridging the gap, and often manifested in terms of the largest scales *determining* features of the smallest scales. Cosmology was thus seen to offer new prospects of uniting quantum and relativity, yielding a novel solution to the problem of quantum gravity.

With this early scientific cosmology came controversy; in particular, the battle over methodologies, instrumental versus realist. In some ways, aspects here paralleled the initial phases of quantum theory in which there was a split over viewing the quantum theory as a formalism about observable entities only or whether it also described some underlying reality. Just as Heisenberg removed reference to anything but the transitions between states, so Edward Milne, for example, would outlaw curved space precisely on the grounds of its unobservability—indeed, Milne viewed general relativity as an "alien" in the realm of physics, preferring to base his theory on invariance under the group of homogeneous Lorentz transformations. But that is only part of the story: Milne also wanted to revise the foundations of scientific reasoning to *rationalize* the process so that one could deduce physical (observable) consequences from general, a priori principles. Eddington would try something superficially similar, though with very different aims and methods, and employing an unusual approach to observation and measurement that transformed the meanings of '*a priori*' and '*a posteriori*'.[62] However, towards the

---

[61] Coming in 1937, one cannot help thinking that the choice of word here is no coincidence...

[62] Eddingtonian *a priori* (which he also refers to as "epistemological knowledge") is "knowledge which we have of the physical universe prior to actual observation of it"—it is, he says, "prior to the carrying out of observations, but not prior to the development of a plan of observation" [Eddington (1939a), p. 24]. Then *a posteriori* simply refers to the *results* of such planned observations. Both, then, involve notions of observation and measurement. As much, if not more, confusion over Eddington comes from his idiosyncratic use of established terms as it does from his introduction of entirely new terms.

end of the 1930s, most physicists were disenchanted with the kind of grand schemes proffered by Eddington and Milne. Herbert Dingle [Dingle (1937), p. 385] famously described such approaches (also including Dirac's cosmological work) as "invertebrate cosmythology," though I think it's clear that he often simply didn't understand the details of any of these approaches if he lumped them together in such a way; for Eddington's methods are a far cry from Milne's, and the former himself viewed the latter's work as a perversion of physics (e.g. [Eddington (1939b), p. 230]).[63]

But one should not discount the impact of these approaches on the history of the problem of quantum gravity. At the very least, Milne's approach offered a contrary voice to the then fairly well established view that gravitation was an aspect of curved space, showing instead how one could recover certain results without invoking this, for him, extravagant concept. We discuss examples that at least have conceptual kinship with Milne's 'flat space' approach in the next chapter.

Eddington wished to unify cosmology with the rest of physics—indeed, cosmology provided the test case for linking gravity and quantum. Schrödinger too sought such a unified field theory, likewise incorporating his wave mechanical ideas in combination with cosmology (the topology and dynamics of the universe).[64] In his correspondence with Einstein from 1939, he excitedly requests information on a story he had heard about Einstein's linking gravitation and matter waves. Schrödinger points out that he had been thinking of the matter waves ("ψ-waves") as "disturbances in the gravitational potential."[65] He viewed these as genuine, energy-transporting gravitational waves to be viewed as "quantized gravitational waves" (ibid.).[66] However, Einstein responds that he simply views the wave picture as an incomplete representation of reality—Schrödinger appears to have dropped this rather Einsteinian line of thinking at this point. Since Eddington and Schrödinger represent the two main leaders of quantum gravity just before, during, and immediately after the Second World War, we spend some time

---

[63] Eddington's primary concern was with Milne's inability to take seriously the "cosmical repulsion" (i.e. the cosmological constant) as the correct explanation of the nebular velocities (ibid., p. 229). Moreover, just as Milne charged general relativity with being cut off from the rest of physics, Eddington returns the favour by pointing out that Milne's kinematical relativity (so called because it focused purely on *motions* of objects from the point of view of observers) must ditch the quantum-specified standards of length and time at large scales (ibid., p. 230). For Eddington, the cosmological constant was the essential link to atomic structure, since from it one could derive the total mass of the universe, which could then be given a quantum mechanical characterization via an appropriate choice of units. One can find an excellent discussion of this controversy over Milne, and other controversial episodes in early cosmology, in [Kragh (1996)].

[64] Schrödinger thought it rather obvious that wave mechanics would have to be substituted for the usual particle-picture used in relativistic cosmology (cf. [Schrödinger (1940b), p. 5]). Notwithstanding the remarks he made to Einstein, mentioned in the next footnote, unlike Einstein, however, Schrödinger saw no way to generate anything like wave mechanics from general relativity alone: it had to be added as a *supplement* to the theory; and, indeed, Schrödinger viewed wave mechanics as a far more "congenial" partner than particle mechanics.

[65] Letter from Schrödinger to Einstein, July 19th, 1939, [Przibram (1934), p. 36].

[66] He further notes that his calculations along these lines, led him to find an error in Eddington's treatment of the origin of the energy operator, in §13.7 of his book *Relativity Theory of Protons and Electrons*, which Schrödinger had reviewed [Schrödinger (1937a)]—this appears to have been Schrödinger's way into Eddington's scheme.

outlining their closely related views, as well as Max Born's competing view, in the remainder of this chapter.

Eddington was perhaps the first person, after Einstein himself, to take very seriously the problem of quantum gravity. His approach was firmly grounded in unpacking the observable content of such a theory, something he was working on in his final (unfinished) work: *Fundamental Theory*.[67] In their treatise on Eddington's physics, Ted Bastin and Clive Kilmister write that:

> Eddington's *Fundamental Theory* embodies an unorthodox attitude to the interpreta-
> tion of physical measurement. This attitude regards physical theories as formulations
> of conditions presupposed by our experimental procedures, rather than determined,
> empirically, by those procedures. Eddington's method was to alter the existing structure
> of relativity theory in accordance with this attitude in order to make it cover quantum
> problems.
>
> [Bastin and Kilmister (1962), p. 59]

Eddington's understanding was rooted in the notion that observables in general relativity must have a relational character, necessarily involving pairs of quantities. Measurements are, therefore, also more complex than often supposed:

> [A]n observable coordinate is measured, not from an abstract geometrical origin, but
> from something which is involved physically in the experiment which furnishes its
> observed value... We must therefore distinguish between the "physical origin" from
> which an observable coordinate is measured, and the "geometrical origin" of the
> mathematical reference frame which is inaccessible to measurement.
>
> [Eddington (1943), p. 3]

This is more orthodox in current treatments of general relativity, yet in his review of Eddington's *Philosophy of Physical Science*, Ernest Nagel [Nagel (1940), p. 163] asks: "what student of relativity will recognize the essence of that theory to consist in the fact that 'we observe only relations between physical entities'?" One wonders what Nagel would have offered in its place. Yet Eddington's analysis was more far reaching than this basic idea and, as I say, now more or less standard picture.[68] He wanted to include quantum theory too. This complicates the analysis considerably: observables are now represented as involving products of wavefunctions ("double wave-functions"): $\phi\psi$. These products are not to be viewed as ordered, in terms of ontological or formal priority;

---

[67] An earlier work, published during Eddington's lifetime, is the pamphlet from his Dublin IAS lectures in 1943: *The Combination of Relativity Theory and Quantum Theory* [Eddington (1943)]. In this, Eddington essentially extracts what he sees as the two basic principles of relativity and quantum mechanics: the principle of relativity (i.e. there is no such thing as absolute motion or locations) and the Heisenberg uncertainty principle. What he sees as being required is a relativistic theory of probability distributions.

[68] I believe this picture is what Bryce DeWitt had in mind when he wrote that "When the two theories are united the result is an operational theory *par excellence*" [DeWitt (1967), p. 1140]. That is, "it will say nothing about time unless a clock to measure time is provided, and it will say nothing about geometry unless a device (either a material object, gravitational waves, or some other form of radiation) is introduced" (ibid.).

both are necessary for the observable to be given an ontological interpretation. Eddington puts the ontological distinction in terms of "potential" and "actual" properties: the components ψ and ϕ, taken in separation, are merely potential, while their conjunction (an invariant), is actual or real. He writes:

> [W]e may say that the so-called physical entities—the relata whose relations furnish the observables—have potential properties only, which must be expressed by a symbol ψ with no observational interpretation; but two such entities in conjunction have actual properties, and $\phi\psi$ has an observational interpretation.
>
> [Eddington (1966), p. 262–3]

The relata are secondary concepts in this scenario, without any direct physical interpretation, while what Eddington calls the "conjunction" is primary.[69] Note that such conjunctions are not yet *measurables*, which will require at a minimum a *pair* of such already complex observables (and therefore four components)—hence, a measurement is a relation that relates a pair of relations. Eddington (ibid.) puts the point thus: "Formally a measurement always refers to just four relata, viz. the terminals of the object relation and the comparison relation.[70] The basis of measurement is therefore a four-point element of world-structure" [Eddington (1936), p. 323]. However, as Eddington again makes plain, the same primacy of the conjunction appears:

> When an object-body is observed in conjunction with a reference body, any measurement that we make determines characteristics which belong, neither to one body nor to the other, but to both jointly. It is, however, customary to allot these mutual characteristics to the object-body, or more defensibly to partition them between the two bodies according to some self-consistent scheme. The conceptual transfer, by which self properties are substituted for mutual properties, is a habit of thought which has been elevated into a convention, and we can scarcely do otherwise than accept in principle.
>
> [Eddington (1943), p. 41]

Inasmuch as we can write down an abstract origin, this is only part of the mathematical framework used to represent systems and their properties. Eddington's theory is, then, one of *real* measurements. Standard accounts of physics, quantum mechanics included,

---

[69] Of course, we will often adopt the practice of supposing that one or other factor is 'really' the one that *has* some property (that *really* moves, or is located, etc.). But this is a *façon de parler*: either could be viewed as the reference factor or standard: "we can no more contemplate an atom without a physical universe to put it in than we can contemplate a mountain without a planet to stand it on" [Eddington (1943), p. 13]. Such composite entities roughly correspond to what Carlo Rovelli [Rovelli (2002)] calls "complete observables" (as opposed to the individual ψ and ϕ, which correspond to gauge-variant "partial observables"). Here, then, Eddington certainly battled with many of the same conceptual issues that have recently re-emerged in the study of quantum gravity, though largely in isolation (often, it has to be said, due to his own reticence to engage in productive dialogue with other physicists).

[70] In *Fundamental Theory* Eddington invokes the cosmological term to function as a "comparison fluid" (an "all-pervading entity"), thus grounding a physical reference frame capable of localizing observables. This is closely related, conceptually if not historically, to what would later be called 'the method of intrinsic coordinates' (or what John Stachel has called an "individuating field" [Stachel (2014), p. 14]).

will usually have no trouble referring to a single electron say, but according to Eddington's account, this is simply wrong-headed since we will have eliminated a necessary standard of comparison. Again:

> We observe only relative positions and relative velocities; consequently an observable coordinate or momentum always involves two physical entities. A measurement involves four physical entities, two to furnish the observable that is said to be measured and two to furnish the comparison observable used as standard.... What is measured is something associated with four entities; we shall call this a *measurable*.
>
> [Eddington (1943), p. 48]

This "something associated with four entities" is the most fundamental measurable possible (in a logical sense). It will only exist (be instantiated) if all four entities are instantiated together. Eddington defines an existence symbol, $M = \mathcal{J}_r\mathcal{J}_s\mathcal{J}_t\mathcal{J}_u$, to represent this idea (and applying to all possible measurables),[71] where M has 16 eigenvalues a_{ij} corresponding to various 'yes/no' value assignments. Hence, $a_{r1}a_{s1}a_{t1}a_{u1}$ is the unique 'yes' eigenvalue (referring to the single way it can exist), and the remaining 15 eigenvalues correspond to 'no existence' (no-eigenvalues). As Eddington puts it, "the characteristic feature of a measurable is that it has 15 different ways of not existing" [Eddington (1943), p. 268]. Measurables in this sense form the building blocks of everything else that follows, with the 'physical entities' being defined in terms of these.[72]

The scheme he goes on to develop is *fundamental* in that it bases *every* piece of physics, including quantum mechanics and gravitation, on these two-entity observables and four-entity measurables (i.e. pairs of observables). As Eddington points out, even positions of entities are not primitives since they require localization, and this requires a material frame. The most primitive entity possible, in a structure, is something characterized by existence or non-existence alone. This property cannot be further reduced and there is nothing more basic.

There was a serious conflict exposed by Eddington in the use of reference frames in the two ingredient theories of quantum gravity: relativity makes use of frames that ignore

---

[71] Here, $\mathcal{J}_r$ is an idempotent existence symbol referring to the rth entity in the structure M, and has two eigenvalues: a_{r1} ('yes,' or existence) and a_{r0} ('no,' non-existence). This was quite explicitly linked to the notion of a binary *choice* by Eddington: it refers to the outcomes of measurements so that if we know the outcome already, then $\mathcal{J}_r$ has eigenvalue a_{r1} (yes). In the absence of a measurement, Eddington speaks of "partial existence" (in the structure). This is a direct ancestor of John Wheeler's much later notion of 'It from Bit,' in which he speaks of "apparatus-elicited answers to Yes/No questions"—since Wheeler's student and long-term friend Peter Putnam was an ardent Eddingtonian (he had written to Dirac expressing his desire to work on Eddington's derivation with him as a graduate student, on August 27th, 1955: Wheeler Papers, Box 22), it is entirely plausible that there was some influence between Eddington's idempotence ideas expressed here and Wheeler's It from Bit ideas.

[72] Physical entities form the "physical universe" which Eddington *defines* as "the systematisation of knowledge obtained by measurement" [Eddington (1943), p. 268]. Given this sleight of hand, it follows that much of what empirical science takes for knowledge of an objective world, is nothing of the sort according to Eddington. Physical science is an elaborate scheme of conventions, referring as much to our capacities for performing measurements (and our inabilities to detect certain quantities and transformations), as to an objective world—this is the meaning of his own view known as "selective subjectivism."

the uncertainty principle, and quantum theory ignores frames that involve spacetime curvature. He sets up the problem as follows:

> Relativity theory begins with a denial of absolute motion. An observed velocity $d\xi/dt$ of a physical entity is necessarily relative to another physical entity. Likewise the coordinate ξ, of which an observed velocity is the time-derivative, is a *relative quantity* of two physical entities.
>
> Quantum theory insists that the connection of a physical object with the geometrical frame of coordinates is governed by Heisenberg's uncertainty principle. A particle is not exactly locatable as a point (or a world-line) in a geometrical frame. It can only be assigned a probability distribution of position and velocity.
>
> In modern physics these two conditions of observability have been applied separately with very far-reaching results. *In relativistic quantum theory they must be applied in combination.* The combined principle is that a coordinate ξ is observable only if it is a relative coordinate of two entities, both of which have uncertainty of position and momentum in the geometrical frame.
>
> [Eddington (1943), pp. 2–3]

In other words, Eddington viewed the problem of quantum gravity as involving a new notion of observables (and measurement) that does justice to the uncertainties of quantum theory, while simultaneously respecting general relativity's relational requirements. We will see in a moment how Eddington chose a case study that could be given a dual description by both quantum theory and general relativity. First let us mention what were seen to be problems with Eddington's proposed solution of the quantum gravity problem. In a review [McCrea (1947)], McCrea directly disputed Eddington's suggested link between quantum theory and relativity, suggesting it tangles him in a logical mess:[73]

> The junction with current quantum theory is made incorrectly, in the opinion of the reviewer. Eddington argues that in the equations of quantum theory the coordinates must refer to a "physical origin" which is the centroid of a large number of particles. However, in quantum theory the states of a physical system are described by points in a Hilbert space of square integrable functions defined over a configuration space. The configuration space is labeled by coordinates which Eddington would call geometrical coordinates and do not refer to any physical origin. Thus current quantum theory deals

---

[73] However, McCrea later linked atomic to cosmological structure in a very different way, based on the continuous creation model (of Bondi, Gold, and Hoyle) in which the creation is understood as "the spontaneous appearance of elementary particles" or "an atomic process" [McCrea (1951), p. 573]. In McCrea's model the creation term is now placed on the right-hand side of the field equations, bundled into the stress-energy tensor $T_{\mu\nu}$, since it is related to the zero-point stress-energy of space that comes from a quantum theory of fields, according to McCrea (ibid., p. 574). Felix Pirani [Pirani (1954)] attempted to further develop this model by providing a specific, physical description of the continual creation process. This he does by postulating particles that come into being at a creation event, and having negative energy-momentum and zero mass—these being responsible for McCrea's zero-point stress. These he calls "gravitinos" and notes that the process cannot be time-invariant because of the expansion of space. Given the close relation to the neutrino (zero charge and mass), he speculates that his gravitinos are simply neutrinos with a change of sign (of the energy). However, he doesn't develop this idea further, and indeed the whole idea fell silent.

with "locatable" systems and not as Eddington insists "non-locatable" ones. It may be
that he is correct and that current quantum theory is wrong, but in that case he cannot
use results of that theory without establishing them from his postulates.

Yet Eddington was not alone in looking towards non-locality as part of the union. Born,
as we will see below, did the same, as did many others coming later, including Wigner,
discussed above. Today, of course, we view some kind of non-locality as an essential
component in any theory of quantum gravity. McCrea's objection to the inability to
recover the particle structure from experimental physics was more on target. McCrea
is not unimpressed by some of the correspondences Eddington achieves between the
calculated and observed values of constants, viewing him as "an amazingly adept numerical manipulator and rationalizer". However, many claims are simply inconsistent with
experiment: the non-Coulomb force between two protons was incorrect, and Eddington
entirely neglects meson physics (and rejected Yukawa's meson theory), claiming he could
make do with nuclear forces without meson-fields.

This is clearly a very different kind of measurement analysis than that carried out
by, e.g., Bohr and Rosenfeld, and looks for the conceptual basis of our modeling and
measuring practices in science.[74] In this respect, we must consider that his later work
was considerably less 'pie-in-the-sky' than many have supposed. Indeed, what one finds
is an account of measurement in quantum gravity that predates the kind of analysis one
finds in recent loop quantum gravity texts by half a century. Indeed, the apparently *a
priori* deductions of empirical features of the world (laws and constants) are all based on
the prior acceptance of his analysis of measurement and observables, which views the
scientific enterprise more like an incredibly complex man-made game than a mirror of
nature. Hence, the calculations of the total number of particles in the universe, or the
fine-structure constant, are understood to be the kinds of deductions that one is *forced* to
make if one follows the ideas of measurement laid out: they are the values *from the point
of view of a physicist acting in accordance with Eddington's principles*.

It is from this focus on measurement that the wilder aspects of his work came to
the surface. Eddington believed that a large part of what we call physics is subjective
in the sense that the operations of human cognition (our makeup as observers and
the constraints we find ourselves subject to) play a large part in the laws we find
and the constants we measure. In this, Eddington was concerned with revealing the
kinds of idealizations in measurements too: we invoke highly simplified backgrounds
('uranoids') for example, that function as the environment for the object-systems we
wish to measure. In this way, he is concerned (somewhat like Poincaré before him) with
exposing *conventions* that we have taken up as if they were objective truths of nature.

---

[74] Eddington was engaged in a kind of measurability analysis for combined quantum theory and general
relativity. As Bergmann and Smith put it, "[m]easurability analysis identifies those dynamic field variables that
are susceptible to observation and measurement ("observables"), and investigates to what extent limitations
inherent in their experimental determination are consistent with the uncertainties predicted by the formal
theory" [Bergmann and Smith (1982), p. 1131]. Additional complexities come from the fact that a physical
reference frame used to localize observables, to build relative quantities such as velocities and positions, in
measurements for example, will be described by a wavefunction.

Unfortunately, this wilder aspect tended to overshadow the more sensible claims about observables and measures.

One can see in relatively simple terms how this rationalist process is supposed to work, and indeed how it might appear somewhat plausible. One can derive the total mass of the universe empirically by measuring the recession of the spiral galaxies. That is, the cosmological constant describing that recession rate furnishes a value for the total mass. One can then change one's picture and view that mass in quantum theoretic terms using atomic units. This is the basis for Eddington's famous number, N, giving the total number of particles (electrons and protons only) in the universe. This number N is the core of his postulated link between the very large and the very small since he takes the atomic scale to be related to the cosmic scale as $\sqrt{N}$. From this link he finds that the radius of the (closed) Einstein universe is of the order $\sqrt{N}$ times the electron radius e^2/mc^2. Eddington claimed that there is an identity between spacetime curvature on the one hand and the wavefunction on the other: both represent the distribution of mass-energy in physical systems. Neither, however, is 'real' in a literal objective sense: they are both 'devices.' When one is doing general relativity one uses curvature (cG), and when doing quantum mechanics one uses Ψ (h). Eddington wanted to find a problem in which both quantum mechanics and general relativity can be used to find a relationship between their respective constants. His example was to do quantum mechanics on a closed space.

In a little more detail, Eddington used two methods to compute the total number of particles in the universe: *a posteriori* and *a priori* (in his sense). Via the former method he computed the number of protons in the universe to be *exactly* $N = 10^{79}$ (from 2.136×2^{256}). He called this "the cosmical number". Initially, he gets the number in a fairly straightforward manner. Firstly, he is dealing with an 'Einstein world' (a finite, unbounded universe). As above, he determines the mass and radius of this world from the recession constant V_0 (i.e., the increase in galactic velocities per unit distance):

$$V_0 = \frac{c}{R_0\sqrt{3}} \tag{4.24}$$

where V_0 is fixed by observation, which allows one to determine the universe's radius R_0. Given R_0 one then solves for the mass with:

$$\frac{\kappa M}{c^2} = \frac{1}{4}\pi R_0. \tag{4.25}$$

Next, given M, and the assumption that there are only protons and electrons, so all mass in the universe is constituted by these protons and electrons[75] (and there is an equal number of them), and given the mass of hydrogen (made up of 1 proton and 1 electron bound by the Coulomb force), one can just divide M by the hydrogen mass and multiply the result by 2 to get the total number of particles.

---

[75] Of course, we now have reason to believe that baryonic matter accounts for only a very small fraction of the total mass of the universe, with the majority comprising dark matter and, more still, dark energy.

But Eddington thought there was a second *a priori* route, starting from qualitative data (rather than quantitative data like V_0). The derivation involves his understanding of measurement and observables in a generally relativistic context according to which observables are represented by two-component entities and measurements by pairs of observables, or four-component entities (a 'measured' observable and a 'standard' observable). A measurable (given both quantum mechanics and general relativity) is represented by a quadruple wavefunction. The number of independent quadruple wavefunctions in an Einstein universe is identical to the number of particles. Eddington calculates $N = \frac{3}{2} \times 136 \times 2^{256}$ which, running this through *Mathematica*, yields 2.36216×10^{79}, close to the *a posteriori* value derived from observational input. Eddington links the force constant and the cosmical number via a square root relation. $F = \frac{2}{\pi}\sqrt{N}$. Given the exactness of his computation of N, he can also compute F exactly. Eddington thus claimed to be able to deduce atomic structure from cosmological structure, in a finite, closed universe. This feature of an Einstein universe makes it act like a box. When one has waves in a closed space one will get countable discrete modes. The exclusion principle (accepted by Eddington as a qualitative generalization) implies that each proper mode cannot be occupied by multiple particles. The sum total of such vibrational modes gives the mass of the universe which, as above, can be used to determine the radius and the number of particles.

Eddington gave similar analyses of other central parts of physics, including other constants (the number of particles in the universe was one such constant). Max Born was especially critical of Eddington's approach with respect to the fine-structure constant:

> When Eddington began his work the experimental value of $hc/2\pi e^2$ [the fine structure constant] was near to 136. Later experiments indicated a larger value, and today it is very near 137. Accordingly Eddington adapted his theory by adding [quite arbitrarily and without proper motivation] a unit. [Born (1943), p. 38] (cited in [Mehra and Rechenberg (2001), p. 936].[76]

Max Born was, then, amongst those with a severe aversion to Eddington's brand of physics, linking as it does the physics of the smallest to the physics of the largest. He proposed an alternative to the view espoused in Eddington's *Relativity Theory of Protons and Electrons*, in his 1938 paper, "A Suggestion for Unifying Quantum Theory and Relativity." This paper suggests that it was common knowledge at the the time that "the difficulties of our present theory of ultimate particles and nuclear phenomena (the infinite values of the self energy, the zero energy, and other quantities) are connected with the problem of merging quantum theory and relativity into a consistent unit" [Born (1938), p. 291], explicitly citing Eddington's book as a case in point. Born, as mentioned

---

[76] The new value of 137 was based on Millikan's data. The apparent fudging led to Eddington being labeled Sir Arthur AddingOne. The values for the constants arise from the number of dimensions of his phase spaces (based on E-numbers), with the generating rule, $f(n) = \frac{1}{2}n^2(n2+1)$, giving 136 (degrees of freedom) for $f(4)$. Eddington argued [Eddington (1929)], not without basis in fact, that he had neglected a degree of freedom introduced by the particle permutation associated with the Fermi–Dirac statistics.

earlier, views general relativity as a purely bulk (or 'molar') phenomenon, not applicable to the ultimate particles. He sees general invariance as the key principle to be merged.

Born's alternative had a conceptual root: at the atomic scale, one simply did not have any way of making sense of distance measurements, since one had no appropriate rods and clocks. On the other hand, while positions and distances in coordinate space made no sense (i.e. was not observable), in momentum space one has a notion of distance that does make sense at the atomic scale. Born found this lack of symmetry puzzling, and proposed that the laws of nature are in fact symmetrical with regard to space-time x^k (where x^k are the coordinates of spacetime events) and energy-momentum p_k (the components of momentum-energy, where $k = 1, 2, 3, 4$) exchanges: $x \to p$ and $p \to -x$. For example, in the context of quantum theory, the commutator, $[x, p] = i\hbar$, was known to be invariant under this duality (called "the postulate of reciprocity"). Likewise, in classical theory, Hamilton's equations were also Born dual. Born traced the apparent breakdown of symmetry to the fact that in general relativity spacetime is curved, while the energy-momentum space is flat. Born was led by his duality to the claim that since general relativity involves curved space, in the coordinate representation, then it must also involve curved phase space, in the momentum representation. That is, there would be a metric in the $p - space$, not directly connected with the standard spacetime metric, so that the notion of a metric is only valid in limiting cases, with the standard metric applying to bulk phenomena ("the molar world"), and the p-space metric applying in the case of nuclear phenomena ("the nuclear world").

Born then utilizes, as did Schrödinger and Eddington, the finite radius of the Einstein universe, which gives an upper limit to the distance between two points. Switching to the $p - space$, reciprocity implies that a curvature must be defined in the same way as with $x - space$—and, given the nature of general relativity, this would have to include a stress-energy tensor. One will end up with solutions (to the $p - space$ analogous Einstein equations) corresponding to a closed momentum space, so that by analogy with the finiteness of the distances in $x - space$, there will be momentum cutoffs determined by the radius of the closed space. This cutoff is used to tame the infinities associated with quantum field theories (and Born gives explicit calculations in the context of quantum electrodynamics), and, Born argues, it does say in a physically motivated manner, based as it is on relativistic considerations. Indeed, the cutoffs amount to the provision of a discrete structure of space (a "granular, or lattice structure" as Born puts it [Born (1938), p. 302]), again grounded in physics rather than a "strange assumption a priori" (ibid.).

However, it can be seen that while Born vehemently argued against a link between the cosmological considerations and atomic considerations, he essentially demonstrates just such a link, albeit more indirectly: his cutoffs are determined by a cosmological parameter, the universe radius. This, then, amounts to an attempt to modify the classical theory in order to resolve thorny problems in the quantum theory.[77]

---

[77] Apart from its deployment by Hartland Snyder [Snyder (1947)], in his paper on quantized spacetime (where curved momentum space was linked to non-commutative spacetime coordinates), the principle was not widely taken up. For example, we find Bryce DeWitt noting, in 1957, that Born's symmetry does not "persist in detail in the dynamical theory of actual physical systems" [DeWitt (1957), p. 378]. However, in

Schrödinger was, however, utterly captivated[78] by Eddington's ideas linking the very small to the very large. He focused his attention precisely on the derivation of atomicity from the closed topology of the universe, believing the latter to be in fact the *only* way of generating the atomicity of matter, linked via wave mechanics.[79] This is, however, somewhat akin to Klein's earlier deduction of the quantization of charge from his topologically closed fifth dimension. In Schrödinger's scheme, there's a bottom to being (described by quantum mechanics), because the universe is finite (and so, qua finite system, possesses discrete proper frequencies), and that is described by general relativity. Hence, general relativity is the more fundamental description in a sense, giving rise to quantum theory. As he puts it very starkly in a 1939 paper:

> Wave mechanics imposes an *a priori* reason for assuming space to be closed; for then and only then are its proper modes discontinuous and provide an adequate description for the observed atomicity of matter and light.
>
> [Schrödinger (1939), p. 899]

That is, the universe must be finite and closed to ensure that the wave equations for matter have a discrete spectrum (cf. [Schrödinger, (1940b)])—these energy levels are interpreted as the proper vibrations of the space, which together give the total mass of the universe. As he expresses it, "[t]he very nature of the matter which forms the Universe is determined by the shape and size of the Universe as a whole" [Schrödinger, (1940b), p. 6]. Moreover, if one considers an *expanding* universe one finds what Schrödinger labels a result of "outstanding importance" (as it has indeed revealed to have been),

---

recent years Born reciprocity has re-emerged as a central component of so-called 'Doubly (or deformed) Special Relativity' [Amelino-Camelia et al. (2011)], in which the non-commutative spacetime coordinates and an energy dependent metric (implying that the events observed in spacetime depend on the energy of the probe) are both reducible to a deeper principle known as *relative locality*. This principle exactly latches onto Born's original motivation, based on the absence of locality at small scales. It has also been suggested that, in the context of string theory, Born duality is essentially just T-duality, linking string theories on large and small radii [Friedel, Leigh, and Minic (2014)]. In both cases, the non-local nature of Born reciprocity plays a key role.

[78] He wrote to Eddington of his work that "there is nothing in the world of knowledge in this moment that interests more" (Schrödinger to Eddington, October 23rd, 1937: Schrödinger Nachlass). This seems to have been prompted by a review Schrödinger was asked to write on Eddington's *Relativity Theory of Protons and Electrons*, which appeared on October 30th, 1937. This showed to Schrödinger how to get out physically motivated discreteness from a closed, finite universe, for this generates waves with discontinuous proper modes, with no ad hoc interventions. Einstein's universe provides a "natural and wall-less box" that necessitates discreteness of its proper vibrational modes (or energy levels) [Schrödinger (1940a), p. 743].

[79] Later work along similar lines came from the desire to probe the Hamiltonian quantization of gravity. Bryce DeWitt used the Hamiltonian formulation in the context of a closed Friedman model universe which was then quantized (essentially 'quantizing the universe'). A similar cosmological approach was used to probe Wheeler's concept of 'superspace.' A curious parallel to the covariant versus canonical split of quantiztion approaches arose in this cosmological context. For example, Charles Misner performed a similar procedure, only using the Lagrangian rather than the Hamiltonian. One can see a trajectory linking this early quantum cosmology all the way to Stephen Hawking's work on, e.g., Euclidean quantum cosmology. We can also find a switch in which theory acts as a resource in taming singularities, with quantum mechanics being proposed as a potential source of resolution of spacetime singularities in cosmological models. The extent to which modern cosmology has benefited from quantum considerations constitutes an interesting open historical problem.

namely that there will exist (for particles) "production or annihilation of matter, merely by the expansion" [Schrödinger, (1939), p. 900]—for light, he notes that there would be light production in the opposite direction. These effects—resulting from the inability to separate out two frequencies of equal but opposite sign (i.e. the positive and negative modes) in a non-static context—he considers to constitute "alarming phenomena." However, he finds that for *linear* functions of time, there is no such problem: only for accelerations would the production phenomena occur, which Schrödinger points out might indeed be the case in critical stages of the universe's evolution, such as when there are turning points (ibid., p. 901).[80]

While Schrödinger is quick to credit Eddington as the source of these ideas, he is equally as quick to distance himself from Eddington's wider epistemological framework for them. Indeed, Schrödinger pulled no punches in his correspondence with Eddington. Writing to Eddington on October 23rd, 1937, he writes that on presenting Eddington's ideas to an audience at a Bologna conference, from which he had just returned, that he had met with "an unvanquishable incredulity of the important group Bohr, Heisenberg, Pauli and their followers."[81] A major problem, he writes, was the consideration of alternative universes in the deduction of his value for $\sqrt{N}$ (the number of particles in the universe). Yet, Schrödinger argued, Eddington's world did not correspond to the actual world, but rather a world with a completely homogenous distributions of his primitive particles. Schrödinger expressed frustration that Eddington does not express very well what might be absolutely self-evident to him.

He had early expressed his frustrations, "teasing" Eddington "over and over again with the same argument".[82] Earlier on he was struggling with the proper vibrations of the world according to Eddington's numbers for the radius R and total number of particles N in the world. Schrödinger pointed out that the number of proper vibrations

---

[80] This is clearly related to what is now called "the Unruh effect," according to which accelerated observers will detect particles, while unaccelerated observers will detect none (and also to the Hawking effect demonstrating particle creation in the gravitational field of a black hole). Note that in this paper, Schrödinger refers to Paul Müller and Wolfgang Hepner as "collaborators" on the *static* universe part of his project. Müller, a fellow Austrian who had just completed his doctoral thesis under Schrödinger was, like Solomon and Bronstein, yet another early quantum gravity researcher whose life was cut short by the war, in this case dying on the Russian front, fighting for the German armed forces, on March 9th, 1942—note, this is the same Müller involved with the German atomic bomb project (the 'Uranmaschine,' together with Carl Friedrich von Weizsäcker and Karl-Heinz Hocker) which he was conscripted into soon after his work with Schrödinger. Müller had considered wave equations on a variety of spacetime backgrounds, including hyperspheres, cylinders, and even on a torus, picking up the additional periodicities associated with the holes and handles (see [Müller (1939), p. 375], the published version of his PhD thesis). Schrödinger possessed an undated manuscript of Hepner's "On the Boundary Conditions in Wave Mechanics" (most likely from 1939, Schrödinger Nachlass), in which Hepner notes that he has done computations for Dirac's equation in an Einstein universe with toroidal coordinates. It seems, from a letter from Hepner (January 17th, 1940: Archives for the History of Quantum Physics, Reel 37, Reel frame 7), that Schrödinger may have been too quick in viewing him as a "collaborator," since Hepner pointed out that he had not intended to look at problems in spherically curved spaces at all.

[81] Schrödinger to Eddington, October 23rd, 1937: Schrödinger Nachlass. His Bologna talk was published as [Schrödinger (1937b)].

[82] Schrödinger to Eddington, September 14th, 1937: Schrödinger Nachlass.

greater than some λ_0, in a volume of space V, is of the order $\frac{V}{\lambda_0^3}$. If R is the radius of this space, then $V = R^3$. Then if this world is to be in the 'Fermi state,'[83] then the number of particles must be of the order $N = \frac{R^3}{\lambda_0^3}$. However, with Eddington's numbers for N and R, namely 10^{79} and 10^{27} cm respectively, the limiting wavelength must come out as as few centimetres; Schrödinger couldn't believe that Eddington could accept this enormous value. Despite Schrödinger's own reputation, even he could not sell others on Eddington's basic scheme unifying the large and small.

Schrödinger finds a similar problem for the radius of the universe if the proton mass is to receive its empirically correct value [Schrödinger (1940b), §6]. He notes that if we assume a closed universe of radius R, then the configuration will be correspondingly restricted by the linear dimensions of R. In which case the momentum p of the ground state will be of order:

$$p = \frac{h}{2\pi R} \tag{4.26}$$

He associates with this p an energy $p^2/2m$ that is to be set equal to the zero-point energy mc^2, giving:

$$\frac{1}{2m}\left(\frac{h}{2\pi R}\right)^2 = mc^2. \tag{4.27}$$

Giving a mass of:

$$m = \frac{h}{2\pi R c \sqrt{2}} \tag{4.28}$$

From this expression for the mass it is relatively straightforward to see that in order to give the correct value for the mass of the proton, the radius R would need to be of the order of the Compton radius, $h/mc \approx 10^{-13}$ cm which, as Schrödinger says, is obviously "pure nonsense."

Once again, Schrödinger expresses his frustrations over Eddington's arguments, which he is not able to follow. However, Schrödinger ends by pointing to the "unusual difficulty of the task" in dealing with the attempted explanation of the laws and features of the smallest scales by the largest scales. As he puts it, since we only have these small-scale laws to go on, by way of explanatory materials, we are in the position of "that worthy baronet [Baron Munchausen–DR] who (said he) dragged himself out of the swamp by pulling his own pigtail" [Schrödinger (1940b), p. 9]. The mist was very deep indeed

---

[83] The Fermi state is the ground state (associated with a static Einstein universe) that results from imposing the exclusion principle (now at the level of the universe), so that every energy level is occupied by at most one particle (with higher energy levels them unoccupied).

in these initial skirmishes with quantum cosmology and, unfortunately, such lines of thought represented the end of an era of somewhat grandiose thinking.

4.5 Conclusion

In this chapter, we showed how the newest developments in quantum mechanics of the late 1920s were very quickly compared with general relativity, with attempts made to demonstrate their mutual coherence. This involved focusing on the basic mathematical structures that formed the first concrete representations of quantum mechanical systems. The aim was structural harmonization, rather than quantization. Likewise, conceptual debates, especially having to do with measurement and the uncertainty relations, as well as new cosmological discoveries (based on applications of general relativity) were also quickly compared, often with surprising results such as explanations of discreteness and predictions of particle production in curved spaces. There were two primary motivations pushing this research forward: coherence (into which the more formal approaches also fit) and *utility* (that is, attempting to gain a better grip on the quantum theory). In the next pair of chapters we see how the development of field quantization methods were incorporated into the quantum gravity problem, with similar underlying motivations.

· ·

REFERENCES

Albers, M., C. Kiefer, and M. Reginatto (2008) Measurement Analysis and Quantum Gravity. *Physical Review D* 78: 064051.

Amelino-Camelia, G., L. Freidel, J. Kowalski-Glikman, and L. Smolin (2011) Relative Locality: A Deepening of the Relativity Principle. *General Relativity and Gravitation* 43: 2547–53.

Anderson, J. L. (1954) Quantum Restrictions on the Measurability of Fields in Gravitational Theory. *Rev. Mexicana Fis.* 3: 176–84.

Bacciagaluppi, G. and A. Valentini (2009) *Quantum Theory at the Crossroads: Reconsidering the 1927 Solvay Conference.* Cambridge: Cambridge University Press.

Bastin, T. and C. Kilmister (1962) *Eddington's Statistical Theory.* Oxford: Clarendon Press.

Bergmann (1956) Quantisierung Allgemein-Kovarianter Feldtheorien. In A. Mercier and M. Kervaire (eds.), *Fünfzig Jahre Relativitätstheorie*, Bern, July 11–16, 1955 (pp. 79–97). Basel: Birkhäuser Verlag.

Bergmann, P. and G. Smith (1982) Measurability Analysis of the Linearized Gravitational Field. *General Relativity and Gravitation* 14(12): 1131–66.

Birrell, N. D. and P. C. W. Davies (1984) *Quantum Fields in Curved Space.* Cambridge: Cambridge University Press.

Blum, A. S. (2018) Where to Start? First Interactions between Wave Mechanics and General Relativity. In A. S. Blum, and D. P. Rickles (eds.), *Quantum Gravity in the*

First Half of the Twentieth Century: A Sourcebook (pp. 49–56). Berlin: Edition Open Access.

Blum, A. S. and D. Salisbury (2018) The Genesis of Canonical Quantum Gravity. In A. S. Blum, and D. P. Rickles (eds.), *Quantum Gravity in the First Half of the Twentieth Century: A Sourcebook* (pp. 455–464). Berlin: Edition Open Access.

Bohr, N. (1931) Space-Time-Continuity and Atomic Physics. In J. Kalckar (ed.), *Niels Bohr Collected Works, Volume 6* (pp. 361–70). Amsterdam: North Holland, 1985.

Bohr, N. and L. Rosenfeld (1933) Zur Frage der Messbarkeit der elektromagnetischen Feldgroössern. *Det Kgl. Danske Videnskabernes Selskab.* **12**(8): 1–65.

Bondi, H., F. A. E. Pirani, and I. Robinson (1959) Gravitational Waves in General Relativity. III. Exact Plane Waves. *Proceedings of the Royal Society of London. Series A* **251**: 519–33.

Born, M. (1943) *Experiment and Theory in Physics*. Cambridge: Cambridge University Press.

Born, M. (1938) A Suggestion for Unifying Quantum Theory and Relativity. *Proceedings of the Royal Society of London. Series A* **165**(921): 291–303.

Born, M. and P. Jordan (1925) Zur Quantenmechanik. *Zeitschrift für Physik* **34**: 858–88.

Brill, D. and J. Wheeler (1957) Interaction of Neutrinos and Gravitational Fields. *Rev. Mod. Phys.* **29**: 465–79.

Bronstein, M. P. (1936) Quantentheorie Schwacher Gravitationsfelder. *Physikalische Zeitschrift der Sowjetunion* **9**: 140–57. (Translated from German by M.A. Kurkov, edited by S. Deser, (2012) Quantum Theory of Weak Gravitational Fields. *General Relativity and Gravitation* **44**(1): 267–83.)

Brown, L. M. (1993) The Legacy of Uncertainty. In D. Han, Y.S. Kim, and V.I. Man'ko (eds.), *The Second International Workshop on Squeezed States and Uncertainty Relations* (pp. 119–32). Washington, DC: NASA Conference Publication [NAS 1.55:3219].

Colin, C. (2010) Jacques Solomon and the Interpretation of Quantum Theory. *Revue d'Histoire des Sciences* **63**: 221–46.

Darwin, C.G. (1928) The Wave Equations of the Electron. *Proc. R. Soc. London, Ser. A* **118** 654–80.

DeWitt, B. (1967) Quantum Theory of Gravity. I. The Canonical Theory. *Physical Review* **160**: 1113–48.

DeWitt, B. (1957) Dynamical Theory in Curved Spaces. I. A Review of the Classical and Quantum Action Principles. *Rev. Mod. Phys.* **29**: 377–97.

Dingle, H. (1937) Modern Aristotelianism. *Nature* **139**: 784–6.

Dirac, P.A.M. (1936) Wave Equations in Conformal Space. *The Annals of Mathematics* **37**(2): 429–42.

Dirac, P.A.M. (1935) The Electron Wave Equation in De-Sitter Space. *The Annals of Mathematics* **36**(3): 657–69.

Dirac, P.A.M. (1928a) The Quantum Theory of the Electron. *Proc. R. Soc. London, Ser. A* **117**: 610–624.

Dirac, P.A.M. (1928b) The Quantum Theory of the Electron. *Proc. R. Soc. London, Ser. A* **118**: 351–61.

Dirac, P. A. M. (1926) The Fundamental Equations of Quantum Mechanics. *Proc. R. Soc. London, Ser. A* **109**: 642–53.

Dowker, J. and Y. Dowker (1966) Particles of Arbitrary Spin in Curved Spaces. *Proceedings of the Physical Society* **87**: 65–78.

Duncan, A. and M. Janssen (2013) (Never) Mind your p's and q's: von Neumann versus Jordan on the Foundations of Quantum Theory. *Eur. Phys. J. H* **38**: 175–259.

Eddington, A. S. (1966) Manuscript on the Transfer Problem. In C. Kilmister (ed.), *Men of Physics: Sir Arthur Eddington* (pp. 248–68). Oxford: Pergamon Press.

Eddington, A. S. (1943) *The Combination of Relativity Theory and Quantum Theory.* Communications of the Dublin Institute for Advanced Studies. Series A, No. 2.

Eddington, A. S. (1943) *Fundamental Theory.* Cambridge: Cambridge University Press.

Eddington, A. S. (1939a) *The Philosophy of Physical Science.* Cambridge: Cambridge University Press.

Eddington, A. S. (1939b) The Cosmological Controversy. *Science Progress* **34**(134): 225–36.

Eddington, A. S. (1936) *Relativity Theory of Protons and Electrons.* Cambridge: Cambridge University Press.

Eddington, A. S. (1929) The Charge of an Electron. *Proceedings of the Royal Society A* **122**(789): 358–369.

Eddington, A. S. (1928) A Symmetrical Treatment of the Wave Equation. *Proceedings of the Royal Society A* **121**(788): 524–42.

Ehrenfest, P. and G. E. Uhlenbeck (1926) Graphische Veranschaulichung der De Broglieschen Phasenwellen in der fünfdimensionalen Welt von O. Klein. *Zeitschrift für Physik* **39**: 495–8.

Flint, H. (1960) Unity in Physics. *Science Progress* **48**: 31–42.

Fock, V.A., (1929a) Geometrisierung der Diracschen Theorie des Elektrons. *Zeitschrift für Physik* **57**: 261–77.

Fock, V.A., (1929b) L'equation d'onde de Dirac et la Géometrie de Riemann. J. Phys. Radium, 10, 392–405. Translated by V.V. Fock as "Dirac Wave Equation and Riemann Geometry" in L.D. Faddeev, L.A. Khalfin, and I.V. Komarov (eds.), *V.A. Fock – Selected Works: Quantum Mechanics and Quantum Field Theory* (pp. 113–131). Boca Raton: CRC, 2019.

Fock, V. (1926) On the Invariant Form of the Wave Equation and of the Equations of Motion for a Charged Massive Point. *Zs. Phys.* **39**(2–3): 226–32. Reprinted and translated in English in L.D. Faddeev, L.A. Khalfin, and I.V. Komarov (eds.), *V.A. Fock – Selected Works: Quantum Mechanics and Quantum Field Theory* (pp. 21–8). Boca Raton: CRC Press, 2019.

Fock, V.A., and D. D. Iwanenko (1929) Géometrie Quantique Linéaire et Déplacement Parallele. *C. R. Acad. Sci.* **188**: 1470–2.

Ford, G. W. (2009) George Eugene Uhlenbeck: 1900–1988. *Biographical Memoir.* Washington, DC: National Academy of Sciences.

Friedel, L., R. Leigh, and D. Minic (2014) Born Reciprocity in String Theory and the Nature of Spacetime. *Physics Letters B* **370**: 302–6.

Gale, G. (1997) The Role of Leibniz and Haldane in Wiener's Cybernetics. *NWCC '94 Proceedings of the Norbert Wiener Centenary Congress* (pp. 247–61). New York: AMS Press, Inc.

Galison, P. (1994) The Ontology of the Enemy: Norbert Wiener and the Cybernetic Vision. *Critical Inquiry* **21**: 228–66.

Goenner, H. F. M. (2004) On the History of Unified Field Theories. *Living Reviews of Relativity* 7: 2. https://doi.org/10.12942/lrr-2004-2.

Goto, K. (1954) Wave Fields in de Sitter Space. *Progress of Theoretical Physics* **12**(3): 311–54.

Goto, K. (1952) On the Meson Wave Equation in de Sitter Space. *Progress of Theoretical Physics* **8**(6): 672–5.

Goto, K. (1951) Wave Equations in de Sitter Space. *Progress of Theoretical Physics* **6**(6): 1013–14.

Gordon, W. (1926) Der Comptoneffekt nach der Schrödingerschen Theorie. *Zeitschrift für Physik* **40**(1–2): 117–33.

Gorelik, G. and V. Y. Frenkel (1994) *Matvei Petrovich Bronstein and Soviet Theoretical Physics in the Thirties.* Basel: Birkhäuser.

Heisenberg, W. (1925) Über Quantentheoretische Umdeutung Kinematischer und Mechanischer Beziehungen. *Zeitschrift für Physik* **33**: 879–93.

Howard, D. (1990) 'Nicht Sein Kann Was Nicht Sein Darf,' or the Prehistory of EPR, 1909–1935: Einstein's Early Worries about the Quantum Mechanics of Composite Systems. In A.I. Miller (ed.), *Sixty-Two Years of Uncertainty* (pp. 61–112) New York: Plenum Press

Ikeda, M. (1953) On a Five Dimensional Representation of the Electromagnetic and Electron Field Equations in a Curved Space-time. *Progress of Theoretical Physics* **10**(5): 483–98.

Jacobsen, A. S. (2012) *Léon Rosenfeld: Physics, Philosophy, and Politics in the Twentieth Century.* Singapore: World Scientific.

Joas, C. and C. Lehner (2009) The Classical Roots of Wave Mechanics: Schrödinger's Transformations of the Optical-mechanical Analogy. *Studies in History and Philosophy of Modern Physics* **40**(4): 338–51.

Kennefick, D. (2007) *Traveling at the Speed of Thought: Einstein and the Quest for Gravitational Waves.* Princeton: Princeton University Press.

Kilmister, C. (1966) *Men of Physics: Sir Arthur Eddington.* Oxford: Pergamon Press.

Kilmister, C. and G. Stephenson (1954) An Axiomatic Criticism of Unified Field Theories—I. *Il Nuovo Cimento* **11**(1): 91–105.

Kilmister, C. and B. Tupper, (1962) *Eddington's Statistical Theory.* Oxford: Oxford University Press.

Klein, O. (1955) Quantum Theory and Relativity. In W. Pauli, L. Rosenfeld, and V. Weisskopf (eds.), *Niels Bohr and the Development of Physics,* (pp. 96–117). London: Pergamon Press.

Klein, O. (1947) Meson Fields and Nuclear Interaction. *Arkiv för Matematik, Astronomi och Fysik* **34A**(1): 1–19.

Klein, O. (1928) Zur fünfdimensionalen Darstellung der Relativitätstheorie. *Zeitschrift für Physik* **46**: 188–208.

Klein, O. (1926a) The Atomicity of Electricity as a Quantum Theory Law. *Nature* **118**: 516.

Klein, O. (1926b) Quantentheorie und fünfdimensionale Relativitätstheorie. *Zeitschrift für Physik* **37**: 895–906.

Kozhevnikov A. B. (2004) *Stalin's Great Science: The Times and Adventures of Soviet Physicists*. London: Imperial College Press.

Kragh, H. (2000) Relativity and Quantum Theory from Sommerfeld to Dirac. *Ann. Phys. (Leipzig)* **9**(11–12): 961–74.

Kragh, H. (1996) *Cosmology and Controversy: The Historical Development of Two Theories of the Universe*. Princeton: Princeton University Press.

Kragh, H. (1990) *Dirac: A Scientific Biography*. Cambridge: Cambridge University Press.

Kragh, H. (1984) Equation with the Many Fathers. The Klein–Gordon Equation in 1926. *American Journal of Physics* **52**: 1024–33.

Kragh, H. (1981) The Genesis of Dirac's Relativistic Theory of Electrons. *Archive for History of Exact Science* **24**: 31–67.

Landau, L. and R. Peierls (1931) Erweiterung des Unbestimmtheitsprinzips für die relativistische Quantentheorie. *Zeitschrift für Physik* **69**: 55–69.

Mandel, H. (1926) Zur Herleitung der Feldgleichungen in der Allgemeinen Relativitätstheorie. *Zeitschrift für Physik* **39**: 136–45.

Martin, D. (1942) On the Methods of Extending Dirac's Equation of the Electron to General Relativity. *Proc. Edinburgh Math. Soc.* **7**(1): 39–50.

McCrea, W. H. (1951) Relativity Theory and the Creation of Matter. *Proceedings of the Royal Society* **206**(1087): 562–75.

McCrea, W. H. (1947) Fundamental Theory. By the late Sir A. S. Eddington. *The Mathematical Gazette* **31**(297): 288–91.

McVittie, G. C. (1932) Dirac's Equation in General Relativity. *Monthly Notices of the Royal Astronomical Society* **92**(9): 868–77.

Mehra, J. and H. Rechenberg (2001) *The Historical Development of Quantum Theory, Volume 6. The Completion of Quantum Mechanics 1926–1941*. Dordrecht: Springer.

Mehra, J. and H. Rechenberg (1982) *The Historical Development of Quantum Theory. The Formulation of Matrix Mechanics and Its Modifications 1925–1926*. Dordrecht: Springer.

Misner, C. (1957) Feynman Quantization of General Relativity. *Reviews of Modern Physics* **29**(3): 497–509.

Moore (1989) W. J. *Schrödinger: Life and Thought*. Cambridge: Cambridge University Press.

Müller, P. (1939) Über die Eigenschwingungen der Zylindrischen und Sphärischen Welt. *Physikalische Zeitschrift* **39**: 366–84.

Nagel, E. (1940) The Philosophy of Physical Science by Arthur Eddington. *The Journal of Philosophy* **37**(6): 161–5.

Ogievetskiĭ, V. and I. Polubarinov (1965) Spinors in Gravitation Theory (translated by G. Volkoff). *Soviet Physics JETP* **21**(6): 1093–100.

Osborne, M. F. M. (1949) Quantum-Theory Restrictions on the General Theory of Relativity. *Phys. Rev.* **75**: 1579–84.

O'Raifeartaigh, L. and N. Straumann (2000) Dirac's Equation in General Relativity. *Reviews of Modern Physics* **72**(1): 1–23.

Pais, A. (1941) The Energy Momentum Tensor in Projective Relativity Theory. *Physica* **8**: 1137–60.

Pauli, W. (1940) Über die Invarianz der Diracschen Wellengleichungen gegenüber Ähnlichkeitstransformationen des Linienelementes im Fall verschwindender Ruhmasse. *Helvetica Phys. Acta* **13**: 204–8.

Pauli, W. (1927) Zur Quantenmechanik des Magnetischen Elektrons *Zeitschrift für Physik* **43**(9–10): 601–23.

Penrose, R. (1962) General Relativity in Spinor Form. In A. Lichnerowicz and M.-A. Tonnelat (eds.), *Les Theories Relativistes de la Gravitation* (pp. 429–32). Paris: CNRS.

Peruzzi, G. and A. Rocci (2018) Tales from the Prehistory of Quantum Gravity. *The European Physical Journal H* **43**(2): 185–241.

Pirani, F. (1954) On the Energy-Momentum Tensor and the Creation of Matter in Relativistic Cosmology. *Proceedings of the Royal Society of London. Series A.* **228**: 455–62.

Pitts, J. B. (2012) The Nontriviality of Trivial General Covariance: How Electrons Restrict 'Time' Coordinates, Spinors (almost) fit into Tensor Calculus, and $\frac{7}{16}$ of a Tetrad is Surplus Structure. *Studies in History and Philosophy of Modern Physics* **43**(1): 1–24.

Przibram, K. (ed.) (1934) *Letters on Wave Mechanics*. New York: Open Road.

Raje, S. A. (1952) Linear Meson Wave Equation in de Sitter Space *Progress of Theoretical Physics* **8**(3): 384–5.

Rosen, N. (1937) Plane Polarised Waves in the General Theory of Relativity. *Physikalische Zeitschrift Der Sowjetunion* **12**: 366–72.

Rosenfeld, L. (1959) Jacques Solomon. In R. S. Cohen and J. Stachel (eds.), *Selected Papers of Léon Rosenfeld* (pp. 297–301). Dordrecht: Springer.

Rovelli, C. (2002) Partial Observables. *Physical Review D* **65**: 124013.

Rowe, D.E. (1989) Interview with Dirk Jan Struik. *The Mathematical Intelligencer* **11** (1): 14–26.

Rüger, A. (1988) Atomism from Cosmology: Erwin Schrödinger's Work on Wave Mechanics and Space-Time Structure. *Historical Studies in the Physical and Biological Sciences* **18**(2): 377–1.

Ruse, H.S. (1938) On the Geometry of Dirac's Equations and their Expression in Tensor Form. *Proceedings of the Royal Society of Edinburgh* **57**: 97–127.

Scholz, E. (2005) Local Spinor Structures in V. Fock's and H. Weyl's work on the Dirac equation (1929). In D. Flament (eds.), *Géométrie au Vingtiéme Siécle, 1930–2000* (pp. 284–301). Paris: Hermann.

Schouten, J.A. (1931a) Dirac Equation in General Relativity. 1. Four Dimensional Theory. *J. Math. Phys. (MIT)* **10**: 239–71.

Schouten, J.A. (1931b) Dirac Equations in General Relativity. 2. Five Dimensional Theory. *J. Math. Phys. (MIT)* **10**: 272–83.

Schrödinger, E. (1940a) Maxwell's and Dirac's Equations in the Expanding Universe *Proceedings of the Royal Irish Academy. Section A: Mathematical and Physical Sciences* **46**: 25–47.

Schrödinger, E. (1940b) The General Theory of Relativity and Wave Mechanics. *Wis-en natuurkundig Tijdschrift* **10**: 2–9.

Schrödinger, E. (1937a) World Structure: Review of *Relativity Theory Of Protons And Electrons*. *Nature* **140**: 742–4.

Schrödinger, E. (1937b) Sur la Theorie du Monde d'Eddington. *Il Nuovo Cimento (N.S.)* **15**: 246–54.

Schrödinger, E. (1939) The Proper Vibrations of the Expanding Universe. *Physica* **6**: 899–12.

Schrödinger, E. (1932) Diracsches Elektron im Schwerefeld. Sitzungsber. *Preuss. Akad. Wiss.* **XI**: 105–28.

Schrödinger, E. (1926) An Undulatory Theory of the Mechanics of Atoms and Molecules. *Physical Review* **28**: 1049–70.

Snyder, H. (1947) Quantized Space-Time. *Phys. Rev.* **71**: 38–41.

Solomon, J. (1938) Gravitation et Quanta. *Journal de Physique et Le Radium* **9**: 479–85.

Stachel, J. (2014) The Hole Argument and Some Physical and Philosophical Implications. *Living Reviews in Relativity* **17**(1): https://doi.org/10.12942/lrr-2014–1.

Synge, J. (1958) Whittaker's Contributions to the Theory of Relativity. *Proceedings of the Edinburgh Mathematical Society* **11**(1): 39–55.

Taub, A. H. (1939) Tensor Equations Equivalent to the Dirac Equations. *Ann. of Math.* **40**: 937–47.

Taub, A. H. (1938) Spin Representation of Inversions. *Bull. Amer. Math. Soc.* **44**(12): 860–65.

Taub, A. H. (1937) Quantum Equations in Cosmological Spaces. *Physical Review* **51**(6): 512–25.

Taub, A. H., O. Veblen, and J. von Neumann (1934) The Dirac Equation in Projective Relativity. *Proceedings of the National Academy of Sciences of the United States of America* **20**(6): 383–88.

Tetrode, H. (1928a) Allgemein-relativistische Quantentheorie des Elektrons. *Zeitschrift für Physik* **50**: 336–46.

Tetrode, H. (1928b) Der Impuls-Energiesatz in der Diracschen Quantentheorie des Elektrons. *Zeitschrift für Physik* **49**: 858–64.

Temple, G. (1929) The Tensorial Form of Dirac's Wave Equation. *Proc. R. Soc. London, Ser. A* **122**: 352–7.

Tonnelat, M.-A. (1941) Sur la Théorie du Photon dans un Espace de Riemann. *Annalen der Physik* **15**: 144–224.

van der Waerden, B. L. (1929) Spinoranalyse. *Nachrichten von der Gesellschaft der Wissenschaften zu Göttingen, Mathematisch-Physikalische Klasse*: 100–109.

Veblen, O. (1935) A Conformal Wave Equation. *Proc. Natl. Acad. Sci. USA* **21**: 484–87.

Veblen, O. (1934) Spinors. *Science* **80**: 415–19.

Veblen, O. (1933) Spinors in Projective Relativity. *Proc. Natl. Acad. Sci. USA* **19**: 979–89.

von Meyenn, K. (1985) *Wissenschaftlicher Briefwechsel mit Bohr, Einstein, Heisenberg u.a. Band II: 1930–1939 / Scientific Correspondence with Bohr, Einstein, Heisenberg, a.o. Volume II: 1930–1939.* Berlin Heidelberg: Springer-Verlag.

Watanabe, S. (1949) Wave Equations in the de Sitter Space. *Physical Review* **76**: 296.

Wessels, L. (1983) Erwin Schrödinger and the Descriptive Tradition. In R. Aris, T. D. Davis, and R. H. Stuewer (eds.), *Springs of Scientific Creativity* (pp. 254–78). Minneapolis: University of Minnesota Press.

Weyl, H. (1929) Gravitation and the Electron. *Proc. Natl. Acad. Sci.* **15**: 323–34.

Whittaker, E. (1937) On the Relations of the Tensor-Calculus to the Spinor-Calculus. *Proc. R. Soc. London, Ser. A* **158**(893): 38–46.

Wiener, N. (1932) Back to Leibniz! (Physics Reoccupies an Abandoned Position). *Technology Review* **34**: 201–3.

Wiener, N. and D. Struik (1927) Quantum Theory and Gravitational Relativity. *Nature* **119**: 853–4.

Wiener, N. and D. Struik (1928) A Relativistic Theory of Quanta. *Studies in Applied Mathematics* **7**(1–4): 1–23.

Wiener, N., and M. S. Vallarta (1929) Unified Field Theory with Electricity and Gravitation *Nature* **123**: 317.

Wigner, E. (1957) Relativistic Invariance and Quantum Phenomena. *Reviews of Modern Physics* **29**(3): 255–68. Reprinted in E. Wigner (ed.), *Symmetries and Reflections* (pp. 51–81). Bloomington: Indiana University Press.

Wigner (1955) Relativistic Invariance of Quantum-Mechanical Equations. In A. Mercier and M. Kervaire (eds.), *Fünfzig Jahre Relativitätstheorie, Bern, July 11–16, 1955* (pp. 210–226). Basel: Birkhaüser.

Wigner, E. (1929) Eine Bemerkung zu Einsteins neuer Formulierung des allgemeinen Relativitätsprinzips. *Zeitschrift für Physik* **53**: 592–6.

Wigner, E. and H. Salecker (1958) Quantum Limitations of the Measurement of Space-Time Distances. *Phys. Rev.* **109**: 571.

Witten, L. (1959) Invariants of General Relativity and the Classification of Spaces. *Phys. Rev.* **113**: 357.

5

Another Field to the Pot

Almost as soon as quantum field theory was invented by Heisenberg, Pauli, Fock, Dirac, and Jordan, attempts were made to apply it to fields other than the electromagnetic field which had given it—and indeed quantum mechanics itself—birth.

Bryce DeWitt[1]

The approaches considered in the previous chapter considered electrons (represented by wavefunctions à la Schrödinger or spinors à la Dirac) in a gravitational field (or a variety of Riemannian manifolds—including static and expanding cosmological scenarios).[2] Once field quantization was established, it was natural to also consider this extended concept from the point of view of the possible relationship, in the quantum context, between electrons, the electromagnetic field, and the gravitational field (and wave-fields more generally). However, in order to get such a theory off the ground, most of these approaches dispensed, at least preliminarily, with curved spacetime and general covariance in order to access the new field-theoretic techniques, which, after all, belonged to the class of special-relativistic theories. Accordingly, such approaches are often called 'covariant,' though often a flat space approach was achieved simply through a linearized theory of gravitation, in which the Minkowski space of special relativity could be viewed as an approximation.

The problem of the formulation of a (specially) relativistic quantum field theory with an electromagnetic interaction ('quantum electrodynamics') was solved by Werner Heisenberg and Wolfgang Pauli in a pair of 1929 papers (from March and September of that year: [Heisenberg and Pauli (1929a)] [Heisenberg and Pauli (1929b)]). In this approach, the electromagnetic field is itself treated as a dynamical system described using Hamiltonian methods. They famously noted, apparently quite casually, at the close of their introduction, that quantizing gravity in a like manner should pose no new problems beyond those already present in the electrodynamic case (their papers

---

[1] [DeWitt (1967a), p. 1113].

[2] Crucially, though, never in a way that coupled the quantized source to the gravitational field through the Einstein field equations. This earlier program focused largely on the harmonization of the basic structures of quantum mechanics and general relativity: the matter waves and the curved spaces.

Covered with Deep Mist: The Development of Quantum Gravity (1916–1956). Dean Rickles, Oxford University Press (2020). © Dean Rickles.
DOI: 10.1093/oso/9780199602957.001.0001

were, of course, supposed to provide a general method for all fields).[3] However, as Alexander Blum ([Blum (2018)], p. 255) has rightly pointed out, we needn't view this as implying that Heisenberg and Pauli believed the gravitational case would be *trivial* by any means: after all, the electrodynamic scenario was still plagued by its own thorny internal problems, not least the problem of the self-energy of the electron, to which Heisenberg and Pauli would turn their attentions immediately afterwards (working with Oppenheimer who was visiting Pauli in Zürich: see, [Blum, Dürr, and Rechenberg, eds. (1993)], p. 4). The point is, in other words, that the difficulties faced in the gravitational context would be on all fours with those faced in the electrodynamic context, qua generic features of quantum field theories. There would be no 'special' problem of quantum gravity, and the gravitational field was to be treated as *just another field*—indeed, the title of this chapter comes from Richard Feynman's take on the problem of quantum gravity, that the gravitational field was in principle no different in fundamental character to the meson field, and all the other fields that go into the big cooking pot that is field theory, so that with quantum gravitation we simply "add another such field to the pot" [Feynman (1995), p. 2]. Feynman reiterated this idea in his contribution to a volume for John Wheeler's 60th birthday: "The questions about making a 'quantum theory of geometry' or other conceptual questions are all evaded by considering the gravitational field as just a spin-2 field nonlinearly coupled to matter and itself...and attempting to quantize this by following the prescriptions of quantum field theory, as one expects to do with any other field" [Feynman (1972), p. 378].[4]

---

[3] The relevant passage is: "Erwähnt sei noch, daß auch eine Quantelung des Gravitationsfeldes, die aus physikalischen Gründen notwendig zu sein scheint, mittels eines zu dem hier verwendeten völlig analogen Formalismus ohne neue Sehwierigkeiten durchführbar sein dürfte" [Heisenberg and Pauli (1929a), p. 3]. (Note that Heisenberg and Pauli explicitly mention, in a footnote attached to this passage, the remark of Einstein's from 1916, suggesting a *necessity* to modify the gravitational theory, along with Klein's 1927 paper on five-dimensional quantum theory in which he also suggests such revisions (himself referring back to Einstein's remark).)

[4] However, it must be said, Feynman's approach is presented as a "pedagogical" one, so his views about the nature of gravity are not so clear-cut; this also holds, as we shall see below, with Suraj Gupta (at least in our chosen time period)—see [Kaiser (1998)] for more on Feynman's pedagogical approach and its sociological significance. Perhaps (briefly jumping ahead to the topic of the next section), however, this gives us a good way of making sense of what split does exist between covariant and canonical approaches in this early phase: the former are *applications*-based and more suited to specific computations and the 'doing' of physics; the latter are *understanding*-based, and concerned more with interpretational issues (i.e. with getting things straight in terms of the correct physical picture). This gets some support from the fact that as soon as one does try to apply the canonical approach, one is forced into covariant-style methods (e.g. perturbation theory, which demands a background around which to perturb). However, if true, this makes it somewhat puzzling as to why there is a sociological split at all, rather than seeing the two approaches as simply complementary descriptions, dependent on the context, e.g. one (covariant) useful for more local features (scattering and so on), the other useful for global and topological features. The existence of a sociological covariant vs canonical split is even more difficult to fathom given the fact that, once the existence of Feynman-style loop diagrams emerged, since both the canonical approach and the covariant approach both achieve correspondence with the classical theory (the latter emerging through the 'tree level graphs = classical field theory' relation)—cf. [Peres (1968), p. 1335]. Moreover, the Feynman functional integral eventually enabled one to establish a kind of duality between canonical and covariant formulations (further enabling, amongst other things, a demonstration of the unitarity of the covariant approach)—cf. [Faddeev and Popov (1973), p. 778]. In fact, DeWitt suggests something along these very lines: "At the present time the two theories play complementary roles, the canonical theory describing the quantum behavior of 3-space regarded as a time-varying geometrical object, and the covariant theory describing the behavior of real and virtual gravitons propagating in this object" [DeWitt (1967a), p. 1115].

This chapter considers the very earliest explorations in quantizing the gravitational field, which were for the most part direct responses to Heisenberg and Pauli's field theory and heavily overlap with parallel explorations in the theory of quantum electrodynamics.

5.1   Exploring the Consequences of the Heisenberg–Pauli Theory

Léon Rosenfeld, after initially having a Copenhagen research visit turned down by an already over-committed Niels Bohr, was accepted by Pauli, in Zürich and ebullient as a result of his success with quantum electrodynamics—he was, as Rosenfeld puts it, "eager to have people brush up the details and explore the consequences."[5] Rosenfeld was charged with exploring the consequences for gravitation of the new method of field quantization. He explored two distinct approaches, rooted in the Heisenberg–Pauli theory and its problems: one involving the invention of new, general canonical methods (with a view to exploring the treatment of gauge degrees of freedom, leading to an early constrained Hamiltonian formalism); the other (directed at a specific computation: the gravitational self-energy of a photon in the lowest order of perturbation theory) using linearization techniques and studying the coupling of the electromagnetic radiation field with the gravitational field. Hence, two rather generic problems were motivating Rosenfeld's study: gauge redundancies and divergences. Though this work is often thought to have instigated two distinct parallel tracks from here on (canonical and covariant respectively[6]), they were really *both* couched in the basic framework of Pauli and Heisenberg's theory. Moreover, Rosenfeld's papers also laid relatively dormant until the two tracks had already been ploughed by others, when they were rediscovered.[7] In fact, as we briefly explain in the next section, the notion of a rigid distinction between 'canonical' and 'covariant' approaches to quantum gravity doesn't quite have the same bite as it now does and emerged only later. The lack of manifest covariance in the Heisenberg–Pauli theory, which lies at the root of the divisions that subsequently emerged between canonical and covariant quantization, led to key, and highly overlapping, foundational developments in both the canonical and covariant approaches to quantum gravity.[8]

---

[5] Interview of Leon Rosenfeld by Thomas S. Kuhn and John L. Heilbron on July 19th, 1963, Niels Bohr Library & Archives, American Institute of Physics, College Park, MD USA, www.aip.org/history-programs/niels-bohr-library/oral-histories/4847–2.

[6] See, e.g. [Rovelli (2002)] for this opinion.

[7] They did not lay completely dormant. Ryōyū Utiyama drew out the consequences in his work on gauge symmetries in 1942, which would be crucial for his later work casting general relativity along the lines of Yang–Mills theories (see footnote 6 in Chapter 6 for more on this). See [Salisbury (2010)] for a discussion of Rosenfeld's work on constrained Hamiltonian systems, and the reasons why it wasn't in fact successful in its aims as regards general relativity—a more detailed, recent treatment is [Salisbury and Sundermeyer (2017)]. We also consider this aspect of Rosenfeld's work further in the next chapter.

[8] I might note that I vacillated considerably over whether to have a single chapter dealing with both early canonical and covariant/flat space approaches, as exemplifications of the same 'just another field' mindset (issuing from the Heisenberg–Pauli theory), or whether to have two separate chapters. I think it is in fact

Though Rosenfeld's work would be strikingly similar to much of the work that was produced over the following couple of decades, Pauli himself later referred to Rosenfeld's work in rather unflattering terms—for example, writing to Oskar Klein[9] who was at that time interested in quantum gravity, he says:

> I would like to draw your attention in this connection to the long work of Rosenfeld, *Annalen der Physik* (4), 5, 113, 1930. He was in Zürich with me when he produced it and was called accordingly 'the man who quantized the Vierbein' (sounds like the title of a Grimm's fairytale, no?). - See Part II of his work, where the 'four-legs' comes to it. The identities between the '*p*' and '*q*' - i.e. canonically conjugated fields - which originate from the very existence of the group of general relativity (coordinate transformations with four arbitrary functions) - were of particular importance at that time. I still remember that Rosenfeld's work was not satisfying in every way, as he had to introduce certain additional conditions that nobody could really understand.

Pauli is here referring to the group-theoretical work of Rosenfeld's [Rosenfeld (1930b)], which we look at in the next chapter. Rosenfeld's work [Rosenfeld (1930a)] on computing the (divergent) gravitational self-energy was rather more impactful, and was followed up by several others, as we see below. Rosenfeld referred to this infinitely large value as a new difficulty for the Heisenberg–Pauli quantum theory of wave fields.

Rosenfeld claims that Heisenberg was the originator of the idea to focus on the self-energy calculation for light, writing that "Heisenberg has raised the question whether analogous relations do not prevail, irrespective of any material influence, in the gravitational effects of light" [Rosenfeld (1930a), p. 589]. Apparently a draft version cited correspondence between Pauli and Heisenberg on the issue:

> In the beginning, you say Heisenberg raised the question of the gravitational self-energy of the light quanta, citing a letter from Heisenberg to me. I can not remember such a letter; but I do believe that I can remember that Heisenberg briefly touched on the question in the Copenhagen discussion of 1929.[10]

That it was Heisenberg's idea makes sense, since he was then searching for ways of bypassing the divergences (including through his 'lattice worlds,' on which see

---

historically more accurate to deal with the canonical and covariant methods in a single chapter, especially since I fully agree with Alexander Blum when he writes: "the two major programs for the quantization of gravity, canonical and covariant quantization, had their origins in the 1930s, in two attempts to get over the quantization difficulty of QED with, at this early stage, only weak and incidental connections to the specific problem of quantum gravity" [Blum (2018), p. 267]. We explore a similar viewpoint in this chapter. However, we purposely deal with these developments in canonical and covariant approaches apart to some extent, for a cleaner picture and one that will be more useful for modern readers. We do, however, point out those areas in which the actual historical developments that occurred turn out to be somewhat convoluted and entangled: it can scarcely be avoided—nor can some degree of overlap and repetition of ideas as we investigate similar ideas across these two chapters.

[9] Letter dated January 25th, 1955: [von Meyenn (2001), p. 64].

[10] Pauli to Rosenfeld, September 19th, 1930: [von Meyenn (1993), p. 742].

Chapter 7), and the classical scenario of electromagnetic radiation in a gravitational field was free of such divergences.[11]

The Heisenberg–Pauli theory was, as mentioned, couched in the Hamiltonian framework. This scheme was of course still embryonic as regards field theory (being the first successful field theory of its kind), and while the Lorentz invariance of the scheme was proven, it was by no means an elegant proof—various versions were eventually given. It was (and still is) widely believed that the canonical approach is defective, relativistically speaking, because it does not treat space and time symmetrically: there is an initial state that evolves with respect to time, a common time variable regardless of any spatial separation between systems. This was part of Heisenberg and Pauli's 1929 theory of quantum electrodynamics, and it bothered many people, not least Heisenberg and Pauli themselves. As Shinichiro Tomonaga, one of the many that formulated new theories in direct reaction to the lack of covariance, so nicely expresses it:

> [in the Heisenberg–Pauli theory] the probability amplitude is given as a function of the field strength at different space points and of one common time variable. However, the concept of a common time at different space points does not have a relativistically covariant meaning.
>
> [Tomonaga (1966), p. 865]

What was sought, then, was a formulation in which this defect of splitting space and time apart was removed: a *manifestly* covariant framework more fitting for a relativistic treatment.

Paul Dirac considered various alternatives, including his Lagrangian-based approach [Dirac (1933a)] dispensing with the Hamiltonian formulation altogether (using coordinates and velocities rather than coordinates and momenta).[12] However, ultimately he returned to the Hamiltonian methods, though with a focus on manifest Lorentz invariance and a firm understanding of the realization of gauge symmetries in Hamiltonian systems—this he did in the late 1940s, at the same time as several others were struggling to quantize gravity beyond flat space approximations. This was also at a time when manifestly covariant quantum electrodynamics was at its peak, with the work of Schwinger, Tomonaga, and Feynman still fresh in the background, and couldn't fail to influence research in quantum gravity. In fact, as we go into more detail below, several physicists applied these new manifestly covariant to Rosenfeld's original problem, finding very different results.

Before we consider the development of these and other covariant approaches in more detail, as well as other related developments, let us briefly discuss this split between

---

[11] Blum [Blum (2018), pp. 259–260] notes that Pauli might be getting mixed up here, since Heisenberg was on a world tour in 1929, returning in November. However, Heisenberg was still in Leipzig in March of 1929, and it seems that Pauli was there too, together with Oppenheimer (see [Monk (2014), p. 695, note 156]).

[12] The Lagrangian method is, of course, centered on the action functional, which is a relativistically invariant object. Of course, Richard Feynman had built on this approach in his own formulation of quantum mechanics in his PhD thesis of 1942, *The Principle of Least Action in Quantum Mechanics*.

'canonical' and 'covariant' that seems to be so rigid and divisive in the quantum gravity community of today. There is a far more complex story underpinning this distinction in the early period than is often made out. There is, in fact, a very close correspondence between canonical and covariant methods in the period before the mid-1950s. It wasn't until the kinds of technology associated with the Feynman loop computation approach came on the scene that such a division really crystallized.[13]

5.2   A Note on the "Canonical versus Covariant" Split

There exist two quite distinct lines of approach to quantum gravity, that emerged fairly early on: that involving quantizing the *full* metric, in all its geometrical and non-linear glory; and that of quantizing a perturbation on a flat spacetime (and so only quantizing part of the metric).[14] Canonical methods tend to be associated with the former, and covariant approaches with the latter. In modern terms this is often couched in terms to the specific kinds of 'violence' one does to Einstein's theory of general relativity. Either one chooses some time-slicing (in the context of a 3+1 split, thus doing violence to the group of spacetime diffeomorphisms and so destroying manifest covariance), or else one splits the role of the metric tensor into its spatiotemporal (i.e. chronogeometric) and gravitational potential parts, leaving the former flat, thus doing violence to Einstein's geometrical vision. Of course, it is precisely this feature (based on what is now called 'background independence') that lies at the root of the contemporary division into distinct quantum gravity 'camps,' with a level of animosity existing between members of these different camps.

As already intimated, these distinctions initially emerged in the context of the development of Heisenberg and Pauli's theory. As also mentioned, this situation prompted Dirac to seek a manifestly covariant alternative based on the Lagrangian (which he in fact found to be "more fundamental," precisely because of its manifest covariance). The Hamiltonian method couldn't fail to break this covariance:

> [T]he Lagrangian method can easily be expressed relativistically, on account of the action function being a relativistic invariant; while the Hamiltonian method is essentially non-relativistic in form, since it marks out a particular time variable as the canonical conjugate of the Hamiltonian function.
>
> [Dirac (1933a), p. 64]

---

[13] Of course, as intimated earlier, divisions into camps like this constitute sociological phenomena, and any account of such divisions is incomplete without also probing these. We already mentioned some elements of this wider issue in Chapter 2.

[14] Of course, the latter is a simplifying move: one separates out a problem into 'hard piece' and an 'easy piece' (which acts as a kind of reference object for the hard piece) and compares the two, looking at the difference. This latter approach was, in the earliest work, coupled with the weak field approximation of Einstein's theory, so that the difference between the easy and hard pieces was minimal. As Bryce DeWitt put it, in a rejected paper intended for *Physical Review*, "[t]he linear (zero order) part of the gravitational Lagrangian function is subtracted from the full Lagrangian [then] the residue ... contains all the self-interaction effects" (CDWA)—see footnote 74. See also [DeWitt (1955)], in which DeWitt explains the philosophy of perturbation theory.

Like many others, on his way to a better relativistically invariant formulation of electrodynamics, he stumbled into tools that could be used in the context of a Hamiltonian
formulation of quantum gravity, as well as the covariant approach. This becomes very
clearly apparent in a comparison of Schwinger and Tomonaga's field theories employing
arbitrary curved surfaces, and the early Bergmann, Pirani–Schild, and DeWitt papers
on canonical quantum gravity, which sit rather uncomfortably between canonical and
covariant.

Tomonaga presented his theory as a generalization of Dirac's 'many-time theory,'
which he called the 'super-many-time theory.' He was focused on the fact that the standard commutation relations of quantum theory give "commutation relations between
the quantities at different points (x, y, z) and (x', y', z') at the same instant of time t"
[Tomonaga (1946), p. 27], which, as he points out, requires a specification of a specific
Lorentz frame. The error was, in Tomonga's mind, that "one has built up this theory
in the way which is too much analogous to the ordinary non-relativistic mechanics"
(ibid., p. 41). Likewise, in addition to the kinematics, the dynamics as supplied by the
Schrödinger equation also firmly distinguishes space and time, so that the notion of a
probability amplitude does not fit with relativity: "the probability amplitude is not a
relativistically invariant concept in the space-time world" (ibid., p. 28). There seems
to be a single plane parallel to the (x, y, z)-plane playing a crucial role. A relativistic generalization of the probability amplitude is therefore sought by Tomonaga. As
mentioned, Dirac's many-time theory provides the basis. Recall that Dirac's approach
replaces functions $\Phi(q_1, q_2, \ldots, q_N, t)$ of many particles at *the same time* with functions of
the form $\Phi(q_1 t_1, q_2 t_2, \ldots, q_N t_N)$, with a time variable for each particle in the system. The
generalization involves extrapolating to infinitely many degrees of freedom, associated
with a field system, so that there are infinitely many time variables too. This amounts to
having a time variable localized to each point of a space-like surface—with a reduction
to the standard many-time theory when the space-like surface equals the *xyz*-plane.
The relevant object is then a functional $\Psi[t_{xyz}]$ which is relativistically invariant by
construction. Tomonaga dissolves the split between equal time commutation relations
giving the kinematics (at a time) and the Schrödinger equation (involving a comparison
between times), but puts another, relativistically invariant one, in its place, involving the
laws for the free fields and the laws for the interactions (amounting to the deviation from
the free case, without interactions). This split leaves a simple free part, and a complicated
interaction part—this is, of course, analogous to Dirac's quantum mechanical interaction
picture (in which the states and observables have time dependence) in the context
of quantum field theory. This covariant field theory is, of course, the root of the
modern approaches to quantum field theory, involving perturbative expansions (such
as Feynman's and Dyson's), including the proper handling of the various infinities
appearing in field theories.[15]

---

[15] Note that the Tomonaga formalism can deal with varying electron number, and so treats electrons in
terms of a quantized field, rather than fundamental particles quite distinct from fields, as Dirac intended. In
this respect, Tomonaga's approach harks back to Heisenberg and Pauli's original 1929 formulation, though
treated covariantly. A good treatment of Tomonaga's path to his theory can be found in [Tomonaga (1966)].

The introduction of the Hamiltonian approach by Heisenberg and Pauli had been somewhat forced upon them by the difficulties they faced in trying to construct a fully covariant, four-dimensional theory of the interacting Maxwell and Dirac fields. Pauli (together with Jordan: [Jordan and Pauli (1928)]) had already succeeded in constructing such a theory for the free Maxwell field, but could not generalize this. The Hamiltonian theory was fraught with difficulties too, primarily relating to the presence of gauge symmetries, expressed through the fact that the Lagrangian is degenerate, leading to problems in satisfying the canonical commutation relations. This problem was eventually resolved in an *ad hoc* manner by adding a term to the Lagrangian $-1/2\epsilon(\partial_\mu A_\mu)^2$, and then eliminating it, once the quantiztion had been achieved, by taking the $\epsilon \to 0$ limit.[16]

The basic idea of a canonical theory of *gravity* is expressed by DeWitt as the theory that "looks at spacetime as a sequence of 3-dimensional slices, each characterized by its intrinsic 3-geometry" [DeWitt (1970), p. 186] (though without a *unique* slicing). As mentioned in the previous chapter, the idea of covariant quantization has much to do with a general distaste for introducing curved spacetime into physics, driving Weinberg's wedge through physics.[17] Early controversies over cosmology included attempts to bring gravitational physics back in line with the rest of physics by making do with flat spacetime.[18] However, in flat space it is harder to establish covariance, and early work introduced the kinds of curved hypersurface we now tend to associate with canonical approaches into what we also now associate with covariant approaches. It is thus hard to separate certain elements in the development of the canonical and covariant approaches.

Thus, one can find glimpses of the 'canonical' curved space concept in both Tomonaga and Schwinger's covariant theories of quantum electrodynamics. Jumping ahead to the next chapter, Schwinger generalized to a plane and then an arbitrary curved surface[19] to establish manifest covariance in exactly the way Paul Weiss, a student of Dirac's and Born's, had done in a bid to bring Heisenberg and Pauli's rather messy proof of invariance into better shape. As Felix Pirani puts it, in his PhD thesis (the first on canonical quantum gravity), "[t]his difficulty [the lack of manifest invariance] was removed by Weiss (1936, 1938) who was able, by the introduction of arbitrary

---

[16] Hence, one can prove Lorentz invariance of the results, but the treatment itself is not covariant. See [Schweber (1994), § 1.5] for a good discussion of the origins and development of Heisenberg and Pauli's theory—though not one that covers any gravitational issues.

[17] The 'covariant versus canonical' split does not necessarily map onto 'flat versus curved.' There are several ways to achieve a covariant approach, including the use of curved spaces. Bryce DeWitt used Rudolph Peierls' (coordinate independent) version of the Poisson bracket to define the commutator in terms of Green's functions. DeWitt was able to covariantly quantize the gravitational field without restricting himself to flat spacetime. The method, known as the 'background field method', invoked physical degrees of freedom (in this case, a stiff elastic medium, like a physical ether, and a field of clocks) to localize points, and thereby allow for the localization of physical quantities. The gravitational field is taken to interact with this background field—see [DeWitt (1967b)]. (Finally, we note that there are formal reasons for doubting a genuine split between covariant and canonical theories, involving the existence of covariant phase space techniques—see, e.g., [Crnković and Witten (1986)].)

[18] Indeed, though this was initially a classical venture, quantum field theory offered new prospects for couching gravitation in flat space, by building up the interaction through graviton-exchange.

[19] Schweber calls the introduction of a state-vector as a functional of a general 3-manifold Schwinger's "chief innovation" [Schweber (1994), p. 319]. However, the next chapter traces this to Paul Weiss.

space-like 3-surfaces, to construct a theory whose invariance was obvious" [Pirani (1951), p. 7].[20] Schwinger directly credits the curved space approach to Dirac, where he refers to Dirac's "Quantum Theory of Localizable Dynamical Systems" in his paper "Quantum Electrodynamics. I. A Covariant Formulation," and quite plainly employs many of the same concepts and notations Dirac uses there. Yet Dirac was using a technique of Weiss's, as he explicitly mentions in the final section of his paper. Tomonaga uses the same technique of establishing manifest covariance by employing arbitrary curved surfaces.[21] Hence, Schwinger, in his explicitly covariant theory, at least indirectly invokes Weiss's theory. Yet this same theory was to become a central piece of the canonical approach to quantum gravity (explicitly credited in the founding papers of Bergmann, Pirani and Schild, and DeWitt) which would, in some sense, make the theory *about* these surfaces, as can be discerned from the DeWitt characterization above, and more so in Bergmann and Brunings' characterization of their approach in terms of a "break down [of] the differences in character between the quantized field variables and the classical space-time continuum" [Bergmann and Brunings (1949), p. 487]. Hence, it is not quite accurate to project our current divisions of 'canonical' and 'covariant' onto earlier work—at least, the story is more convoluted that has been suggested.

It probably makes more sense to say there simply was no canonical or covariant quantization *program* before the 1950s. The canonical program emerged with a flurry of mid-century activity around Pirani and Schild, Peter Bergmann, Bryce DeWitt, and Paul Dirac. The covariant program also has origins with DeWitt, and then Suraj Gupta, though at this stage there is still a close formal link to the canonical approach through parallel attempts to deal with the problems of the Heisenberg–Pauli theory. In particular, the first so-called canonical approaches in the late 1940s and early 1950s, now credited with being the founding papers in this line of attack, were in fact couched in a four-dimensional, manifestly covariant formulation known as the parameter formalism.[22] The key element unifying both methods in this early period appears to have been a consistent application of the electrodynamical analogy, to which we now turn.

---

[20] Indeed, Weiss might well have anticipated the kinds of Hamiltonian approach to gravity of Pirani and Schild, and Bergmann, had he treated also the situation of theories with constraints, in which solutions of Hamilton's equations for the generalized velocities can't be assumed (which would be violated in generally covariant theories, for example): that required Dirac's framework presented in his Vancouver lectures in 1949—an extrapolation of his earlier 1933 work on homogeneous velocities: [Dirac (1933b)]. Indeed, Weiss's parameter formalism in the context of field theory is a generalization of Dirac's earlier particle mechanics version.

[21] Indeed, the links to Weiss were acknowledged in the aftermath of the Schwinger–Tomonaga formalism, in e.g. [Matthews (1949)]. Keith Roberts was also engaged in a project to derive a classical version of the Schwinger–Tomonaga theory, in the interaction representation, using Weiss's theory as a covariant basis [Roberts (1950)] [Roberts (1951)]—see also [de Wet (1950)]. However, it must be borne in mind that Weiss's approach was itself heavily grounded in Dirac's transformation theory, describing quantum mechanics in terms of transition amplitudes ('transformation functions') from the state at one time to the state at a later time. Dirac had already suggested the possibility of generalizing this to the states of quantum fields on arbitrary surfaces—Tomonaga used this generalized notion in his own covariant theory; as did Weiss, though generalizing to the canonical theory. In both cases, the aim is a formalism that is relativistically invariant throughout, which is achieved by dispensing with a space fixed once and for all.

[22] We take a look at the introduction of parameterized hypersurfaces in this context and its role in canonical quantum gravity in the next chapter.

5.3 The Electrodynamical Analogy

Oliver Darrigol has written that "the early history of meson theory can be viewed principally as a history of the use of the quantum electrodynamical analogy" [Darrigol (1998), p. 226]. We might say the same thing about quantum gravity.[23] This makes sense: electrodynamics was, of course, the prototype for field quantization. As Bryce DeWitt nicely put it, at the time of the first studies of quantum gravity "[q]uantum field theory had...scarcely been born, and its umbilical cord to electrodynamics had not yet been cut" ([DeWitt (1970)], p. 182). Such formal analogies between general relativity and Maxwell's theory of electromagnetism misled quantum gravity researchers for many years. However, as in the case of mesons, often the analogies were necessary to get a foothold on general relativity and its quantization.[24]

The analogy between quantum gravity and quantum electrodynamics was surprisingly persistent, despite the many early voices of dissent pointing to very clear disanalogies. It certainly formed the basis of most of the earliest work, including Rosenfeld, Fierz and Pauli, Kraichnan, DeWitt, and Gupta. Solomon and Bronstein also utilized the analogy, though they of course identified *obstructions* (relating to, amongst other things, the inadmissibility of direct generalizations of the Bohr–Rosenfeld analysis of measurements from electrodynamics to gravitation). It is also responsible for the mutual interactions of canonical and covariant methods in the earliest period.

Among the very earliest studies was the paper "Über die Gravitationswirkungen des Lichtes" by Léon Rosenfeld [Rosenfeld (1930a)] in which he undertook a (tree-level, lowest perturbative order[25]) computation of the gravitational self-energy of a photon.

---

[23] The analogy to electrodynamics even pervaded the early naming of the theory of quantum gravity, which was labeled 'quantum gravidynamics' by Bryce DeWitt in his PhD thesis (a name long since discarded, perhaps in recognition of the many *disanalogies*).

[24] For example, in the context of gravitational radiation the notion of electromagnetic radiation offered many essential clues. Felix Pirani is unequivocal about the utility of this radiation analogy (partial though it is): "*Some analogy has to be sought, because the concept of radiation is until now largely familiar through electromagnetic theory, and one cannot define gravitational radiation sensibly without some appeal to electromagnetic theory for guidance*" ([Pirani (1962)], p. 91). However, as we see in Chapter 9, the analogy also genuinely fooled funding agencies (including the military) into thinking that one might have the same potential to *manipulate* gravity as with electromagnetism, to generate spectacular new forms of power and propulsion. In fact, it was known at the birth of general relativity that there was a disanalogy between the kinds of waves that would be generated in gravitating systems, as compared to electromagnetic waves; namely, that there is no dipole ("the one-sidedness of the sign of scalar T") in the case of gravity (see, e.g., Einstein letter to Schwarzschild, February 19th, 1916: [Einstein (1998), p. 196]).

[25] In field theory one seeks to calculate the *amplitude* for occurrence of processes. The perturbative approach, where one expands some quantity in powers of the coupling constant of the theory, offers the standard methodology, and usually involves perturbing around a flat space. Later, Feynman developed a fairly mechanical diagrammatic notation for doing these computations. There are two types of diagram: those with (one or more) closed loops and those without closed loops. The latter are known as tree diagrams, and they are the simplest to evaluate since the (4-momenta of the) external lines determine the (4-momenta of the) internal lines, with no need to perform integration over the internal momentum variables. By contrast, diagrams with closed loops have internal lines that are not determined by the external ones. For each loop there is a 4-fold integration to be performed (each involving integration over the independent momentum variables). This new technology would reveal flaws in much earlier work undertaken along broadly covariant lines.

It was known that the electron's interaction with its own field (the electron's electromagnetic 'self-energy') suffered from divergences (attributed to its non-vanishing mass), and, as was to be expected, Rosenfeld's calculation revealed (quadratic) divergences for the gravitational case too.[26] In Rosenfeld's mind, this pointed to a generic problem facing any field theory.

Let us delve further into Rosenfeld's work thought to have founded the so-called covariant approach, but really constitutes the first *linearized, perturbative* approach. It must first be said that the paper commonly thought to have started this approach on its way was not primarily concerned with quantizing gravity per se, but was more of a study of singularities in the context of a theory of quantum fields. As Rosenfeld himself describes the motivation (my emphasis):

> [A]t that time, the calculation of the self energy of the light quantum arising from its gravitation field *was done with a very definite purpose*. One knew that there was an infinite self energy for particles like electrons which had finite mass, and it was unclear whether this self-energy was not due to the idealization of a mass point. Of course, one knew that the quantum self energy was different from the classical one, it was not just the $\frac{e}{2r}$ for $r = 0$, but we knew that there were other contributions from quantum theory. One did not know that it was reduced to a logarithmic expression, but anyhow, one knew that there were other quantum effects. But one did not know whether those divergences were not due to this original divergence which was there in the model itself, in the Hamiltonian, actually. So it was thought interesting to study the self-energy of things like the light quantum whose Hamiltonian did not contain anything, with the zero rest mass and no singular field; but then the only field that one thought of at that time, because electron pairs were not imagined, was the gravitation field.[27]

In other words, Rosenfeld's work here falls firmly within the 'quantum gravity as a resource' camp, rather than in the 'peaceful coexistence' camp: it was an exploratory exercise intended to further probe the first work on quantum electrodynamics, directly suggested to Rosenfeld by Pauli, as we saw. The concern then was to show that the divergences that, for example, the Heisenberg–Pauli work had revealed, did not reduce to the classical singularity. Instead they were of a purely quantum nature. Indeed, the perturbative approach remained associated with the presence and treatment of infinities beyond Rosenfeld's first work, but it is certainly clear that the general problem of separating out divergences was the primary task:

---

[26] Though Rosenfeld was one of the earliest quantizers of gravity, by 1966 he was less convinced that there *was* a problem of quantum gravity [Rosenfeld (1966)]. Or at least, in his mind the problem had been radically misconceived as a mathematical one (involving the necessity of unification on the basis of formal inconsistencies) instead of an empirical one. For Rosenfeld, the absence of empirical clues meant that one could only ever probe quantum gravity from the "epistemological side," which implied that the considerations could not establish the conformity of any such investigations to the world of phenomena: "no logical compulsion exists for quantizing the gravitational field" (p. 606).

[27] Interview of Leon Rosenfeld by Thomas S. Kuhn and John L. Heilbron on July 19th, 1963, Niels Bohr Library & Archives, American Institute of Physics, College Park, MD USA, www.aip.org/history-programs/niels-bohr-library/oral-histories/4847-2. Subsequent quotations refer to this interview, unless otherwise specified.

One did not know how to separate those infinities. We knew that the classical model contained by itself an infinity, and if it had been possible to trace the other infinities to that one then that would have been fine. But then this paper on the gravitation energy showed that this was not the case, that one got an infinity in that case also, so one had the impression that there was something deeply wrong with the whole theory. Then it was not clear whether the difficulty was a methodical one, or was in the mathematical procedure or whether it was in the physics; and this uncertainty prevailed until one had this manifestly co-variant formulation which showed that the difficulties were not in the mathematics, but in the physics. Probably many people guessed at that time that it was in the physics and not just a matter of the formalisms.

Here, "manifestly covariant" refers to the later Schwinger–Tomonaga formalism. Bryce DeWitt would later re-perform Rosenfeld's computation of the gravitational self-energy of the photon, using just these manifestly gauge-invariant, Lorentz covariant techniques that had, in 1950, only just become available (using the tools of renormalization that had only recently been showcased at the first Shelter Island conference). Whereas Rosenfeld's own calculation showed quadratic divergence, DeWitt found a vanishing value—DeWitt showed that contra Rosenfeld, the analysis revealed the necessity of charge renormalization, rather than a non-vanishing mass).[28] His diagnosis of Rosenfeld's error (more of a guess) was that the method used was not Lorentz-invariant (let alone generally covariant), nor was the correct interaction Hamiltonian used, and only the first order terms in this case too (see [DeWitt (1950), p. 129]). DeWitt's approach in his thesis was nonetheless itself firmly in the weak-field approximation (an approximation method "thoroughly amalgamated with a standard perturbation method"), though he flags that it is intended only as "a temporary expedient" [DeWitt (1950), p. 3], and indeed the only one available at that time for providing a physical interpretation of the quantized gravitational field. He does in fact also mention that he has already performed calculations using a "feedback principle" (going beyond the weak-field approximation) to evade some of the problems associated with the full non-linear theory.[29]

Suraj Gupta, often credited with the first genuine, exact covariant quantization of gravitation (including a treatment of the non-linearities),[30] was also building on this manifestly covariant foundation, and used the fact that, in this new scheme, the supplementary condition for the electromagnetic field is modified when one passes from the Heisenberg representation to the interaction representation. Gupta applies this feature to the case of general relativity, thereby deriving supplementary conditions (i.e. the coordinate conditions) for the gravitational field in the interaction representation.[31] Gupta also redid Rosenfeld's calculation (and, indeed, thanks Rosenfeld, who submitted

---

[28] See [DeWitt (2009)] for DeWitt's own brief discussion of this thesis work.

[29] DeWitt credits this non-linear feedback concept to Robert Kraichnan (in his "unpublished researches").

[30] However, given DeWitt's citation of Kriachnan *before* Gupta's papers had appeared, it seems clear that Kraichnan was the first to attempt such a feat.

[31] Alexander Blum locates the birth of the covariant quantization of gravity program here, in the generalization of the interaction representation from QED to gravitation [Blum (2018), p. 267]. This seems broadly correct, so long as we insert several steps on the way, which we cover in this chapter and the next.

Gupta's paper, for discussions at the end of his paper), using Dyson and Feynman's techniques, and also finds a vanishing photon self-energy (from gauge-invariance). He goes on to compute electron self-energy, finding a situation worse than for QED, showing that "the gravitational field involves many more divergencies besides those occurring in quantum electrodynamics" [Gupta (1952b), p. 619].[32]

As with the divergence paper, *mutatis mutandis*, Rosenfeld's other 1930 work on the constrained Hamiltonian formalism was not strictly an exercise in the quantization of gravity *per se*, but a study of ambiguities issuing from the symmetry structure of gauge theories more generally, which had been dealt with in, what appeared to most, to be an unsatisfactory manner (e.g. fixing the gauge, and thereby breaking symmetry). One of his chief innovations was to link the gauge invariance of the action to the presence of constraints in the Hamiltonian formalism. This implies that not all momenta are independent functions of the velocities. As he explains:

> There was this point in their proof in which the invariance of the Hamiltonian seemed to depend on a special structure of the Hamiltonian, and that looked very suspicious. In fact, I said that directly to Pauli when I had read the proofs, and he said, "Yes, I understand that, but we have not been able to find any mistake in our calculation and we do not understand what this means; we suspect that it must be wrong, but we don't know." Then the thing came to a crisis through the fact that I then tried to make a more general formulation of this field quantization, starting from Lagrangian, defining the momenta through the variational derivatives, and then I saw that I could do that without specializing the Hamiltonian. It was a purely abstract scheme which worked in a completely general way with only this complication of the accessory conditions, but that was, at any rate, not due to any special structure but only to the existence of invariance with respect to group. So at that stage I was convinced that there must be a mistake in the original paper; it was quite clear, and then I set myself to finding this mistake. By going very carefully through every step several times I spotted it; it was an extremely subtle point about two points which had been exchanged in the delta function which was not legitimate. But then it came out all right, so I wrote up a little note about it and I started the note with the sentence, "In ihrer grundlegenden Abhandlung uber Quantenelektrodynamik," and so on. When Pauli saw that, he said, "Ja, das ist ein ziemlich morscher Grund, den wir da gelegt haben." [so that was struck out] Then I suggested "In ihrer ersten Abhandlung," and Pauli agreed, "Ja, das ist sinnvoll." Then he told me that obviously, the general method must be right, that is, we have to quantize the fields; since there is a classical Hamiltonian, the Hamiltonian method must be good enough, but since it leads to those consequences, it must be wrong somewhere. Probably we have not yet got the right variables, he said at that time. That was the first reaction to it; very early in '29.

Thus, neither of the two papers usually taken to be the origination of the covariant and canonical approaches, then, were primarily focused on providing theories of quantum gravity. Nor did they provide the spearhead for future developments in any strong way.

---

[32] In 1950, Finnish physicists, Nilsson and Laurikainen [Nilsson and Laurikainen (1950)] also used Schwinger's invariant techniques to redo Rosenfeld's calculation; using covariant perturbation theory they too find the gravitational self-energy of a photon can be made to vanish, by implementing a regularization procedure.

Rosenfeld was, however, followed in his divergence explorations by his colleague Jacques Solomon [Solomon (1931)] (written May 24th, 1931), who developed Rosenfeld's work by investigating the role of zero-point energy in the difficulties faced in analyzing the energy of the gravitational field generated by electromagnetic waves. Solomon shows that the infinite self-energy of the photon was not due to the zero-point energy by showing how the interaction energy of a single photon in the gravitational field generated by it remains infinite in models in which the zero-point has been eliminated.[33] Hence, we can perhaps view Solomon's later pessimistic remarks about the clash between quantum and gravitational principles in the context of measurability and non-linearity (discussed in the previous chapter) as having a well founded prior basis in this work. A natural method of quantizing the theory gives physically impossible results. Of course, this was part of a rather generic problem of field theories that would not be properly understood until the 1940s, with the development of renormalization techniques. Solomon concludes that one requires an alternative conceptualization of the nature of the interaction itself, though he wasn't forthcoming about what this would involve.

Such difficulties eventually led to the idea that gravity had features that could not be treated along the same lines as electromagnetism.[34] *Forcing* the analogy certainly led to a theory along the lines of quantum electrodynamics, but only by ignoring non-linearities which are part and parcel of general relativity. Thus, in 1962, we find Christian Møller writing:

> During the last ten years, this very difficult question has been treated in a large number of important papers whose aim has been to perform the quantization on the lines of the usual quantum mechanics of fields. To my knowledge, however, no one has yet succeeded in carrying through this program consistently for the exact non-linear gravitational-field equations. Only on substitution of the linear approximations for the exact equations has it been possible to perform the quantization in a simple way according to the usual rules of quantum mechanics, and in this way we are led to the notion of gravitons on the analogy

---

[33] Solomon credits Rosenfeld with kind advice in the acknowledgements of this paper. He remarks that the model which eliminates the zero-point contribution (by altering the classical Hamiltonian function) was a collaboration with Rosenfeld, in close connection with Landau and Peierls. The results formed part of his doctoral thesis.

[34] However, in the 1960s, the electromagnetic analogy did bear some fruit in the construction of quantities of the 'right sort' for classical and quantum gravity; namely, coordinate independent, gauge-invariant observables. If one can find these observables, then one could in principle turn them into operators and construct a Hamiltonian. In the electromagnetic case, one has the electric and magnetic fields at one's disposal. These are gauge invariant entities. However, if one then considers the behaviour of charges interacting with the fields, then operators associated with the particles are not gauge invariant: gauge transformations alter the phase of the particles. The solution is to consider loop-like, path-dependent quantities known as 'holonomies' (see [Mandelstam (1962)], p. 347): $\Phi(x, P) = \phi(x)e^{-ie\int_P^x dz_m A_m(z)}$. Hence, rather than working with local particle operators, one works with these 'spread out' quantities (and the electromagnetic field). This approach retains the covariance of the original theory. Mandelstam argues that, in this case at least, "there is a close analogy between the electromagnetic field and the gravitational field" ([Mandelstam (1962)], p. 353), and both can be cast in the same intrinsic form. (Mandelstam's idea of formulating gravity, and gauge theories, in terms of observables of this form, was recently taken up again in [Gambini and Pullin (2018)].)

of photons and mesons in the case of electromagnetic and nuclear fields. However . . . the linear theory is an entirely different theory in which characteristic features of the general theory of relativity are lost. On the other hand, we cannot resign ourselves to the situation in the classical theory of relativity with the argument that gravitational quantum effects would anyway be unmeasurably small. Somehow we must account for the simultaneous existence of Planck's quantum of action and gravitational fields in our universe.

[Møller (1962), pp. 21–22]

There was, then, around the end point of the period this book focuses on clearly a dawning recognition[35] that gravity has some special qualities that render the electrodynamical analogy inert, or at least incomplete. As Bryce DeWitt would put it, again in 1962:

Although the quantum description of the gravitational field has many points of similarity to conventional quantum field theory, it nevertheless seems incapable—or capable only with difficulty—of incorporating certain conventionally accepted notions.

[DeWitt (1962), p. 330]

Along with this goes a disenchantment with the power of linearization techniques in a gravitational context. The field equations of general relativity simply are non-linear, and this has immediate consequences for the superposition principle, as Bronstein and Solomon had surmised early on. Any quantization of the full theory would result in quanta that interacted with each other, and *self*-interacted! This is quite unlike the situation in quantum electrodynamics, which did itself turn out to be special with respect to its linearity: photons are not charged, so the theory is linear and obeys the superposition principle in a straightforward way. These blatant differences did not deter early researchers, for the case of quantized Maxwell theory was one of the few available tools to guide theory construction, in spite of its imperfect fit. Later, in the mid-century, it was suggested that the now-acknowledged important non-linearities of general relativity might in fact somehow 'cure' the divergences of quantum field theory.

---

[35] However, still in 1981, at the first Marcel Grossman meeting, one finds Peter van Neuwenhuizen writing that "The strategy of particle physicists has been to ignore these problems [of non-linearities] for the time being, in the hope that they will ultimately be resolved in the final theory," after which he proceeds to ignore them himself and follow suit: "particles (including gravitons) are always in flat Minkowski space and move *as if* they followed their geodesics in curved spacetime because of the dynamics of multiple graviton exchange" [van Niuwenhuizen (1981), p. 2]. Carlo Rovelli, who himself cites this passage of van Neuwenhuizen's, uses it to highlight the divide we mentioned in Chapter 2: "the divide reflects the different understanding of the world that the particle physics community on the one hand and the relativity community on the other hand, have. The two communities have made repeated and sincere efforts to talk to each other and understanding each other. But the divide remains, and, with the divide, the feeling, on both sides, that the other side is incapable of appreciating something basic and essential: On the one side, the structure of quantum field theory as it has been understood in half a century of investigation; on the other side, the novel physical understanding of space and time that has appeared with general relativity. Both sides expect that the point of the other will turn out, at the end of the day, to be not very relevant. One side because all the experience with quantum field theory is on a fixed metric spacetime, and thus is irrelevant in a genuinely background independent context. The other because GR is only a low energy limit of a much more complex theory, and thus cannot be taken too seriously as an indication on the deep structure of Nature. Hopefully, the recent successes of both lines will force the two sides, finally, to face the problems that the other side considers priority: background independence on the one hand, control of a perturbation expansion on the other" [Rovelli (2002), pp. 760–1].

The development of Yang–Mills theory saw the development of a new analogy. Because of the closer similarities between Yang–Mills fields (which also have non-linear field equations) and the gravitational field (both of which also have infinite dimensional gauge groups), Murray Gell-Man (in the late 1950s) suggested to Feynman, who was becoming interested, that he attempt to quantize Yang–Mills theory first, as a preparatory exercise ([Feynman (1972), p. 378]). This led to groundbreaking work in the construction of quantum gauge field theories (of the kind that make up the current standard model of particle physics). However, this new analogy began to be probed within our timeframe. In 1955–6, while visiting the Institute for Advanced Study, in Princeton, Ryōyū Utiyama [Utiyama (1956)] worked out the details of Yang–Mills theories and general relativity qua gauge theories, bringing out important similarities not shared by electromagnetism. The gravitational field was formulated as a gauge field of the Lorentz group $SO(1,3)$—this was on the back of Yang and Mills' 1954 paper focussing on non-Abelian, $SU(2)$ symmetries. Utiyama generalized Yang and Mills'[36] work to spacetime transformations (diffeomorphisms), so that the gravitational field is generated through Lorentz invariance by a gauge argument in a similar way to the electromagnetic potentials A_μ coming from the invariance of the Lagrangian under $U(1)$ transformations. In other words, one is *forced* to introduce a field to maintain invariance of the Lagrangian under the specified symmetries (so long as the original Lagrangian is invariant under Lorentz transformations). Utiyama's method proceeded by first introducing a system of tetrads $h_a^\mu(x)$ at each point and extending Lorentz transformations at those points to a larger group of transformations depending on six arbitrary functions of x.[37] In some ways, this blunts the case for the specialness of gravity, since it provides a common framework for all forces, with both gravity and Yang–Mills theories becoming a species of non-Abelian gauge theory, with the gauge principle playing a unifying role. It also appears to advocate a flat space (particle physicists) viewpoint, since in this approach gravity is associated with the Poincaré group acting in Minkowski space, from which general relativity is derived.

5.4 β-Decay and the Neutrino–Graviton Hypothesis

A very short-lived flurry of activity in the 1930s centered on a different analogy, the possibility of a connection between neutrinos and gravitons—indeed, the neutrino hypothesis was born amidst these gravitational connections and indicates that gravitation played a fairly central role in quantum issues, despite its weakness.[38] Pauli had publicly

---

[36] As Yang later noted, his work with Mills "was completely divorced from general relativity and we did not appreciate that gauge fields and general relativity are somehow related" [Yang (1986), p. 17].

[37] Though coming after our period under study, Feynman, in 1963, would figure out the quantum perturbation theory for such non-Abelian gauge fields. As DeWitt notes, however, Yang–Mills theories yield renormalizable theories while the gravitational field does not [DeWitt (1978), p. 275].

[38] A useful discussion of this episode is [Jensen (2000), §6.5]. See also [Gorelik and Frenkel (1994), §4.2.3] for a different, Soviet, perspective.

proposed the existence of the neutrino[39] in 1930, in response to Bohr's own proposal that there was energy (and charge) non-conservation in so-called β-rays (that is, β-decay).[40] In a subsequent letter to Bohr from George Gamow,[41] Gamow points out that in a small impromptu conference, at the Institute in Kharkov, with Paul Ehrenfest, Lev Landau, and others, they concluded that Bohr's proposed violation of energy conservation would contradict general relativity—there would be no solutions since the sources are constrained to conserve energy. Using the simple diagram in fig. 5.1, Gamow notes (ibid.):

> It looks as if non-conservation of energy is in contradiction to the equations of gravitation for the vacuum. If the gravitational equations are correct for region B, this implies that the total mass in region A (where we do not know the laws) must be constant. If in A we have, for example, an RaE [Radium E—DR] nucleus, and alter its total mass with a jump in a transmutation process, we can no longer apply the usual gravitational equations in region B.

However, Bohr simply agreed in his response to Gamow that there would be "equally sweeping consequences for Einstein's theory of gravitation" (and indeed, so will the

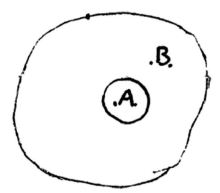

Fig. 5.1 *Image source, [Jensen (2000), p. 175].*

---

[39] As laid out in a letter addressed to the "radioactive ladies and gentlemen" (an open letter to the radioactivity group at the Gauvereins meeting in Tübingen, December 4th: [von Meyenn (1985), pp. 39–40]). Pauli's original name for the particle was 'neutron,' and he referred to his proposal as "a desperate remedy" for the apparent energy (and angular momentum) imbalance in nuclear beta decay, with the neutron able to "exist in the nuclei" and "which have the spin 1/2 and follow the principle of exclusion". In a later letter to Oskar Klein (December 12th, 1930) he claimed to not much believe in the hypothesis himself [von Meyenn (1985), p. 45], and even in the original open letter he notes that it might seem "unlikely" ("but only those who dare win").

[40] The problem that led to Pauli and Bohr's desperate acts stemmed from features of Rutherford's new experiments on the expulsion of protons under the bombardment of atomic nuclei with 'α-rays'.

[41] December 31st, 1932: [Peierls (1986), p. 569].

charge conservation problem pose equal problems for Maxwell's theory).[42] Bohr went on to speculate, at the 1933 Solvay Conference, whether this non-conservation might have observable consequences. In fact, Pauli had been expressing the worry that there would be a conflict with gravitational physics nearly two years before Gamow's letter. He wrote to Oskar Klein on January 8th, 1931, asking him to ask Bohr "how he thinks of the balance of weight and the gravitational effect on the outside of a closed box, inside of which radioactive β-decay processes take place" [von Meyenn (1985), p. 51].[43]

The links between the new neutrino and gravitation did not end with the unrealized possibility of damaging general relativity. Bohr, still antagonistic to the neutrino, further speculated, likely in a desperate bid to save the correspondence principle, that perhaps the neutrino is really in some sense the manifestation of a quantized gravitational wave. As Gamow put it in a letter to Goudsmit (March 8th, 1934): "Bohr ... (well, you know that he absolutely does not like this chargeless, massless little thing!) thinks that continuous β-spectra is compensated by the emmition [sic] of gravitational waves (!!!) which play the role of the neutrino, but are much more physical things" (cited in [Pais (1994)], p. 369). Just a week later, perhaps as a result of Fermi's theory, as well as the derision of Gamow and company, Bohr expressed this idea to Pauli:[44]

> It also pleased me that you understood the basic mood of my final remarks on energy conservation. However, I have since become more skeptical with regard to the hope that implicitly lies in these remarks, namely to use the gravitational theory for a correspondence-derivation of the Law of β-Radiation. The idea was that a Neutrino, for which one assumes a rest mass of 0, may not be anything but a gravitational wave with appropriate quantization. I have convinced myself, however, that the Gravity constant is far too small to justify such a conception and am therefore fully prepared for us to really have a new atomic trait for us here that could mean the real existence of the Neutrino.

The idea is a close cousin of Louis de Broglie's [de Broglie (1934)], then only recently proposed, neutrino theory of light,[45] according to which a photon is understood to be a bound state (a "fusion") consisting of a (complementary) neutrino and anti-neutrino pair (the two spin-1/2 particles, then thought to possess zero rest mass and charge, combining to produce a spin-1 particle).[46] In other words, any process involving what

---

[42] Bohr, letter to Gamow, January 21st, 1933: [Peierls (1986), p. 571].

[43] He had already skirted this problem with gravitational measurements in the context of β-decay, again in a letter to Klein (as mediator for Bohr), in December 1930: [von Meyenn (1985), pp. 45–6].

[44] Bohr to Pauli, March 15th, 1934, [von Meyenn (1985), p. 307].

[45] Note, however, that de Broglie's fusion theory of light was developed independently of the neutrino idea, and was developed two years prior [de Broglie (1932)], as part of a rigid adherence to the light-matter analogy, using the general notion of a pair of Dirac particles, and so a photon would be represented by a 16-component object transforming under Lorentz transformations as would the product of two Dirac spinors (see [Darrigol (1998), p. 291–2]). Marie-Antoinette Tonnelat [Tonnelat (1943)] generalized this scheme to include gravitons, finding, as we have seen before, a correspondence with general relativity for the weak-field case—she also looked at the situation for a non-vanishing cosmological constant, in de Sitter space, and found that this involved a non-vanishing graviton mass (the case with matter was considered in [Tonnelat (1944)]).

[46] This was proposed to account for the creation and annihilation property of photons but suffered from an obvious discrepancy in the statistical behavior of photons (qua bosons) and neutrinos (qua fermions): how does

looks like a photon (an elementary particle) should be understood as a complex system involving a pair of fermions. Pauli had discussed this photon theory, with Heisenberg, in the context of Bohr's idea of treating the neutrinos as gravitational quanta.[47] Pauli is more favourably disposed to the idea of a photon rather than a gravitational reduction to neutrinos, though he wonders "why the light quantum emission always occurs in such a way that two neutrinos are emitted in the same direction". Interestingly, he notes also that Fermi had suggested connecting the neutrinos with "half gravitational quanta" (see [von Meyenn (1985), p. 278]).[48]

Following on from such issues, and Dirac's recent excursion into wave equations for higher spin theories [Dirac (1936)],[49] Pauli had already by 1937[50] set his graduate student Markus Fierz to work on the problem of particles of arbitrary spin, including spins higher than 1 that would of course become associated with gravitational quanta. Peierls had his own doctoral student, Fred Hoyle, working on this same topic of wave equations for higher-spin particles ([Peierls (1986)], pp. 561–2), and in fact there was a regular back and forth between Pauli and Peierls on the progress of their students, with Hoyle up against the problem of a non-positive-definite spin 3/2 particle which was ultimately shown by Fierz and Pauli to rest on a fallacy. Hoyle's work was unsuccessful. Indeed, Pauli remarked (in a letter to Uhlenbeck, July 9th, 1938) that "Hoyle's reflections…were quite wrong, and he had to retract his work just before printing (a disgrace more for Peierls than for Hoyle)" [von Meyenn (1985), p. 586]. (Cf. [Blum (2014), §4], in which he argues that the introduction of new nuclear particles and the study of arbitrary, higher-spins were in fact closely linked). Pauli does not seem to have been particularly impressed by the 'spin-2=gravitational quanta' result, since he first mentions it somewhat casually to Heisenberg in figuring out his talk at the Solvay

---

Bose–Einstein behavior of photons emerge from the Fermi–Dirac statistics of the neutrinos? In fact, Pascual Jordan [Jordan (1935)] resolved this issue by introducing an interaction with electric charges that provides a mechanism for the photonic behaviour. Cf. [Darrigol (1998), p. 229] for the analogy between β-decay and photon emission.

[47] See Heisenberg to Pauli (February 5th, 1934); Pauli's reply (February 6th, 1934): [von Meyenn (1985), pp. 272–8].

[48] Pauli in fact, in 1936, returned to the issue of the relationship between β-radioactivity and gravitation in a popular level review article on Space, Time, and Causality in Modern Physics. Here he speculates that this radiation points to a "deeper level" of physical reality since they seem to involve a new constant of nature, lying beyond those known. He writes: "In this connection it is of interest to point out that present-day classical field theories, including the relativistic theory of gravitation, do not give a satisfactory interpretation of the essentially positive character of the constant κ, which is responsible for the fact that gravitation manifests itself as an attraction and not as repulsion of gravitating masses. Such an interpretation could consist only in the reduction of the constant κ to the square of another constant of nature. This suggests looking for phenomena in which the square root of the constant κ plays a part. While hitherto it has been regarded as almost certain that gravitational phenomena play practically no part in nuclear physics, it now appears that the possibility that the phenomena of β-radioactivity might be connected with the square root of κ can no longer be rejected out of hand" [Pauli (1936), pp. 104–5].

[49] As Dirac put it: "it is desirable to have the equation ready for a possible future discovery of an elementary particle with a spin greater than a half, or for approximate application to composite particles" (ibid., p. 448).

[50] See, e.g., Pauli's request for information along these lines to Rudolph Peierls, in a letter dated October 15th, 1937: [von Meyenn (1985), p. 538].

meeting.[51] Pauli's focus seems to have been the more general spin-statistics link for arbitrary spins.[52]

Bohr had completely abandoned this energy non-conservation idea by 1936[53] not only due to his thoughts about the magnitude of the gravitational constant, but primarily thanks to Fermi's theory of β-decay [Fermi (1934)][54] which was based on Pauli's neutrino hypothesis and is best encapsulated in the process:

$$\text{neutron} \longrightarrow \text{proton} + \text{electron} + \text{neutrino} \qquad (5.1)$$

That is, the transition from neutron to proton is accompanied by the creation of an electron and a neutrino (and annihilation thereof for the inverse process). Bohr's reticence to accept Pauli's neutrino no doubt had much to do with the inability to use the correspondence principle to figure out the neutrino's behaviour, yet the approach of Fermi, firmly grounded in experimental data (using the "comparison with experience" as Fermi put it), was enough to dislodge this prejudice.

However, though the neutrino–graviton link did not become established, it was still mentioned for some time afterwards. Gorelik and Frenkel refer to an article, "The Neutrino Hypothesis and the Conservation of Energy," by Dmitri Blokhintsev and his teacher Fedor Gal'perin ([Blokhintsev and Gal'perin (1934)]), in which they suggest, like Bohr, that β-decay would render graviton radiation observable if the neutrino–graviton

---

[51] See letter from Pauli to Heisenberg, April 27th, 1939: [von Meyenn (1985), p. 638]. It seems that it was Fierz who had initially mentioned it to Heisenberg, who was evidently far more enthusiastic about it than Pauli (see Heisenberg to Pauli, April 23rd, 1939: ibid., p. 635; cf. also letter from Heisenberg to Pauli, May 19th, 1939: [von Meyenn (1985), p. 659]. Pauli later wrote to Heisenberg that "I only write about gravitational quanta because of your special wish" (June 10th, 1939: ibid., p. 662). Heisenberg wrote back glowingly that "[t]he gravitational quanta of spin 2 made me happy" (Heisenberg to Pauli, August 14th, 1939: [von Meyenn (1985), p. 673)].

[52] A parallel, largely independent (initially, at least) spin-statistics classificatory scheme was developed from de Broglie's fusion theory, in Tonnelat's work, such as that already cited in footnote 45, and also by Gérard Petiau (e.g. [Petiau (1941a)], [Petiau (1941b)]). Odd fusions of Dirac particles gave fermions, and even fusions gave bosons, with a simple additive scheme whereby if the fusion of two Dirac particles gives a photon of spin 1, then a fusion of two photons gives a particle of spin-2: the graviton (consisting of four Dirac particles). Both had been students in de Broglie's seminar, and in fact, Petiau had already begun working on this classification in his PhD thesis ("Contribution á la Théorie des Equations d'Ondes Corpusculaires"), under de Broglie's supervision, at the University of Paris, which he completed in 1936. This was separated from the quantum field theoretic (though they eventually made use of Fierz and Pauli's link between the spin-2 case and gravitation) work by the war (as Tonnelat points out: [Tonnelat (1976), p. 228)], but also, perhaps, by its grounding in de Broglie's idiosyncratic theory—see [Pitts (2010)] for a philosophical analysis of this work, including the theories of massive gravity that stemmed from it; [Rocci (2015), §4.3.3] provides an historical overview; Tonnelat's own perspective can be found in [Tonnelat (1976)].

[53] See Bohr's short note on "Conservation Laws in Quantum Theory" in *Nature* (1936) 138: 25–6. Bohr attempted to link this energy non-conservation to stellar energy production—for an explanation of this idea, see: Interview of George Gamow by Charles Weiner on April 25th, 1968, Niels Bohr Library & Archives, American Institute of Physics: www.aip.org/history-programs/niels-bohr-library/oral-histories/4325.

[54] As Darrigol puts it: "the first successful transposition of quantum electrodynamics to nuclear processes": [Darrigol (1998), p. 229].

link is true. However, they note that the mismatch of spin values makes de Broglie's fusion theory more likely [Gorelik and Frenkel (1994), pp. 96–7].[55]

Rudolph Peierls was still referring to the neutrino–graviton link, in a review article about fundamental particles, in 1946, though in rather dismissive terms, now drawing attention to the clash of spins, in the wake of Fierz and Pauli's result:

> Lastly, it has been pointed out that, just as the wave equations of the electromagnetic field give rise to photons, and those of the nuclear force field to mesons, the gravitational field should be similarly quantized. Since gravitational forces are extremely weak if expressed in units appropriate to elementary particles, the quanta constituting the gravitational field would have only an extremely weak interaction with other particles. It is tempting to relate this to neutrinos, which are indeed extremely elusive particles; but the idea is somewhat less attractive when it is remembered that the symmetry properties of the gravitational field are such as would correspond to particles of spin 2, and that therefore the only way of introducing neutrinos would be in terms of an elementary process consisting of the simultaneous emission or absorption of four neutrinos. It is likely that this problem will remain in the realm of speculation for a considerable time.
>
> [Peierls (1946), p. 773]

Still later, in a general review of neutrino physics in 1948, Gamow, in considering the problem of neutrinos' physical nature, returned to the idea that they might be quanta of the gravitational field [Gamow (1948)].[56] The idea is put here in terms strikingly close to that of Bohr's, about which he was so dismissive; namely as a neutrino annihilation process leading to the emission of a "gravitational quantum". Neutrinos were to stand to the gravitational field as electrons stand to the electromagnetic field. However, all of this is clearly supposed to be taken with a grain of salt, or, as Gamow puts it, "with a due amount of phantasy" (ibid., p. 31), and was clearly intended to whet the *Physics Today* readers' appetite for an open puzzle.[57]

---

[55] It is clear from the cited excerpts given by Gorelik and Frenkel that Blokhintsev and Gal'perin were doing little more than rehashing what Bohr, Pauli, Heisenberg, and others were saying at the time (as indeed Gorelik and Frenkel suggest). They also accept Einstein's earlier claim that gravitational radiation would occur in atoms according to the theory, resulting in instability, and therefore suggesting the existence of gravitational quanta, in the same way as photons. Gorelik and Frenkel also claim that this article also contains the first ever usage of the term 'graviton' to describe gravitational quanta—this might well be so, since it was more usual at the time to use the full expression "gravitational quanta" to describe them, though they refer to it as the "so-called *graviton*," so presumably it was already in currency to some extent. (Albeit in a slightly different context, there is in fact another, earlier, usage of the term from 1935, by one Sir Shah Sulaiman (a mathematician and high-court judge), who put forward a competing theory to Einstein's which postulates the existence of gravitons on which the pull of gravity depends (*Science News Letter,* November 16th, 1935, p. 309).) The Russian-Italian physicist Gleb Wataghin added the twist of suggesting that it is gravitational quanta that should have spin of one half, thus securing the identification with the neutrino [Wataghin (1937)]—he was perhaps following Fermi's idea here; however, he claims to derive this result from the analysis of Dirac's equation in curved space (following Tetrode's methods, presumably associating this with the presence of Dirac particles of spin 1/2). See [Hagar (2014)] for a detailed treatment of Wataghin's work, including in §5.3.2, his graviton-neutrino speculations.

[56] Gamov rather amusingly gave his famous Mr Tompkins (with the first book appearing in 1939, as *Mr Tompkins in Wonderland*) the initials "C.G.H." (after the three fundamental constants).

[57] Not long after Gamow's discussion, the fermionic neutrino would, in John Wheeler's mind, furnish a fatal obstacle to his geometrodynamical program, in which he attempted to recover particle physics from aspects of

5.5 From Flat Space Theories to Covariant Quantization

Simplification often characterizes the earliest work in some field. One might find toy models, for example. Or, in cases were one has a non-linear theory, the use of linearization techniques, as we have saw earlier in this chapter. General relativity is a non-linear theory: gravity couples to energy-momentum, and the gravitational field has energy-momentum, therefore gravity gravitates. The non-linearities led to many (but by no means all) of the complications that are faced in quantum gravity. Rosenfeld, in his 1930 work, attempted to quantize the linear theory only. This, as most acknowledge, is a preliminary exercise. One would attempt to account for the non-linearities by adding quantum corrections.

In a comparative review of Lorentz-invariant theories of gravitation, Whitrow and Morduch write:

> Despite the great intellectual appeal of Einstein's general theory of relativity, various alternative theories of gravitational motion have been formulated during the past fifty years. All these theories have one feature in common: like Einstein's special theory of relativity they retain the principle of Lorentz-invariance, but unlike general relativity they do not admit the general covariance of all frames of reference in all possible states of motion. The incentive to devise Lorentz-invariant theories of gravitation, which has persisted until the present day, appears to be mainly due to the great difficulty in solving the non-linear equations that occur in general relativity. Lorentz-invariant theories are, in general, linear theories, but even those that are not strictly linear tend to be mathematically more tractable than general relativity, at least to order $1/c^2$.
>
> [Whitrow and Morduch (1965), p. 15]

We mentioned in the previous chapter that Milne's alternative theory of gravitation, based on flat spacetime and a strict focus on observable structure only, offered an important contrary voice to the prevailing view that gravitation was simply an aspect of curved space. By showing how one could recover certain key results without invoking this additional geometrical superstructure, he pointed to the possibility of an approach to gravitation in line with the rest of physics. This mindset was attractive to many others,[58] including several early quantum gravity researchers who would try and quantize such 'flat space' theories as a more practical and applicable approach to the problem, though often treating the flat space as a zeroth-order approximation to the curved space of Einstein's theory. Let us begin by looking at a purely classical venture.[59]

---

spacetime geometry and topology: while electric charge was no problem, he was never able to find a satisfactory geometrodynamical analog for the neutrino (see, e.g., [Klauder and Wheeler (1957)]; and Chapter 8 of this book).

[58] In a sense, general relativity was bad for the study of gravitation, cocooning it in unfamiliar garb.

[59] In another earlier review (covering scalar, vector, and tensor based Lorentz-invariant alternatives to general relativity), Whitrow and Morduch argued that "no Lorentz-invariant theory of gravitation predicts all the effects considered here in complete agreement with general relativity" [Whitrow and Morduch (1960), p. 794]. However, they do briefly consider quantum gravity theories in a later review [Whitrow and Morduch (1965)], from which the above quotation was taken, though their focus is primarily on the issue of positive-energy in such theories—again, quantum gravity theories are not included in their claim that Lorentz-invariant

George Birkhoff, one of the many drawn into the links between Schrödinger's wave equation and general relativity, presented a theory, in 1927 [Birkhoff (1927)], in which matter was described by a 'perfect fluid' and atomic potential ψ (constant along the worldlines of the perfect fluid[60]), initially defined relative to Einstein's curved spacetime, from which he obtains an equation like the Schrödinger equation. This idea did not take, but later, in 1942,[61] Birkhoff returned to the theory of gravitation, only this time with a theory based on flat Minkowski space with gravitational tensor h_{ij} satisfying $\Box h_{ij} = 0$ in empty space. Birkhoff's primary motivation in his flat space alternative to Einstein's geometrical theory was the fact that no new tests, beyond the three classical tests, were forthcoming and gravity had simply languished in isolation from the rest of physics. Indeed, he writes, "because of its complicated mathematical character, [it] seems to be essentially unworkable" [Birkhoff (1944), p. 325]—see also [Birkhoff (1943)].

His theory had some interesting results. For example, it managed to account for the red shift by invoking something akin to Fritz Zwicky's 'tired light' hypothesis, related to the energy change of a photon as it travels from an emitting body (see [Barajas, *et al.* (1944), p. 138]). The equivalence principle is de-geometrized and interpreted as the claim that "certain equations in the Einstein theory are linear homogeneous in the density of matter" [Birkhoff (1944), p. 333]. It was also explicitly promoted as a 'dualistic' alternative to Einstein's monism: in addition to flat spacetime, there is the perfect fluid introduced as a fundamental bearer of mass and charge, with disturbance velocity of c. It was set up explicitly as an absolutist alternative to Einstein's theory, in which there exists a set of independent variables, t, x, y, z, valid for the whole of spacetime, thus violating general covariance. However, the theory was scrutinized by Hermann Weyl [Weyl (1944a)] who found that Birkhoff had simply reproduced something along the lines of Einstein's own weak-field approximation from 1916. That is, one works with $g_{\mu\nu}$ close to $\delta_{\mu\nu}$ (i.e. flat space), and splits $g_{\mu\nu}$ according to $g_{\mu\nu} = \delta_{\mu\nu} + h_{\mu\nu}$, ignoring higher powers of the $h_{\mu\nu}$ piece that differs from $\delta_{\mu\nu}$. Weyl was also critical of the introduction of a perfect fluid to model matter, which Weyl viewed as a "primitive irreducible physical entity," though it is unclear what exactly is wrong with having a fundamental level such as this. There was also the problem that the proportionality between inertial and gravitational mass is unaccounted for, putting physics back into a pre-Einsteinian mystery [Weyl (1944b), pp. 205–6]. Weyl notes, however, that this is a problem for any such flat spacetime view (what Weyl calls "the degenerate Einstein theory") if viewed as something other than an approximation, for they will always involve the split: "$g_{ik} = \delta_{ik} + \epsilon h_{ik}$ (δ_{ik} = inertia, h_{ik} = gravitation)" (ibid., p. 206).

Weyl's low opinion seems to have been shared by Schrödinger who, in a letter to Frederik Belinfante, responding to a request for an assessment of Birkhoff's theory, wrote back that he has no doubt that "the waste paper basket is the right place for it".[62] In his letter, Belinfante characterizes Birkhoff's theory as treating the gravitational

---

theories of gravitation cannot be in complete agreement with general relativity. A good lengthy treatment of flat space gravitational theories is [Pitts (2001)].

[60] This potential is also supposed to lead to an atomic frequency equation along the lines of Schrödinger's.

[61] First presenting his ideas at the Astrophysical Congress, in Puebla Mexico, in February of that year.

[62] March 16th, 1951: Schrödinger Nachlass, University of Vienna.

field "as just another field".[63] However, in his reply, Belinfante points out that he will spend some more time with Birkhoff's papers before he throws them in the wastepaper basket. Indeed, he pursued something very close to Birkhoff's theory in his own "phenomenological linear theory of gravity" (which he also quantized).[64] Belinfante's own linear theory was meant to be a self-subsistent theory, not an approximation or preliminary exercise until the 'real work' with the non-linearities is done. Given that Maxwell's theory in linear form (together with its quantization) "gives a fair description of" electromagnetism, it could be supposed, according to Belinfante, that the same could be said of a linear theory of gravity (with quantization) [Belinfante and Swihart (1957a), p. 169]—hence, this approach too falls within the scope of the electrodynamical analogy. In order to stand alone, it had to differ from the standard linearizations, which all failed to reproduce the classic tests of general relativity. It achieved this by inserting in the field equations *ad hoc* constants tailored using measurement results to yield the same results as general relativity for the three classic tests (for the one-body problem).

A little earlier, Marcos Moshinsky (a student of Wigner's at Princeton), was one of the few to take seriously and directly extend Birkhoff's theory along the "just another field" line, including other fields (meson, electromagnetic, and pair fields) by adding extra terms to the Lagrangian function originally used by Birkhoff [Moshinsky (1950)]. His reasoning was, again, to study gravitation from the point of view of field theory. Moshinsky argued that for various choices of these additional terms, one could derive the correct result for the deflection of light around the Sun, as well as the gravitational red shift, in complete agreement with the general theory of relativity. Moshinsky ends up constructing a Dirac equation for an electron in a combined gravitational and electromagnetic field (à la Birkhoff), to describe the effect of a gravitational field on atoms and electrons. While he notes that the gravitational interactions between particles are too weak to observe (being 10^{40} times smaller than the electromagnetic), he argues that large masses can have observable effects at the atomic level. For example, a hydrogen atom in a large gravitational field (such as near the surface of a star) will be in a portion of space with an altered dielectric constant (given Birkhoff's theory)—Belinfante borrowed this explanation in his own linear theory [Belinfante and Swihart (1957b)]. Moshinsky manages to derive the gravitational red shift of the wavelength of light emitted in this case. He also derives a gravitational correction to the Zeeman effect, again couched in terms of the interaction of Birkhoff's gravitational field with other fields (ibid., pp. 518–9).[65]

---

[63] Letter from Belinfante to Schrödinger, April 2nd, 1951: Schrödinger Nachlass, University of Vienna.

[64] This project was pursued over several papers with his student James Swihart, who produced a PhD thesis on the topic at Purdue in 1955 (see appendix)—Swihart later turned his attentions to condensed matter physics.

[65] This 'gravitational Zeeman effect' would later be studied by Zel'dovich, in 1965 [Zel'dovich (1965)], in the context of the gravitational field of a rotating star, much as Moshinsky had also suggested—of course, this phenomenon is related to the Lense–Thirring effect, though the interpretation of Moshinsky (based on Birkhoff's theory) is radically different, involving field interactions (which change rods and clocks through the coupling, giving results that mimic curved spacetime) rather than metric dragging on the surface surrounding a body—a similar mimicry is involved in the more well-known particle physics/flat space accounts of gravity.

Birkhoff's theory was curiously detached from other work on the topic from the time.[66] For example, in 1940, Nathan Rosen had already showed that Einstein's field equations can be given a flat space interpretation by introducing a second metric tensor for flat space, γ_{ik} (the Minkowski tensor) at each point of space, in addition to the usual Riemannian metric g_{ik} which now is only to represent the gravitational potential [Rosen (1940a)][Rosen (1940b)].[67] This was originally a purely classical endeavour. Papapetrou [Papapetrou (1948)] had taken Rosen's approach and rendered the de Donder gauge condition in a bi-metric fashion. In an earlier project to that considered above, Belinfante attempted to integrate all of this with Suraj Gupta's covariant quantization approach (which we turn to shortly) to come up with a flat theory useful "in visualizing space, as a help in calculations, and as a basis for a simple quantum theory of fields" [Belinfante (1955b), p. 800]. Though the theory is labelled 'flat,' however. Belinfante also finds a curious feature of this bi-metric theory (which he labels the "swiss cheese" property), which is that when electrically neutral point particles are present, these will manifest as 'holes' (values of coordinates in the flat Lorentz frame that are not filled with 'real' points) in the flat space, along worldlines of elementary particles—in this regard, he also speculates that there is a "natural cut-off radius for the fields around the hole" that can function to suppress high-momentum states (or short-distance states) in their expansion, which of course could be helpful in eliminating some of the divergences of quantum field theory (the topic of Chapter 7). However, given the supposedly auxiliary nature of the flat metric in this approach, it is difficult to see how this is supposed to work physically.

As mentioned, Belinfante (and Swihart) simultaneously developed a flat space theory of gravitation intended to stand alone, without the curved space (for which they claim priority over Gupta: [Belinfante and Swihart (1954), p. 2]; see also [Belinfante (1955a)]). They initially attempted a quantization of Birkhoff's theory of gravity but were unable to find a Lagrangian that gave Birkhoff's equations of motion—a fact they interpreted as the inability of the theory to satisfy the reciprocity (i.e. action-reaction) principle linking gravity and matter. The theory is of course, not generally covariant, which Belinfante views as somewhat positive, since they evade such problems as the constraints that plague canonical approaches. However, this fact (manifesting in terms of the existence of a canonically conjugate momentum for each gravitational potential component), also implies that their theory is not a valid approximation to Einstein's theory [Belinfante (1955a), p. 195].[68] Gupta had already argued that Einstein's theory could in fact be fully

---

[66] However, it was utilized in early Soviet work on the calculation of cross-sections involving gravitons and other particles—see, e.g., [Piir (1957)]. Of course, by the time of Birkhoff's later flat space theory and the discussions by Weyl, Fierz and Pauli had already shown how one can associate the weak field approximation of general relativity with the quantum theory of a massless spin-2 particle. What remained to be done was really to include the non-linearities, to step away from the weak-field limit. Indeed, it would take the observation of black holes to distinguish these approaches, since the flat space approaches do not possess Schwarzschild horizons.

[67] Again, as with Birkhoff's, this approach distinguishes between an inertial field and a gravitational field.

[68] Gupta had reviewed Belinfante's paper "Quantization of the interacting fields of electrons, electromagnetism, and gravity" [Belinfante, Caplan, and Kennedy (1957)] and found it "extremely long and laborious" using work "freely borrowed from the work of the reviewer" and for which the "resulting formalism is rather unintelligible" (*Mathematical Reviews*, MR0090400).

recovered from a flat space theory derived from a Lagrangian with an infinite number of terms.

Suraj Gupta was responsible for one of the first approaches to what is now known as 'covariant quantum gravity,' involving a perturbative, Lorentz-invariant formulation of the general theory of relativity.[69] His innovation, beyond Belinfante and Swihart, was to break general covariance during the quantization procedure, fixing the coordinates by adopting a gauge condition (namely the so-called de Donder coordinate condition: $\partial g^{\mu\nu}/\partial x^{\nu} = 0$).[70] Gupta's approach is conceptually very distinct from the Hamiltonian methods of Bergmann, Pirani and Schild, and the DeWitts of around the same time, in which the metric tensor retains its geometrical significance (albeit post-quantization). Gupta by contrast completely denies the metric tensor the status of a Riemannian space, arguing instead that $g^{\mu\nu}$ has the status of an orthodox quantum field, so that it makes sense to consider the geometrical interpretation as a purely macroscopic feature only. One might consider the approach to involve the original Riemannian, curved space conception by viewing the theory in flat Minkowski space as a kind of zeroth order approximation. The field quantities in the original theory are recovered in terms of expansions in powers of the coupling constant κ. Hence, general relativity is recast as a theory in flat spacetime, with the Lagrangian containing an infinite number of terms. It is, then, very much along the lines of the electromagnetic field analogy. One first casts general relativity as a flat space theory, and then one quantizes it almost exactly along the same lines as in the electromagnetic case.[71]

Gupta gave two versions of his approach,[72] at varying levels of generality, beginning with a version based on the linearized field equations. In this initial skirmish, Gupta invokes an indefinite metric in Hilbert space to quantize the theory—this effort grew out of his earlier work on quantum electrodynamics using an indefinite metric and what is now called the "Gupta–Bleuler method" (after both Gupta and Konrad Bleuler independently devised the scheme, in 1950: [Gupta (1950)], [Bleuler (1950)]). This

---

[69] In their review of covariant quantum gravity, Faddeev and Popov identify Gupta as the originator of the scheme, following the electrodynamic analogy including, as we see below, an analogous version of the 'Fermi trick' (i.e. $\frac{\partial A'_{\mu}}{\partial x_{\mu}} \Psi' = 0$) to deal with the singularity of the Lagrangian (from the gauge freedom). However, they note that Gupta's direct transfer of Fermi's method led to a violation of unitarity, as discovered by Feynman in 1963, and resolved by themselves and DeWitt in 1967 using functional integration [Faddeev and Popov (1973), pp. 777–8].

[70] Having a flat, fixed spacetime background brings with it the need to impose coordinate conditions to get rid of the gauge degrees of freedom related to general covariance, thereby picking out an individual coordinate system (or privileged family of such) in which it is satisfied. Gupta adopts an exactly analogous approach to that used in his quantization of the radiation field, using an indefinite metric. The indefinite metric introduces photons (with negative commutator and states with negative probabilities), and so Gupta utilizes a supplementary condition (similar to the Fermi trick in which the Lorentz condition is imposed to stop the longitudinal and scalar photons from being observable—i.e. appearing as free particles) to eliminate them, rendering them unobservable. The coordinate conditions here, then, play a role in carving out the physically realistic parts of the theory.

[71] Hence, by this stage, we have a distinctive split in the approaches, despite their common ancestry.

[72] In fact, Gupta had earlier mentioned the relevance of his approach to the Fermi supplementary condition in his paper (written in Cambridge in 1951) to the case of coordinate conditions in general relativity.

results in a picture involving two types of free graviton, which Gupta shows must be spin-2 particles. The full (non-linear) theory (splitting a field into a free and an interacting part) amounts to the interaction picture, and clearly maps quite directly onto the splitting of the metric into flat and residual terms. This suggests a straightforward treatment of quantum gravity in which the residual parts are viewed in terms of interactions between the quanta associated with the free field (coming from the flat part of the metric). Gupta's approach essentially springs from the generalization of a program for getting supplementary conditions from a switch to the interaction representation; in this case, the supplementary condition has the form of a coordinate condition. That is, Gupta's quantization of gravity amounts to couching the theory in the interaction representation (based on Schwinger's formulation) and following the consequences. As he describes it his theory amounts to: "expressing the field quantities in the Riemannian space as expansions in the flat space, and then splitting the gravitational field into the linear and the non-linear parts. The linear part of the gravitational field is regarded as the free gravitational field, while the non-linearity is treated as a direct interaction between the gravitons."[73]

In fact, Bryce DeWitt[74] had already followed a similar path. DeWitt and Gupta both write down the gravitational potentials in a form in which : $g^{\mu\nu} = \epsilon^{\mu\nu} - \kappa\gamma^{\mu\nu}$ (with ϵ the Minkowski tensor)—in fact DeWitt uses a '+'. They both employ the interaction representation in the quantized theory. The field operators $\gamma^{\mu\nu}$ are made to obey the field equations and commutation relations of the linearized theory with a free Lagrangian L_0, and the wavefunction then satisfies a Schrödinger equation in which the "Hamiltonian" is just the operator $\kappa L'$ expressed as a function of the $\gamma^{\mu\nu}$ so defined. The effect of $\kappa L'$ is to produce an apparent interaction between two quanta of the gravitational field, corresponding to the non-linear character of the original field equations. The quantized theory so constructed is an exact one and not merely a weak-field approximation. It is however obviously well adapted to calculations in which a weak-field approximation is made by treating $\kappa L'$ as a small perturbation.

[73] Gupta returned to quantum gravity in 1954, with a kind of inversion of his previous work [Gupta (1954)]. In this case, he begins with a field of gravitons zero-mass, rather than deriving them. Allowing the field to interact with other fields, implies that its source must be a symmetric tensor with zero divergence.

[74] Bryce DeWitt (then Seligman) wrote a doctoral thesis under Julian Schwinger devoted entirely to quantum gravity—this seems to be only the second such thesis solely on this topic, after Bronstein's. DeWitt later described this thesis as "both mediocre and huge" [DeWitt (1996), p. 30]. DeWitt evidently attempted to have the thesis published in the *Physical Review* (curiously, at the same time Felix Pirani and Alfred Schild's paper on the canonical approach to quantum gravity—on which see the next chapter), with a paper entitled "On the Application of Quantum Perturbation Theory to Gravitational Interactions," in 1950, but the editor, E. J. Hill, had trouble finding reviewers willing to take on so long a paper (a common theme with DeWitt's papers). Howard Robertson originally agreed to carry out the review (of both papers), but later backed out of a proper review of DeWitt's, not willing to take any responsibility for accepting or rejecting it, and noting that he was "extremely loathe to review" it. He points out, showing a clear misunderstanding of the methods involved in the respective papers, that DeWitt should "carry out his work in terms of their [Pirani and Schild's—DR] theory"! Given that DeWitt was promising a second paper Robertson saw the whole venture as an "inordinate drain on our publishing facilities" (May 17th, 1950, Robertson Papers, Caltech, Box 7)—see [Lalli (2016)] for a study of the early reviewing practices of *The Physical Review*. Were this paper published, DeWitt would have beaten Gupta to the first published approach to quantum gravity based on manifestly covariant techniques.

Curiously, in his correspondence with John Wheeler,[75] following the publication of his "Gravitation and Electromagnetism," Gupta points out that he believes that Einstein's original theory is "the only sensible theory for the description of the gravitational field and its interaction with other fields". His flat space approach is motivated by expedience rather than as constituting a closer fit to reality. That is, "for all practical purposes" one can work in the Lorentz-invariant framework.[76] The idea is that the reduced theory provides a far more practical framework, given its closer affinity to other field theories. Hence, this is very much in line with the kind of reasoning offered by Milne and Birkhoff for their flat space approaches. However, it seems that Gupta's view altered over the years, so that by 1960 he was convinced that his method's apparent success (over curved space approaches) was evidence for a flat spacetime picture.[77]

Robert Kraichnan also wrote a pair of papers [Kraichnan (1955), Kraichnan (1956)] bearing much similarity to Gupta's, including a similar usage of non-covariant auxiliary conditions to fix the coordinate system.[78] Kraichnan had been a postdoctoral fellow at the Institute for Advanced Study to work on part of the paper, during 1949–1950— indeed, he credits Einstein as having worked with him on the paper (Kraichnan was Einstein's research assistant, working with him to find non-linear particle-like solutions to Einstein unified field theory). Curiously, though he cites Gupta's later review of his earlier work on quantum gravity, he does not directly cite Gupta's 1952 papers ([Gupta 1952a], [Gupta 1952b]).

Ernesto Corinaldesi defended Gupta's approach to the quantization problem at the 1955 Berne conference.[79] He aimed to apply the procedure to the gravitational two-body problem, as treated by Einstein, Infeld, and Hoffmann, showing that their equations hold

[75] We will see in Chapter 8 that once John Wheeler focused on gravitational physics, he was quick to repurpose his student Richard Feynman's covariant, path-integral approach in the service of his geometrodynamical vision, to generate all of physics from geometry and topology.

[76] Suraj Gupta, letter to John Wheeler, February 22nd, 1955: Wheeler Papers, Box 11. Writing a little later, on April 18th, 1955, Gupta mentions that he has been studying Wheeler's then-recent geons, but foresaw considerable complications in providing a quantum mechanical treatment, given that it would not only involve computing the exchange of virtual gravitons, but would also require consideration of the electromagnetic potential between photons.

[77] See his report from an invited address to the American Physical Society, 29 January 1960, reprinted in [Gupta (1962)].

[78] Note also that Kraichnan seemingly developed his approach independently of Gupta, writing an MIT S.B. thesis on the subject in 1947, aged only 18, with the title "Quantum Theory of the Linear Gravitational Field"—I have not been able to find a copy of this thesis to check its contents. However, as mentioned, it seems clear that there is also an iterative 'feedback' principle at work in this early work too, since Bryce DeWitt credits Kraichnan in 1949. Blum and Hartz chased up the issue of Kraichnan's thesis, and the extent to which he considered non-linearities therein, but were not able to find it in M.I.T.'s library [Blum and Hartz (2017), pp. 109–10]. (Peter Bergmann would also later refer to the style of approach employed by Gupta (and also by Kraichnan and Thirring) as the "feed-back" approach—letter to Wheeler, August 18th, 1971: JAW, Box 4).

[79] Corinaldesi also submitted an extended version of this paper to *Proceedings of the Physical Society A* [Corinaldesi (1956)]. The paper was written while he was a visiting researcher at the Dublin Institute for Advanced Study in 1955—this paper, as with Gupta's, was communicated by Léon Rosenfeld. Note too that Suraj Gupta was registered as a visiting student at the Dublin IAS the same time as Corinaldesi was there as a scholar (Annual Report of the work of the Institute, 1949–1950.). Leopold Infeld also gave a lecture on the problem of motion in general relativity on the 29th of April. 1949, during Corinaldesi's tenure, which no doubt influenced Corinaldesi's choice of the EIH theory as an application of Gupta's theory.

for wave-packets and so, in the classical limit, for macroscopic bodies too. He wished to deduce the equations by means of Gupta's flat space re-interpretation. Hence, this wasn't a quantization of gravity for its own sake, but an attempt to understand the two-body problem from a different angle. The idea is that, in the approximation used by EIH (using an expansion in only the first few powers of gravitational constant, at κ^4, and velocities of the bodies), identical equations of motion could have been derived from the linear theory instead, which Corinaldesi argued was easier to manipulate—indeed, he states that using the theory of the quantized gravitational field is done for "reasons of convenience rather than necessity" [Corinaldesi (1956), p. 190]. Indeed, at the level of approximation used, Corinaldesi encounters none of the usual quantum field theoretic difficulties (divergences and so on); these would arise at the next approximation, with the divergences that appeared making it likely that renormalization would prove unsuccessful (cf. ibid., p. 195). Interestingly, Corinaldesi concludes noting that there is an underdetermination that exists up to κ^4: there would be no way to distinguish between the traditional, geometrical picture of general relativity and the linearized, flat space, quantum field theoretic version. It is exactly at the higher powers that one would expect to see distinguishing features.

Recall that much of the work related to quantum gravity immediately before, during, and immediately after the war years was couched in broadly cosmological terms, most often having links to unified theories. In his postwar thesis, Bryce DeWitt aimed to detach quantum gravity from such considerations, and place it squarely in the realm of more orthodox particle physics. He states quite plainly in the opening of his thesis:

> It should be stated in passing that the point of view of the present procedure, in which the description of matter is accomplished through the addition of extra fields, is directly opposed to the philosophy of the "unified field theories," in which the attempt is made to describe not only gravitation, but all phenomena in terms of the geometry of space or in terms of a single field in some abstract covariant formalism. The singular lack of success of the unified field theories, in particular in attaching any kinds of world-geometric significance to spinor field, leaves, at the present time, no other alternative but to add extra fields. It is the present author's belief that no other alternative is actually possible.
>
> [DeWitt (1950), p.1]

In this sense, DeWitt too was methodologically closer to the likes of Birkhoff and Milne, than Eddington and Schrödinger. The earliest approaches to quantum gravity in this sense (i.e. in the sense of quantization of the gravitational field) were, quite naturally, pursued by those who had skills in quantum field theory. However, even as early as 1938 Jacques Solomon argued that the standard field quantizations methods would fail for *strong* gravitational fields, for then the approach ceases to give a good approximation to Einstein gravity ([Solomon (1938)], p. 484). Alternative approaches were suggested as the extent of the problems facing flat space quantum field theory became ever more apparent. The linearization approach was recognized by many to be a provisional step up the ladder, of course. With the feedback principle, some believed that the limitations of flat space theories had been overcome; however, problems mounted in the period beyond our timeframe. Canonical approaches (after Rosenfeld at least) were developed precisely to incorporate the full non-linear theory.

We note that Bryce DeWitt did in fact attempt a reconciliation between the covariant and the canonical, introducing a 'background field' that evades the usual objection 'flat space' to the covariant approach. DeWitt felt confident enough to state at this stage that "quantum geometrodynamics now exists that is as well-founded as any other quantum field theory in which infinities occur" [DeWitt (1972), p. 436]. Ultimately, at least gravitation could be modelled by analogy with other fields (albeit, in this case, the Yang–Mills field) and with the addition of a background field which allows a freedom in choosing the background spacetime geometry. DeWitt's approach involved recycling Rudolf Peierls' [Peierls (1952)] manifestly covariant redefinition of Poisson brackets (doing without canonical variables) in the service of quantum gravity. Difficulties with standard quantization methods led to the idea that one should keep the gravitational field purely classical. We finish this chapter with a brief look at some early examples.

5.6   Great Expectations

In his brief historical account of the early years of the canonical formulation of general-relativistic theories, Peter Bergmann writes that "[w]ith the birth of quantum field theory in the late 1920s, physicists decided that nature could not be half classical and half quantum, and that the gravitational field ought to be quantized, just as the electromagnetic field had been" ([Bergmann (1989)], p. 293). However, this is projection of a more modern mindset onto the past: there was no such debate in the earliest period.[80] An important idea, that took time to emerge and *led* to such debate, and in fact would be rehashed over and over again (and still appears today, in the form of 'semiclassical theory') is that the gravitational field is *sui generis*, and must be kept classical.[81] That is, it should *not* be treated like any other field.

By 1955, Peter Bergmann believed that this approach, in which the gravitational field is "exempt ... from the approach valid for all other physical phenomena," suffered from

---

[80] Of course, Einstein's 1916 remarks (see § 3.3) concerning the necessity to modify general relativity involved a kind of stability argument whereby gravitational radiation must also be quantized in order to avoid the kinds of problem that would face continuous electromagnetic radiation (i.e. a spiralling of electrons into the nucleus). However, this is distinct from the kinds of compatibility issues being presented here, which concern whether the quantization of one field must thereby lead to the quantization of some other field that it couples to.

[81] As we saw in Chapter 1, Roger Penrose argues that there is a fundamental principle at work that forbids coherent superpositions of gravitational fields corresponding to distinct geometries (i.e. associated with a superposition of states placing a mass at different locations, making the geometries differ by an amount of the order of the Planck length), and in fact this is responsible for wavefunction collapse. That is, the non-linearity of the gravitational field leads to the breakdown of linearity (the superposition principle) in a fairly intuitive physical manner. As we see below, Iwao Sato had, in 1948, something akin to the breakdown of the superposition principle as a result of coupling quantized fields to the spacetime metric via the (semiclassical) Einstein equations.

"internal contradictions in the foundations of the theory" [Bergmann (1955), p. 492], though he gave no actual argument. Later too Dirac, speaking at the Trieste Symposium in 1968, writes that, while there is no experimental evidence for the quantization of the gravitational field, "we believe that quantization should apply to all the fields of physics." After all, he writes, "[t]hey all interact with one another, and it is difficult to see how some could be quantized and others not" [Dirac (1968), p. 539]. There was indeed, as Bergmann put it, "a certain measure of contradiction between the general theory of relativity and quantum theory" [Bergmann (1955), p. 491] but, according to his viewpoint, a careful treatment of the classical theory (isolating the truly physical degrees of freedom to be quantized) would lead to the solution, not some radical move of leaving gravity unquantized. However, we find that the gravity as 'just another field' approach was directly rejected in several early approaches.

Iwanenko and Sokolov [Iwanenko and Sokolow (1947)] gave an early analysis of a quantized scalar field acting as the source of the gravitational field. This is of course, a crucial step. While we saw in the previous chapter examples of electrons and so on in a variety of backgrounds, these did not involve this kind of coupling through the field equations. Schrödinger considered the effect of gravity on waves, but not vice versa. Iwanenko and Sokolov discuss particle annihilation/production processes (what they call the "transmutations of scalar particles into gravitons"): for an emission of two gravitons there is an annihilation of a scalar particle and its anti-particle (with an effective cross-section of the order of the square of the gravitational radius of the scalar particle). However, for much of the paper, they simply re-derive the Pauli–Fierz equations for a spin-2 field, showing how this lines up with general relativity in the weak-field limit (and how gravitons emerge). In other words, they revert to a linearized theory. They do, however, perform an actual computation to reveal the weak-field limit correspondence, using the intensity of gravitational radiation from a source in motion.[82] The authors seem not to be aware of the prior work of Schrödinger, and take themselves to have originated the study of pair creation in cosmological contexts.

The first explicit example of a 'half-and-half' world, with a quantized source and a classical gravitational field,[83] is Iwao Sato's rather remarkable paper, "An Attempt to Unite the Quantum Theory of Wave Field with the Theory of General Relativity" in 1949:[84] [Sato (1949)]. Sato begins by formulating a generally covariant quantum theory of a 'wave field.' This he does along lines we have already seen, namely in curvilinear coordinates with the Schrödinger equation then written in a coordinate independent form. This is simply establishing the coherence of a quantum system in a curved space. However, he then considers the equations of the gravitational field from this

---

[82] Iwanenko gives an excellent review of early Soviet work on gravitation in [Iwanenko (1967)].

[83] Of course, I am bracketing those approaches discussed in the previous section (e.g. Belinfante's 'phenomenological' theory) in which the gravitational field is treated as a classical field, though in a flat spacetime, in interaction with quantized fields, in which case there is none of the kind of coupling that occurs in Einstein's theory. Note that Stephen Hawking's paper on particle creation in black hole backgrounds employs exactly this setup, where it is viewed as an approximation: [Hawking (1975)].

[84] Submitted September 13th, 1948; talk read on May 21st, 1948 at the Annual Meeting of the Physical Society of Japan.

well-established foundation. This is the Einstein equation, only now with the energy-momentum tensor replaced by its expectation value (of a scalar field U; switching to the electromagnetic field in his second paper on the topic), and so taking the tensor $g_{\mu\nu}$ to be described by c-numbers. Sato's reason for using this approach is precisely to evade the by then very well known difficulties facing the direct quantization of the gravitational field. In other words, there is no sense that this is a 'correct' theory.

Interestingly, Sato gets remarkably close to some basic aspects of the first versions of the Hamiltonian formulation to quantum gravity given by Pirani and Schild, Bergmann and his students, and the DeWitts—to be discussed in the next chapter.[85] He uses a parameter formalism similar to that due to Born's and Dirac's student, Paul Weiss, in which a $3+1$ slicing of spacetime is made, according to which the points of three-dimensional surfaces are characterized by a set of parameters ξ_1, ξ_2, ξ_3 (with the hypersurface, S, determined by $x_4 = x_4'$), such that the states (or the observables) vary with S. There is a notion of displacement from one surface S to another ($x_4 = x_4' + \delta x_4$) in terms of the normal vector of S. He writes a Schrödinger equation for the scalar field in terms of these parameterized surfaces and arrives at covariant commutation relations along with field equations for the quantized scalar field U, with parameter dependence. (Note that Sato gives his theory in both the Schrödinger and the Heisenberg representation.) However, the real innovation, as mentioned above, is in his field equations, which he writes as (where f is the Newton constant):

$$G_{\mu\nu} - \frac{1}{2}g_{\mu\nu}G = \frac{-8\pi f}{c^4} \times \Psi^*[S]T_{\mu\nu,S}\Psi[S] \tag{5.2}$$

Here, the second multiplicand in the right-hand side is the expectation value of the energy tensor of the scalar field U, given the quantum state Ψ; hence, here we see the reaction of the quantized scalar field on the metric (even if it is only the expectation-value of the energy-momentum tensor of the quantized field). The fundamental equations of motion of his scheme are non-linear since they involve the metric tensor; Sato points out that the superposition principle must fail in such cases. That is, when one integrates the equations, to get some 'later' surface S from S_0 (i.e. given data on S_0) then if $\Psi_1[S]$ and $\Psi_2[S]$ are solutions, it does not follow that $\Psi_1[S] + \Psi_2[S]$ is a solution.[86]

Freeman Dyson reviewed Sato's work in *Mathematical Reviews*, in which he draws attention to the fact that there appears to be a novel argument for quantizing gravity buried in there, that wasn't in fact pursued. Again, the focus is on the breakdown of the superposition principle in a situation in which the classical $g_{\mu\nu}$ are genuinely coupled to

---

[85] It is a bridge in many ways, since it provides a semiclassical version. Unfortunately, Sato gives no references at all, so it is difficult to get a sense of how he developed his theory. There are hints of Tomonoga's work on electrodynamics using general surfaces too, so this is a likely link.

[86] A second paper ([Sato (1950)]) introduces an electromagnetic field, alongside the scalar field, now with these both quantized with the spacetime metric remaining classical. The same principles apply again, with the quantized fields again acting on the classical metric via their expectation values. Sadly, there is no interpretive discussion of what this actually means physically. Sato promises a third paper in his series, focusing on the treatment of divergences in the context of his semiclassical theory; however, this failed to appear.

the quantized source. Felix Pirani also discussed Sato's paper (and Dyson's review) in his PhD thesis, finding it to be rather a more general problem for hybrid theories. He frames his own quantization scheme as a reaction to this problem (see [Pirani (1951), p. 6]), though initially the $g_{\mu\nu}$'s are treated independently from a geometrical interpretation, and viewed as ordinary field variables in an 'amorphous space' to be reinterpreted later. The total system is, on this view, represented by a state functional Ψ, satisfying: $\partial\Psi/\partial t = H\Psi$ (with the Hamiltonian H dependent on the $g_{\mu\nu}$). This restores linearity, preserving the superposition principle, according to Pirani, which leads him to the idea that the $g_{\mu\nu}$ must become non-commuting q-numbers.

The idea that the world might actually *be* half classical (the gravitational part) and half quantum (the rest) was not a natural one in the early phase studied in this book. It required several blind alleys and wrong turnings before people speculated that quantum gravity (understood as involving quantum properties of the gravitational field) might simply not exist.[87] Present day researchers usually associate this view with the older Rosenfeld, arguing (e.g. during an exchange at the Chapel Hill Conference on the Role of Gravitation in Physics: [DeWitt and Rickles (2011), p. 257]) that there is simply no empirical evidence for the quantization of gravitation and therefore the problem of quantum gravity becomes a purely epistemological problem, not a logical one. The exchange was perhaps triggered by Peter Bergmann's opening of the quantum gravity session of the conference with the question: "Why quantize?" [DeWitt and Rickles (2011), p. 165].[88] Rosenfeld returned to semiclassical gravity in the 1960s, more firmly stating his view that there is no empirical evidence one way or the other for the necessity of quantization of gravity [Rosenfeld (1963)], as did Christian Møller [Møller (1962)] (speculating on whether the semiclassical equations could be exact, applying also to single atoms), which is where the origin of the semiclassical idea is often wrongly placed.

Another early version (before either Møller or Rosenfeld) of semiclassical gravity was given (in French) by Robert Potier [Potier (1956)]. Potier couples a quantized spinor field and an unquantized gravitational field $g_{\mu\nu}$ through the semiclassical field equations. He chose the simplest form for this, using the expectation value (of the

---

[87] Though, as we saw in the case of Born and van Dantzig in the previous chapter, there were those who believed that gravity was essentially macroscopic (or 'molar') and not subject to quantization. However, they don't pursue this in any detail.

[88] However, by this point Bergmann notes that what is needed is a "a study should be made of the extent to which arguments based on the uncertainty principle force one to the conclusion that the gravitational field must be subject to quantum laws," with a focus on two questions: "(a) Can quantized elementary particles serve as sources for a classical field? (b) If the metric is unquantized, would this not in principle allow a precise determination of both the positions and velocities of the Schwarzschild singularities of these particles?" (ibid.). A large part of the remainder of the conference after this involved thought experiments trying to reveal inconsistency with the uncertainty relations unless the gravitational field is quantized. A secondary part had to do with establishing gravitational waves, which were associated with the problem of the reality of their quanta too. Indeed, Pirani would later express this point very explicitly: "In my opinion ... a direct observation of classical gravitational radiation is not necessary or sufficient as a justification for the gravitational radiation theory. It is my view that the primary motivation for the study of this theory is to prepare for quantization of the gravitational field. The classical theory has to be untangled first, but unless it is eventually brought into the contemporary domain by quantization, the theory of gravitational radiation cannot have much to do with physics" [Pirani (1965), p. 368].

stress-energy-momentum density of the spinor field ψ) as the source of the gravitational field. The coupled system is then described by three equations, broadly similar to those derived by Sato:

$$\gamma^\mu \psi_\mu = \chi \psi \tag{5.3}$$

$$i\frac{\delta \Psi}{\delta \sigma(x)} = \mathfrak{h}(x) \Psi \tag{5.4}$$

$$R_{\mu\nu}(x) - \frac{1}{2}g_{\mu\nu}(x)R(x) = \chi(\Psi, T_{\mu\nu}(x)\Psi) \tag{5.5}$$

Where the first equation is just the generally relativistic Dirac equation; the second equation describes the time dependence of Ψ (the total state vector) through a Tomonaga–Schwinger equation; and $\mathfrak{h}(x)$ is the interaction Hamiltonian. Then there are the semiclassical Einstein equations, as with Sato coupling the gravitational field to the expectation value of the quantized source, this time a spinor field. Potier examines the logical relationship between these equations, but once again he does not consider conceptual implications of this kind of coupling vis-à-vis the superposition principle and so on.

Despite the strong intuition that quantized sources coupled to a (universal) gravitational field leads to a quantized gravitational field, there was no direct proof for this link, certainly not before 1956. As mentioned in the first chapter of this book, it was more often an intuition or feeling (for unity and harmony) that quantization anywhere means quantization everywhere.

5.7   Conclusion

I don't think it is that much of an exaggeration to state that much of the work carried out on quantum gravity after 1930 until mid-century was an exercise in 'exploring the consequences' of the Heisenberg–Pauli theory: understanding the symmetries and the divergences, and attempting to find ways of dealing with both. The goal was very much to treat all fields in much the same way, and so one could also envisage learning about one field from another. However, there was a separate track, superficially similar, though issuing from a desire to have a theory of gravitation more in line with the rest of physics, and in particular one not involving the difficulties of curved, dynamical spacetime. The interaction representation and a desire for a manifestly covariant description played a crucial role in the development of such approaches, and involved a curious borrowing of concepts often associated with canonical approaches. An apparently orthogonal approach developed alongside these later manifestly covariant approaches, involving a hybrid approach retaining a classical gravitational field, albeit still coupled to quantized sources through the Einstein field equations. These were done largely to avoid complications, however, and the conceptual consequences, though hinted at, were not further explored. Subsequent work, until the present day (though with notable exceptions) on quantum gravity has been largely dominated by the methodology of treating gravity on a par with other fields. The next chapter focuses on canonical quantization methods, which have a rather different flavor to those in the present chapter, with a more rigorous focus on establishing the classical theory in the correct Hamiltonian form to then enable the well-known canonical quantization algorithm.

....

REFERENCES

Barajas, A., G. D. Birkhoff, C. Graef, and M. Vallarta (1944) On Birkhoff's New Theory of Gravitation. *Physical Review* **66**(5–6): 138–43.

Belinfante, F. J. (1955a) Attempts at Quantization of Gravitational Field. *Revista Mexicana de Fisica* **4**(4): 192–206.

Belinfante, F. J. (1955b) Use of the Flat-Space Metric in Einstein's Curved Universe, and the "Swiss-Cheese" Model of Space. *Physical Review* **98**(3): 793–800.

Belinfante, F. J., D. I. Caplan, and W. L. Kennedy (1957) Quantization of the interacting fields of electrons, eletromagnetism, and gravity. *Rev. Mod. Phys.* **29**: 518–46.

Belinfante, F. J. and J. C. Swihart (1957a) Phenomenological Linear Theory of Gravitation: Part I. Classical Mechanics. *Annals of Physics* **1**(2): 168–95.

Belinfante, F. J. and J. C. Swihart (1957b) Phenomenological Linear Theory of Gravitation: Part III: Interaction with the Spinning Electron. *Annals of Physics* **2**(1): 81–99.

Belinfante, F. J. and J. C. Swihart (1954) *A Theory of Gravitation and its Quantization*. Research Report, Purdue University.

Bergmann, P. G. (1989) The Canonical Formulation of General-Relativistic Theories: The Early Years, 1930–1959. In D. Howard and J. Stachel (eds.), *Einstein and the History of General Relativity* (pp. 293–9). Boston: Birkhaüser.

Bergmann, P. G. (1955) Fifty Years of Relativity. *Science* **123**: 486–94.

Bergmann, P. G., and J. H. M. Brunings (1949) Non-Linear Field Theories II. Canonical Equations and Quantization. *Reviews of Modern Physics* **21**: 480–7.

Birkhoff, G. D. (1944) Flat Space-time and Gravitation. *Proceedings of the National Academy of Sciences of the United States of America* **30**: 324–34.

Birkhoff, G. D. (1943) Matter, Electricity and Gravitation in Flat Space-Time. *Proceedings of the National Academy of Sciences of the United States of America* **29**(8): 231–9.

Birkhoff, G. D. (1927) A Theory of Matter and Electricity. *Proceedings of the National Academy of Sciences of the United States of America* **13**(3): 160–5.

Bleuler, K. (1950) Eine neue Methode zur Behandlung der Longitudinalen und Skalaren Photonen. *Helv. Phys. Acta* **23**: 567–86.

Blokhintsev, D. I. and F. M. Gal'perin (1934) Gipoteza Neitrino i Zakon Sokhraneniya Energii. *Pod Znamenem Marxisma* **6**: 147–57.

Blum, A. S. (2018) Without New Difficulties: Quantum Gravity and the Crisis of the Quantum Field Theory Program. In A. S. Blum and D. P. Rickles (eds.), *Quantum Gravity in the First Half of the Twentieth Century: A Sourcebook* (pp. 255–69). Berlin: Edition Open Access.

Blum, A. (2014) From the Necessary to the Possible: The Genesis of the Spin-statistics Theorem *The European Physical Journal H* **39**(5): 543–74.

Blum, W., D. Dürr, and H. Rechenberg (eds.) (1993) *Werner Heisenberg: Gesammelte Werke, Series A/Part III*. Berlin Heidelberg: Springer-Verlag.

Blum, A. and T. Hartz (2017) The 1957 Quantum Gravity Meeting in Copenhagen: An Analysis of Bryce S. DeWitt's Report. *The European Physical Journal H* J42(2): 107–157.

Corinaldesi, E. (1956) The Two-body Problem in the Theory of the Quantized Gravitational Field. *Proc. Phys. Soc. Sect. A* **69**: 189–95.

Crnković, C. and E. Witten (1986) Covariant Description Of Canonical Formalism In Geometrical Theories. In S.W. Hawking and W. Israel (eds.), *Three Hundred Years of Gravitation* (pp. 676–84). Cambridge: Cambridge University Press.

Darrigol, O. (1998) The Quantum Electrodynamical Analogy in Early Nuclear Theory or the Roots of Yukawa's Theory. *Revue d'histoire des Sciences* **41**(3–4): 225–97.

de Broglie, L. (1934) Sur la Nature du Photon. *Comptes rendus de l'Académie des Sciences* **198**: 135–8.

de Broglie, L. (1932) Sur une Analogie entre l'électron de Dirac et l'onde électromagnétique. *Comptes rendus de l'Académie des Sciences* **195**: 536–53.

de Wet, J. (1950) The Interaction Representation in the Quantum Theory of Fields. *Proceedings of the Royal Society of London. Series A* **201**(1065): 284–96.

DeWitt, B. S. (2009) Quantum Gravity: Yesterday and Today. *General Relativity and Gravitation* **41**: 413–19.

DeWitt, B. S. (1996) The Uses and Implications of Curved Spacetime Propagators. In Y. Jack Ng (ed.), *Julian Schwinger: The Physicist, the Teacher, and the Man* (pp. 29–60). Singapore: World Scientific.

DeWitt, B. S. (1978) The Formal Structure of Quantum Gravity. In M. Levy and S. Deser (eds.), *Recent Developments in Gravitation* (pp. 275–322). New York: Plenum Press.

DeWitt, B. S. (1972) Covariant Quantum Geometrodynamics. In J. R. Klauder (ed.), *Magic without Magic* (pp. 409–40). New York: W. H. Freeman.

DeWitt, B.S. (1970) Quantum Theories of Gravity. *General Relativity and Gravitation* **1**(2): 181–9.

DeWitt, B. S. (1967a) Quantum Theory of Gravity: I. The Canonical Theory. *Phys. Rev.* **160**: 1113–48.

DeWitt, B. S. (1967b) Quantum Theory of Gravity: II. The Manifestly Covariant Theory. *Phys. Rev.* **162**: 1195–239.

DeWitt, B. S. (1962) The Quantization of Geometry. In L. Witten (ed.), *Gravitation: An Introduction to Current Research* (pp. 305–18). New York: John Wiley and Sons.

DeWitt, B. S. (1955) *The Operator Formalism in Quantum Perturbation Theory*. University of California Radiation Laboratory Pub. No. 2884. Berkeley: University of California Radiation Laboratory.

DeWitt, B. S. (1950) *Part I. The Theory of Gravitational Interactions. Part II. The Interaction of Gravitation with Light*. PhD thesis, Harvard University.

DeWitt, C. and D. Rickles, eds. (2011) *The Role of Gravitation in Physics: Report from the 1957 Chapel Hill Conference.* Berlin: Edition Open Access.

Dirac, P. A. M. (1968) The Quantization of the Gravitational Field. In A. Salam (ed.), *Contemporary Physics: Trieste Symposium 1968: Volume I* (pp. 539–43). Vienna: International Atomic Energy Agency.

Dirac, P. A. M. (1936) Relativistic Wave Equations. *Proceedings of the Royal Society of London. Series A* **155**: 447–59.

Dirac, P. A. M. (1933a) The Lagrangian in Quantum Mechanics. *Phys. Z. Sowjetunion* **3**: 64–72.

Dirac, P. A. M. (1933b) Homogeneous Variables in Classical Dynamics. *Mathematical Proceedings of the Cambridge Philosophical Society* **29**(3): 389–400.

Einstein, A. (1998) *The Collected Papers of Albert Einstein, Volume 8 (English). The Berlin Years: Correspondence, 1914–1918.* (English supplement translation, translated by A. M. Hentschel). Princeton: Princeton University Press.

Faddeev, L. D. and V. N. Popov (1973) Covariant Quantization of the Gravitational Field. *Sov. Phys. Usp.* **16**: 777–88.

Fermi, E. (1934) Tentativo di una Teoria Dei Raggi β. *Il Nuovo Cimento* **11**: 1–19.

Feynman, R. P. (1972) Closed Loop And Tree Diagrams. In J. R. Klauder (ed.), *Magic Without Magic* (pp. 355–75). New York: W. H. Freeman.

Feynman, R. P. (1995) *Lectures on Gravitation.* Boston: Addison-Wesley.

Gambini, R. and J. Pullin (2018) Gravitation in Terms of Observables. *Classical and Quantum Gravity* **35**(21): 215008.

Gamow, G. (1948) The Reality of Neutrinos. *Physics Today* **1**(3): 4.

Gorelik, G. and V. Frenkel (1994) *Matvei Petrovich Bronstein and Soviet Theoretical Physics in the Thirties.* Basel: Birkhäuser.

Gupta, S. (1962) Quantum Theory of Gravitation. In *Recent Developments in General Relativity: A Collection of Papers Dedicated to Leopold Infeld* (pp. 251–8). Oxford: Pergamon Press.

Gupta, S. (1954) Gravitation and Electromagnetism. *Phys. Rev.* **96**: 1683–5.

Gupta, S. (1952a) Quantization of Einstein's Gravitational Field: Linear Approximation. *Proceedings of the Physical Society. Section A* . **65**(8): 161–9.

Gupta, S. (1952b) Quantization of Einstein's Gravitational Field: General Treatment. *Proceedings of the Physical Society. Section A* . **65**(8): 608–19.

Gupta, S. (1950) Theory of Longitudinal Photons in Quantum Electrodynamics. *Proc. Phys. Soc. A* **63**: 681–91.

Hagar, A. (2014) *Discrete or Continuous? The Quest for Fundamental Length in Modern Physics.* Cambridge: Cambridge University Press.

Hawking, S. W. (1975) Particle Creation by Black Holes. *Communications in Mathematical Physics* **43**: 199–220.

Heisenberg, W. and W. Pauli (1929a) Zur Quantendynamik Der Wellenfelder. *Zeitschrift Für Physik* **56**: 1–61.

Heisenberg, W. and W. Pauli (1929b) Zur Quantentheorie Der Wellenfelder. Band II. *Zeitschrift Für Physik* **59**: 168–90.

Iwanenko, D. (1967) Fifty Years of Soviet Work on Gravitation. *Soviet Physics Journal* **10**(10): 16–21.

Iwanenko, D. and A. Sokolow (1947) Kvantovaia Teoria Gravitatsii. *Vestnik Moskovskogo Universiteta Seriya 3 Fizika Astronomiya* **8**: 103–16.

Jensen, C. (2000) *Controversy and Consensus: Nuclear Beta Decay 1911–1934.*

Jordan, P. (1935) Zur Neutrinotheorie des Lichtes. *Zeitschrift Für Physik* **93**: 464–72.

Jordan, P. and W. Pauli. (1928) Zur Quantenelektrodynamik Ladungsfreier Felder. *Zeitschrift Für Physik* **47**: 151–73.

Kaiser, D. (1998) A ψ is just a ψ? Pedagogy, Practice, and the Reconstitution of General Relativity, 1942–1975. *Studies in History and Philosophy of Modern Physics* **29**: 321–38.

Klauder, J. and J. Wheeler (1957) On the Question of a Neutrino Analog to Electric Charge. *Reviews of Modern Physics* **29**: 516–17.

Klein, O. (1927) Zur Fünfdimensionalen Darstellung der Relativitätstheorie. Zeitschrift für Physik **46**: 188–208.

Kraichnan, R. (1956) Possibility of Unequal Gravitational and Inertial Masses. *Physical Review* **101**: 482–88.

Kraichnan, R. (1955) Special-relativistic Derivation of Generally Covariant Gravitation Theory. *Physical Review* **98**: 1118–22.

Lalli, R. (2016) 'Dirty work', but someone has to do it: Howard P. Robertson and the refereeing practices of *Physical Review* in the 1930s. *Notes Rec R Soc Lond* **70**(2): 151–74.

Mandelstam, S. (1962) Quantization of the Gravitational Field [and Discussion]. *Proceedings of the Royal Society of London, Series A* **270**(1342): 346–53.

Matthews, P. T. (1949) The Generalized Schrödinger Equation in the Interaction Representation. *Physical Review* **75**: 1270.

Møller, C. (1962) The Energy-Momentum Complex in General Relativity and Related Problems. In A. Lichnerowicz and M. A. Tonnelat (eds.), *Les Theories Relativistes de la Gravitation* (pp. 2–29). Paris: Centre National de la Recherche Scientifique.

Monk, R. (2014) *Inside The Centre: The Life of J. Robert Oppenheimer.* New York: Anchor.

Moshinsky, M. (1950) On the Interactions of Birkhoff's Gravitational Field with the Electromagnetic and Pair Fields. *Physical Review* **80**(4) 514–19.

Nilsson, S. B. and K. V. Laurikainen (1950) On the Gravitational Self-energy of Light. *Physical Rev.* **2**(80): 291–2.

Pais, A. (1994) *Niels Bohr's Times: In Physics, Philosophy, and Polity.* Oxford: Oxford University Press.

Papapetrou, A. (1948) Einstein's theory of gravitation and flat space. *Proc. Roy. Irish Acad. (Sect.A)* **52**: 11–23.

Pauli, W. (1936) Space, Time and Causality in Modem Physics. In C. P. Enz and K. von Meyenn (eds.), *Wolfgang Pauli: Writings on Physics and Philosophy* (pp. 95–105). Berlin Heidelberg: Springer-Verlag.

Peierls, R. ed. (1986) *Niels Bohr – Collected Works, Volume 9: Nuclear Physics (1929–1952).* Amsterdam: Elsevier.

Peierls, R. (1952) The Commutation Laws of Relativistic Field Theory. *Proc. R. Soc. Lond. A* **214**(1117): 143–57.

Peierls, R. (1946) Fundamental Particles. *Nature* **158**: 773–5.

Peres, A. (1968) Canonical Quantization of Gravitational Field. *Physical Review* **171**(5): 1335–44.

Petiau, G. (1941a) Sur la théorie du Corpuscule de Spin 2. *Comptes rendus hebdomadaires des séances de l'Académie des Sciences* **212**: 292–5.

Petiau, G. (1941b) Sur une Représentation du Corpuscule de Spin 2. *Comptes rendus hebdomadaires des séances de l'Académie des sciences* **212**: 47–50.

Piir, I. (1957) O Probleme Vzaimodeistvia Kvantovannykh Elektromagnitnogo i Gravitatsionnogo Polei. *Trudy Instituta Fiziki i Astronomii Akademii nauk Estonskoi S.S.R.* **5**: 41–52.

Pirani, F. (1965) Introduction to Gravitational Radiation Theory. *Lectures on General Relativity, Lectures from Brandeis Summer Institute in Theoretical Physics 1964. Volume 1* (pp. 249–73). Englewood Cliffs, NJ: Prentice-Hall, Inc.

Pirani, F. (1962) Survey of Gravitational Radiation Theory. In *Recent Developments in General Relativity* (pp. 89–105). New York: Pergamon Press.

Pirani, F. (1951) *On the Quantization of the Gravitational Field of General Relativity*. PhD Thesis, Carnegie Institute of Technology.

Pitts, J. B. (2010) Permanent Underdetermination from Approximate Empirical Equivalence in Field Theory: Massless and Massive Scalar Gravity, Neutrino, Electromagnetic, Yang-Mills and Gravitational Theories. *The British Journal for the Philosophy of Science* **62**(2): 259–99.

Pitts, J. B. (2001) *Gravitation with a Flat Background Metric*. PhD Thesis, University of Austin at Texas.

Potier, R. (1956) Sur la théorie des Champs Quantifiés en Relativité Généralisée. *C. R. Acad. Sci. Paris* **243**: 939–42.

Roberts, K. V. (1951) On the Quantum Theory of the Elementary Particles. II. Quantum Field Dynamics. *Proc. R. Soc. Lond. A* **207**: 228–51.

Roberts, K. V. (1950) On the Quantum Theory of the Elementary Particles - I. Introduction and Classical Field Dynamics. *Proc. R. Soc. Lond. A* **204**: 123–44.

Rocci, A. (2015) *Dalla Nascita Della Relativitaá Generale al Secondo Dopoguerra (1915–1945)*. PhD Thesis, Universitá Degli Studi di Padova.

Rosen, N. (1940a) General Relativity and Flat Space. I. *Phys. Rev.* **57**: 147.

Rosen, N. (1940b) General Relativity and Flat Space. II. *Phys. Rev.* **57**: 150.

Rosenfeld, L. (1966) Quantum Theory and Gravitation. In R. S. Cohen and J. Stachel (eds.), *Selected papers of Léon Rosenfeld* (pp. 599–608). Reidel: Dordrecht, 1979.

Rosenfeld, L. (1963) On the Quantization of Fields. *Nuclear Physics* **40**: 353–56.

Rosenfeld, L. (1930a) Über die Gravitationswirkungen des Lichtes. *Zeitschrift für Physik* **65**(9–10): 589–99.

Rosenfeld, L. (1930b) Zur Quantelung der Wellenfelder. *Annalen der Physik* **397**(1): 113–52.

Rovelli, C. (2002) Notes for a Brief History of Quantum Gravity. In R. T. Jantzen et al. (eds.), *Proceedings of the Ninth Marcel Grossmann Meeting on General Relativity* (pp. 742–768). World Scientific: Singapore.

Salisbury D. (2010) Léon Rosenfeld's Pioneering Steps Toward a Quantum Theory of Gravity. *J. Phys.: Conf. Ser.* **222**: 012052.

Salisbury, D. and K. Sundermeyer (2017) Léon Rosenfeld's General Theory of Constrained Hamiltonian Dynamics. *Eur. Phys. J. H* **42**: 23–61.

Sato, I. (1950) General-relativistic quantum theory of wave field. II. *Sci. Rep. Tohoku Univ., Ser. 1.* **33**: 136–3.

Sato, I. (1949) An Attempt to Unite the Quantum Theory of Wave Field with the Theory of General Relativity. *Sci. Rep. Tohoku Univ., Ser. 1* **33**: 30–7.

Schweber, S. (1994) *QED and the Men Who Made It.* princeton: Princeton University Press.

Solomon, J. (1938) Gravitation et Quanta. *Journal de Physique et Le Radium* **9**: 479–85.

Solomon, J. (1931) Nullpunktsenergie der Strahlung und Quantentheorie der Gravitation. *Zeitschrift für Physik* **71**: 162–70.

Tomonaga, S. (1966) Development of Quantum Electrodynamics: Personal recollections. *Science* **154**(3751): 864–8.

Tomonaga, S. (1946) On a Relativistically Invariant Formulation of the Quantum Theory of Wave Fields. *Progress of Theoretical Physics* **1**(2): 27–42.

Tonnelat M.-A. (1976) From the Photon to the Graviton and to a General Theory of Corpuscular Waves. In M. Flato, Z. Maric, A. Milojevic, D. Sternheimer, and J. P. Vigier (eds.), *Quantum Mechanics, Determinism, Causality, and Particles.* Dordrecht: Springer.

Tonnelat, M.-A. (1944) La Particule de spin 2 et la loi de Gravitation d'Einstein dans le cas de Presence de Matière. *Comptes rendus de l'Académie des Sciences* **218**: 305–8.

Tonnelat, M.-A. (1943) Théorie Euclidienne de l'électromagnétisme et de la Gravitation. *Disquisit. Math. Phys.* **3**: 249–74.

Utiyama, R. (1956) Invariant Theoretical Interpretation of Interaction. *Phys. Rev.* **101**: 1597.

van Niuwenhuizen, P. (1981) An Introduction to Covariant Quantization of Gravitation. In R. Ruffini (ed.), *Proceedings of the First Marcel Grossman Meeting on General Relativity* (pp. 1–26). Singapore: World Scientific.

von Meyenn, K. (2001) *Wissenschaftlicher Briefwechsel mit Bohr, Einstein, Heisenberg u.a. Band IV, Teil III: 1955–1956 / Scientific Correspondence with Bohr, Einstein, Heisenberg, a.o. Volume IV, Part III: 1955–1956.* Berlin Heidelberg: Springer-Verlag.

von Meyenn, K. (1993) *Wissenschaftlicher Briefwechsel mit Bohr, Einstein, Heisenberg u.a. Band III: 1940–1949 / Scientific Correspondence with Bohr, Einstein, Heisenberg, a.o. Volume IVII: 1940–1949.* Berlin Heidelberg: Springer-Verlag.

von Meyenn, K. (1985) *Wissenschaftlicher Briefwechsel mit Bohr, Einstein, Heisenberg u.a. Band II: 1930–1939.* Berlin Heidelberg: Springer-Verlag.

Wataghin, G. (1937) Sulla Teoria Quantica della Gravitazione. *La Ricerca Scientifica* **8**(2): 1–2.

Weyl, H. (1944a) How Far Can One Get With a Linear Field Theory of Gravitation in Flat Space-Time? *American Journal of Mathematics* **66**(4): 591–604.

Weyl, H. (1944b) Comparison of a Degenerate form of Einstein's with Birkhoff's Theory of Gravitation. *Proceedings of the National Academy of Sciences of the United States of America* **30**(8): 205–10.

Whitrow, G. J. and G. E. Morduch (1965) Relativistic Theories of Gravitation: A Comparative Analysis with Particular Reference to Astronomical Tests. *Vistas in Astronomy* **6**: 1–67.

Whitrow, G. J. and G. E. Morduch (1960) General Relativity and Lorentz-Invariant Theories of Gravitation. *Nature* **188**: 790–94.

Yang, C. N. (1986) Hermann Weyl's Contribution to Physics. In K. Chandrasekharan (ed.), *Hermann Weyl: 1885–1985: Centenary Lectures* (pp. 7–22). Berlin Heidelberg: Springer-Verlag.

Zel'dovich, Y. B. (1965) Analog of the Zeeman Effect in the Gravitational Field of a Rotating Star. In J. P. Ostriker (ed.), *Selected Works of Yakov Borisovich Zeldovich, Volume II* (pp. 193–5). Princeton: Princeton University Press, 2017.

6

Forming the Canon

General-relativistic theories differ in important respects from other field theories commonly dealt with in theoretical physics. Not only are they invariant with respect to a transformation group whose members represent a function space whose functions are defined on the four-dimensional spacetime manifold (they share this property with gauge-invariant theories), but the resulting system of first-class constraints includes the Hamiltonian constraint, that is to say the invariance group includes the transformations that correspond to a mapping of three-dimensional space-like hypersurfaces ("instants in time") on each other; these mappings correspond to what ordinarily would be called the unfolding of the physical situation in time.

Peter Bergmann and Arthur Komar[1]

In this chapter, I chart the early development of the canonical quantum gravity (that is, the quantization of the gravitational field in Hamiltonian form).[2] The concern is more with the *genesis* of canonical quantum gravity and with filling in what I take to be a troubling incompleteness in the emergence of canonical approaches in the 1950s, which seem to amount to a case of spontaneous generation. There are equally puzzling and related issues concerning the correlations between the first approaches which apparently sprang up independently of one another.

The first inroads on this approach were made early on in 1930, almost as soon as the method had been applied to electrodynamics, by Léon Rosenfeld and at Wolfgang Pauli's behest, as we saw in the previous chapter. However, after this initial, impressive

---

[1] [Bergmann and Komar (1962), p. 32].

[2] We fall just shy of the full completion of the *classical* Hamiltonian formulation of GR (by Dirac in 1958: [Dirac (1958a)]) whose roots are thoroughly grounded in this earlier period. What we find in this period include: the establishment of a procedure for quantizing in curved spaces; the first expressions for the Hamiltonian of general relativity; recognition of the existence and importance of constraints (i.e. the generators of infinitesimal coordinate transformations); a focus on the problem of observables (and the realization of conceptual implications in defining these for generally relativistic theories), and a (template of a) method for quantizing the theory. This template (often called "formal quantization") would be much criticized on account of not dealing with the difficult machinery of the full theory: constraints and so on. James Anderson, who had worked under Peter Bergmann, who spearheaded this approach, was soon disenchanted for this reason. As he put it: "I tried to do physics and not the formalism. In particular, he [Bergmann] and Art Komar had very elaborate programs…but I felt the physics was missing" (Interview of James Anderson, March 19th, 2011, conducted by Donald Salisbury and Dean Rickles).

Covered with Deep Mist: The Development of Quantum Gravity (1916–1956). Dean Rickles, Oxford University Press (2020). © Dean Rickles.
DOI: 10.1093/oso/9780199602957.001.0001

effort, it is often thought that progress stalled until 1949, when Peter Bergmann first began thinking along these lines. Felix Pirani and Alfred Schild also tackled the problem at around the same time, pipping Bergmann to the post in writing down an expression for the Hamiltonian. DeWitt also studied the canonical approach in his PhD thesis. Then, the story goes, Paul Dirac settled matters having to do with the treatment of the symmetries and constraints of general relativity within the (constrained) Hamiltonian context. From here, Arnowitt, Deser, and Misner presented the framework for the dynamics of general relativity, and then considerations of the meaning of the quantum version of the theory, including the domain space, and the Wheeler–DeWitt equation followed.

However, this is incomplete. There was work being carried out just prior to the Second World War, by Paul Weiss (a former PhD student of Paul Dirac's), that was pivotal in the application of canonical methods to the gravitational field (even if not directly tackling that topic, it was incorporated into it later), and in Dirac's theory of constrained Hamiltonian dynamics. The first work on canonical quantum gravity is essentially a direct combination of Weiss's and Dirac's insights (and Bergmann's insights) in the area of manifestly covariant field theories applied to generally covariant theories. Yet much of this work has been all but forgotten, and so a large part of this chapter will be devoted to correcting this and revealing a hidden layer of continuity in the history of canonical quantum gravity.

However, Rosenfeld certainly deserves credit for his initial explorations in this area, especially as regards the treatment of the gauge symmetries, and so we begin with him. Indeed, his initial exploration is part of a more general, wider project to understand symmetry in field theories, and to construct invariant formulations of such theories, and so we find the dark age between Rosenfeld and the 1950s work to be a little lighter than is usually supposed. But let us first make some general remarks about the nature of the canonical approach.

6.1 On Hamiltonian Formulations of Gravity

There are several reasons for wishing to have a Hamiltonian (canonical) formulation of general relativity, but the chief motivation for much of the work in the 1950s was certainly its subsequent quantization—Rosenfeld notwithstanding. There were two approaches to setting up the Hamiltonian formulation, one preserving the four-dimensional spacetime symmetry (the 'parameter formalism'), the other splitting spacetime into space and time (the 3+1 formalism) in such a way that this symmetry is broken. Thus, the modern day idea of partitioning approaches to quantum gravity via a 'covariant versus canonical' split doesn't map so simply onto the earliest approaches, as discussed in the previous chapter. Though Hamiltonian (canonical) approaches in general involve specifying an instant of time (since the canonical variables refer to states at an instant of time), one does not have to break covariance to preserve this feature. The very earliest Hamiltonian approaches were covariant in the sense that they preserved the four-dimensional symmetry by making the spacelike hypersurface (on which the Hamiltonian dynamical variables are defined) independent of the spacetime coordinates, x^μ. This was dispensed with fairly quickly, though it provided a necessary leg up. Dirac's approach, abandoning manifest covariance, would eventually prove more successful, especially in extracting the physical content of the theory (i.e. in separating out which degrees of freedom are physical, in the sense of coordinate independent, and which are artifacts). However, breaking covariance in this manner would prove to be too much for some of the more ardent advocates of general relativity.

A key problem faced in the Hamiltonian formulation, already detected by Rosenfeld, was the presence of constraints in the theory (showing that not all apparent degrees of freedom were *physical*) arising from the general covariance of the four-dimensional formulation. The treatment of these manifestations of symmetry, and their impact on the definition of the observables of theory, accounted for much of the work on the problem of quantum gravity in the 1950s, and simultaneously gave the subject a more conceptual flavour. In certain importance senses, these were simply manifestations of an old problem known to Einstein, and in modern parlance known as the 'hole argument.' The basic idea is that the field equations do not determine the propagation of the metric components uniquely, but only up to diffeomorphism. This non-uniqueness makes itself known in the canonical approach through the Hamiltonian itself possessing some ambiguity.

The first phase of canonical quantization is to establish the relevant dynamical variables describing the instantaneous states (which must form a Poisson algebra, given as a Poisson bracket structure), which will then be used to build the Hamiltonian from which the dynamics is extracted. The standard field variables for general relativity are the 3-metric on a spatial hypersurface and its conjugate momenta, interpreted as something like the time derivative of the 3-metric and related to the extrinsic curvature of the hypersurface—for this reason, this approach is often seen to concern the 'dynamics of geometry.'[3] To quantize the theory there is a standard algorithm according to which one converts the elements of this Poisson algebra to operators acting on a space of wavefunctions, with the bracket structure converted to a commutator structure—this algorithm (canonical quantization) is of course one of the major benefits of a canonical approach.

The difficult part comes from any gauge symmetries present, which manifest themselves canonically as constraints, or what Dirac called "weak equations" (equations between variables that vanish in the classical theory and that arise from the definition of momentum variables) on the initial data. In the case of general relativity (using the 3-metric), there are 4 constraints, which specify a set of relationships between the canonical variables at an instant of time (thus breaking the manifest covariance of the full four-dimensional theory): three of these form a vector quantity and correspond, roughly, to spatial diffeomorphisms of the hypersurface, while the remaining constraint

---

[3] Rosenfeld's 1930 treatment [Rosenfeld (1930b)] was couched in the tetrad formalism used for considering the Dirac equation in curved spaces—Rosenfeld's choice reflects his desire to have a unified framework for all then-known fundamental interactions, which therefore involved a Lagrangian for the gravitational field interacting with an electromagnetic field and with spinors. This radically alters the structure of the formalism, involving 16 field variables (and 16 corresponding conjugate momenta), as opposed to the 10 field variables (along with 10 conjugate momenta) of $g_{\mu\nu}$ of the metric-variable formulation.

corresponds to diffeomorphisms off the hypersurface.[4] These constraints must be treated carefully in the quantum context: they annihilate quantum wavefunctions (or, in other words, they vanish as operators)—for example, the vector constraint just means that wavefunctions must be spatially diffeormophism invariant (depending only on the invariant content), which is easy to state but hard to describe explicitly; the other constraint is famously fiendishly difficult and remains a problem to this day (not least the problem of its regularization). The inner product structure must be introduced to respect this constrained formulation. The observables too must be defined in relation to the constraints, and amount to quantities that have vanishing Poisson brackets with respect to the constraints. Viewing the theory in terms of a slicing of spacetime into spaces (space-like hypersurfaces) and imposing Cauchy data, for the gravitational field, on this slice, it is well known that the Einstein equations cannot uniquely determine the 'next' slice along but do so only up to a diffeomorphism (i.e. an infinitesimal displacement of the coordinates). Since the Hamiltonian generates time evolution (pushing data from one slice to the next), if there is such a non-uniqueness then the Hamiltonian (as the canonical conjugate of the Hamiltonian) must reflect this to match the features of the theory. In other words, the Hamiltonian has to contain terms responsible for the above infinitesimal displacement of the coordinates.

A curiosity that emerges from the work in the 1950s is that the Hamiltonian itself is a sum of constraints, and so is itself a vanishing quantity—this is a result of the fact that the covariance-breaking selection of a time variable was arbitrary, rather than physically preferred.[5]

Much of the earliest work on this approach to *quantum* gravity in fact tackled the initial *classical* problem of finding the variables and the Hamiltonian, which was complicated by the symmetry of general relativity. In this way, the ultimate pursuit of the quantum problem led to considerable developments in the understanding of the classical theory. This journey began with Rosenfeld in 1930 and ends[6] with a complete formulation with Dirac in 1958 [Dirac (1958b)] (though, of course, many new developments extend beyond

---

[4] It is more accurate to speak of *families* of constraints, since they each form infinite sets.

[5] In 1958, Dirac [Dirac (1958b)] would show how one can eradicate the problem of constraints by fixing a coordinate system (i.e. fixing the gauge), since the appearance of the constraints in the first place emerges from the arbitrariness of the coordinate system.

[6] There was another notable earlier skirmish with a kind of linearized canonical quantum gravity by Ryōyū Utiyama, who presented a pair of papers on the interaction of vector mesons with the gravitational field (using the canonical approach introduced by Heisenberg and Pauli in their 1929 paper on quantum electrodynamics, as well as Rosenfeld's developments), [Utiyama (1947)] and [Utiyama (1948)], in 1941 and 1942 (at the meeting of the Physico-mathematical Society of Japan)—though final publication in international journals was curtailed by the war, so they didn't appear until until 1947 and 1948. This pair of papers were the first, after Rosenfeld's 1930 work, to explicitly treat the complications brought about by the general freedom of the functions of the gravitational field in general relativity, though nothing like the constraint framework of Dirac appeared and he seems not to have gotten very far. Utiyama recovered Rosenfeld's results for photons interacting with the gravitational field by setting the meson mass to zero. In the weak approximation, Utiyama showed that the ten Einstein equations are not all mutually independent, but that four arbitrary functions exist in the solutions, stemming from the arbitrariness of coordinates in general relativity—these, of course, correspond to the constraints in later work, and it can be seen that Utiyama got to this fundamental aspect prior to Bergmann and his group; however, Rosenfeld too found four identities from the general covariance group [Rosenfeld (1930a), p. 90].

Dirac)—much complex work followed, in fact, trying to get to grips with the nature of the domain space of the wavefunctions, where one considers wavefunctionals of the metric.[7]

6.2 Rosenfeld Again

We saw in the previous chapter that Léon Rosenfeld was the first to explicitly apply the new techniques developed within quantum field theory of electromagnetism to the gravitational field. In his first paper on the subject, "Zur Quantelung der Wellenfelder" [Rosenfeld (1930b)] ('On the Quantization of Wave Fields'[8]), he deals with casting the theory of general relativity into canonical form, selecting canonical field variables for the gravitational field suitable for quantization. The key obstacle that Rosenfeld identified was the presence of constraints—though there were also constraints in the electromagnetic case too. Such constraints arise from the fact that in theories with (gauge) symmetries the canonical momentum densities (i.e. the partial derivatives of the Lagrangian density divided by the velocities) are not independent, but instead stand in time-independent relationships (the constraints). As mentioned, in the case of general relativity the constraints arise from the diffeomorphism invariance of the theory.[9]

Rosenfeld identified the greater difficulties with respect to gravitation in a later paper [Rosenfeld (1932)], in which he noted that the relativity group does not submit to the Fermi method, or Heisenberg and Pauli's methods for demonstrating invariance. Rosenfeld seeks instead an approach that emphasizes the importance of groups of (local symmetry) transformations attached to the Lagrangian. This method was the birth of constrained Hamiltonian dynamics, in which a Hamiltonian scheme is developed that is compatible with the existence of identities coming from the arbitrary parameters of a gauge theory.[10] Rosenfeld had written to Dirac in 1932 pointing out that the aim of his paper was "to give an alternative proof of the invariance of this scheme [the Heisenberg–Pauli theory; Heisenberg and Pauli (1929a, 1929b)] for the whole gauge and (general) relativity group".[11]

---

[7] For a good history dealing with Rosenfeld, Bergmann, and Dirac that goes further than the treatment here, see [Salisbury (2008)].

[8] For a fine translation of Rosenfeld's paper into English, see [Rosenfeld (1930a)].

[9] These difficulties no doubt account for the fact that Rosenfeld himself chose not to utilize this general method in the context of his quantization of gravity in which he was computing the gravitational self-energy of light. Had he used his framework, involving a sophisticated treatment of the symmetries and gauge invariance, he might have generated better results. However, he made the task more difficult for himself with his choice of Lagrangian, which involved invariance with respect to four-dimensional diffeomorphisms, local Lorentz transformations of the tetrad field, and U(1) gauge transformations.

[10] Note that Rosenfeld had also identified the essential bosonic nature of gravity (and tensor fields in general), with Fermi statistics only possible for "material fields" (electrons and protons)—he returns to this point in [Rosenfeld (1932), §6] ("Les statistiques"), revealing that Bose quantization (i.e. quantization via commutators) is required for quantum gravity. As he expresses it: "as regards gravitation, it is not possible to quantize the corresponding field quantities [tetrads] with Fermi statistics" [Rosenfeld (1930a), p. 91].

[11] Dirac Papers, Churchill College: DRAC 3/4. As Donald Salisbury [Salisbury (2009)] makes clear, Dirac undoubtedly 'stored' this discussion with Rosenfeld for a couple of decades until he turned to the issue of constrained dynamics in the context of Lorentz invariant field theories, in his 1950 Vancouver lectures, using many of the same concepts, and even many of the same notations.

The idea, in simple terms, is to attempt to find a way to get around a problem with the Heisenberg–Pauli theory, in that in going to the Hamiltonian formulation (which, recall, they felt obligated to use), one loses the symmetries of the original theory (as given in the Lagrangian). All of the methods Heisenberg and Pauli had tried (which Rosenfeld reviews), were cumbersome or *ad hoc*, and in fact did not preserve the symmetry in any nice way, often explicitly destroying it with a choice of gauge or frame. Moreover, it is difficult to see how the methods would apply in the case of gravitation. Rosenfeld managed to describe in a mathematically rigorous way the apparent breakdown in symmetries as a result of the 'canonizing' of a theory—namely, some canonical momenta are not independent. Putting things in this way also allows for a general treatment that covers both gravitation and electromagnetism, with the specific symmetry problems arising from the same source. This project was, as with Rosenfeld's study of divergences, prompted directly by Pauli:

> it was immediately after I came to Zürich that I got the proof of this paper by Heisenberg and Pauli to study. Then Pauli told me that he was not at all pleased with those longitudinal waves, so he wanted to have them treated another way, which I did, but that was not more enlightening, far from it.[12]

Hence, this paper on constrained Hamiltonian systems is also grounded in a dissatisfaction with the Heisenberg–Pauli theory by its architects. There were unphysical degrees of freedom in the theory (the longitudinal waves), which appeared to be generated in the transformation to a Hamiltonian formulation, and it wasn't clear how to deal with them. Rosenfeld certainly shared this confusion about the strange properties of the theory, and the proof of invariance it employed:

> There was this point in their proof in which the invariance of the Hamiltonian seemed to depend on a special structure of the Hamiltonian, and that looked very suspicious. In fact, I said that directly to Pauli when I had read the proofs, and he said, "Yes, I understand that, but we have not been able to find any mistake in our calculation and we do not understand what this means; we suspect that it must be wrong, but we don't know." Then the thing came to a crisis through the fact that I then tried to make a more general formulation of this field quantization, starting from Lagrangian, defining the momenta through the variational derivatives, and then I saw that I could do that without specializing the Hamiltonian. It was a purely abstract scheme which worked in a completely general way with only this complication of the accessory[13] conditions, but that was, at any rate, not due to any special structure but only to the existence of invariance with respect to group. So at that stage I was convinced that there must be a mistake in the original paper; it was quite clear, and then I set myself to finding this mistake. By going very carefully through every step several times I spotted it; it was an extremely subtle point about two points which had been exchanged in the delta function which was not legitimate. But

---

[12] Interviewed by Thomas S. Kuhn and John L. Heilbron. Location: Carlsberg. Interview date: Friday, July 19th, 1963 https://www.aip.org/history-programs/niels-bohr-library/oral-histories/4847–2.
[13] I suspect this should be "auxiliary" conditions.

then it came out all right, so I wrote up a little note about it and I started the note with the sentence, "In ihrer grundlegenden Abhandlung uber Quantenelektrodynamik," and so on. When Pauli saw that, he said, "Ja, das ist ein ziemlich morscher Grund, den wir da gelegt haben." [so that was struck out] Then I suggested "In ihrer ersten Abhandlung," and Pauli agreed, "Ja, das ist sinnvoll." Then he told me that obviously, the general method must be right, that is, we have to quantize the fields; since there is a classical Hamiltonian, the Hamiltonian method must be good enough, but since it leads to those consequences, it must be wrong somewhere. Probably we have not yet got the right variables, he said at that time. That was the first reaction to it; very early in '29.

(ibid.)

We should not split off Rosenfeld's paper on gravitational self-energy of light from this paper on gauge symmetries. When asked by Thomas Kuhn, following the previous quoted statement of Rosenfeld's, "Which of the consequences of that paper do you suppose most troubled him [Pauli]: the zero point energy, the self energy?", Rosenfeld replied:

Yes, all those things which were in it. One did not know how to separate those infinities. We knew that the classical model contained by itself an infinity, and if it had been possible to trace the other infinities to that one then that would have been fine. But then this paper on the gravitation energy showed that this was not the case, that one got an infinity in that case also, so one had the impression that there was something deeply wrong with the whole theory. Then it was not clear whether the difficulty was a methodical one, or was in the mathematical procedure or whether it was in the physics; and this uncertainty prevailed until one had this manifestly co-variant formulation which showed that the difficulties were not in the mathematics, but in the physics. Probably many people guessed at that time that it was in the physics and not just a matter of the formalisms.

(ibid.)

Here we see, in fact, how the two projects are intimately linked, by a need to know how one's representations are mapping onto reality. No result is certain unless one can be certain that one is dealing with the physical degrees of freedom. Without a good grip on the symmetries, one couldn't be sure what was 'real physics' and what was 'representation'. Rosenfeld is suggesting that these issues were not clear until the emergence of the manifestly covariant formulation of Schwinger and Tomonaga. The later architects of the new canonical formulation of general relativity would face identical problems, with much of the effort placed exactly on this preliminary stage of isolating the true physical structure.

Although one might think that Pauli's pushing of the problem of the quantization of the gravitational field was simply a matter of making good on his and Heisenberg's remark at the conclusion of their 1929 paper on the quantization of the electromagnetic field, as mentioned in the previous chapter, we can really see that it was rather used as an investigative probe to study the nature of infinities and singularities, as well as symmetries appearing in field theories. The first investigation of the quantization of gravity, then, revealed something more general about quantum theories and the special problems they generate: it was more of a by-product of learning about gauge degrees of freedom in

QED. In this work something of the specialness of the symmetries of general relativity were recognized early on; however, the recognition that this would pose special difficulties *vis-à-vis* quantization was slow to come. What Rosenfeld did in his symmetries paper was construct a quantum theoretical Hamiltonian for the *linearized* theory only. To a certain extent, this of course dodges various difficulties having to do with non-linearities, but it still had to face some of the difficulty of the symmetries of the full theory. Yet, to a certain extent, other features of general relativity which had not yet been developed in 1930[14] prompted the decision to study the theory of general relativity in its full, non-linear form.

6.3   Paul Weiss and the Parameter Formalism

As mentioned, the history of the canonical approach to the quantization of the gravitational field usually begins with Bergmann and, simultaneously, Pirani and Schild (who built directly on Dirac's Canadian lectures, which they both attended[15]) establishing, at the end of the 1940s, the *classical* theory along canonical lines—with earlier work of Rosenfeld from 1930 duly acknowledged.[16] It is perfectly true that many of the standard concepts of the modern approach (such as 'true observables' and the notion of constraints related to the formulation of these via the latter's elimination) can be found to have their origin here.

However, I argue that this view of the origins of the modern approach omits the curious mixing of key ideas of both Weiss (especially his parameter formalism) and Dirac, which were then seized upon by Bergmann, and Pirani and Schild (and later Dirac himself), to be put to work in quantizing the gravitational field—which then flowed into the later work and persists today. In particular, Weiss's rejection of the single-time (scalar) τ formalism of relativistic quantum mechanics and field theory in favour of a picture of temporal evolution as a 'slicing' of n-dimensional spacetime S into $n-1$-dimensional spacelike hypersurfaces, now commonly labeled by Σ.[17]

Here I might point out that John Stachel's [Stachel (1992)] excellent (pre-)history of the Cauchy problem in general relativity, which features an almost 'discrete leap' from the more mathematical work carried out in the 1920s (by Hadamard, Cartan, and others, primarily from France) to the work by Bergmann's group in the 1950s, is smoothed out by the inclusion of Weiss's contributions, in which he can be seen to function as a conduit between the disciplines (from the mathematical theory of partial differential equations to quantum field theories).[18] Weiss, being a Göttingen-trained mathematical physicist, also

---

[14] Such as the Einstein–Infeld–Hoffmann result which offered a glimpse of some of the power of non-linearities.

[15] Schild was teaching a course on tensor calculus at Carnegie Tech in the fall of 1949, where he had just moved as a result of tensions in Toronto, where he was previously. He took Felix Pirani along with him.

[16] For example, Peter Bergmann himself paints the development in this way: [Bergmann (1992)].

[17] This section is based on on work carried out together with Alexander Blum—see [Rickles and Blum (2015)].

[18] We should not forget André Lichnérowicz here, who certainly provides a more obvious link between the earlier mathematical work and that in relativity research in the 1950s—Lichnérowicz's supervisor was

knew of Hilbert's work on the Cauchy problem in general relativity and his slicing into hypersurfaces on which Gaussian coordinates are defined. Indeed, Weiss refers directly to the work of Cartan, Hadamard, and Hilbert (precisely those mentioned by Stachel) in his [Weiss (1938a)] (see pp. 103–4)—the Cauchy problem was vital for his approach, as he makes clear in the introduction to [Weiss (1938a)].

Weiss in effect takes the mathematical notion of "state" from Cartan (involving the notion of initial data: "correctly set data"), showing that it will be relativistically invariant and can ground a description of motion of a dynamical continuum through spacetime. The advance is that he transfers this idea to quantum mechanics, by linking it to 'state' in Dirac's sense: "the maximum information of non-contradictory data relating to the dynamical system" [Weiss (1938a), p. 103]. The spacelike surfaces then become 'carriers' of data in both classical and quantum scenarios. The motion for a field (a dynamical continuum) is then understood in terms of the "continuous unfolding of canonical transformations, starting from some initial data on some space-like region" [Weiss (1938a), p. 114]. Weiss's method is general enough to apply to any spacelike region (with its associated normal) and for this reason was seen to be a clear advantage in establishing a quantization of gravity for, prima facie, one can treat the metric variables in this framework as if they were any other kind of field variable.

Weiss's importance to the development of canonical quantum gravity rests on the method he devised for quantizing field theories. The method was employed directly and with due attribution in the founding papers of the field, by Dirac, Bergmann, Pirani and Schild, and DeWitt. The crucial element is the usage of an arbitrary, parameterized 3-surface (embedded in spacetime) along with normals to the surface that can be given a background-metric-independent definition. By forming momenta conjugate to the (generalized) coordinates, one can then write the Einstein field equations in a Hamiltonian form. The approach naturally leads to what many working on non-geometrical approaches to quantum gravity referred to as 'formal quantization'. The quantization is carried out in a metric-free (amorphous) space thanks to Weiss's method, and it is then expected (rather than demonstrated) that the standard geometrical interpretation of the metric can be reinstated post-quantization.

Weiss's story is an interesting one, and, since it overlaps with relevant individuals and wider events, we explore it in some detail here, before returning to more quantum gravity specific aspects.

Weiss was initially a student of Max Born, and later Paul Dirac, at Cambridge University—indeed, becoming Dirac's *first* doctoral student as a result of Born's leaving Cambridge to take up the Tait Chair of Natural Philosophy at Edinburgh. Born wrote of him that "he is one of the most learned students I have ever met in my field" and that

George Darmoir who had also worked on the Cauchy problem in general relativity, though much earlier in 1927 (see [Stachel (1992)] and [Stachel (2014)] for more details). However, unlike Lichnérowicz, Weiss pressed the Cauchy problem into the service of canonical quantization, and in a way that made use of general hypersurfaces, and so it was his approach that most directly connects to the problem of quantum gravity.

"[h]is knowledge and the thoroughness of it is amazing, in physics and mathematics".[19] In this recommendation letter, Born goes on to point out that Weiss is so modest that it is difficult to find this out without knowing him a long time. He also notes, in the same letter, that he will be delivering a series of lectures at the Institut Poincaré (in April 1937), in which the "main source...will be this dissertation of Weiss" (ibid.). Born also briefly mentions Weiss in his autobiography, pointing out that he had earlier been a research assistant to Fritz Haber [Born (1975), p. 267]. Dirac wrote that Weiss "has an unusual ability for research" (letter to SPSL dated November 29th, 1936: MS. SPSL 286/2).

Despite this high praise, then, Weiss's is not a name known to most physicists, nor to historians of physics.[20] However, as we shall aim to demonstrate in this section, he was responsible for some of the pivotal concepts involved in the canonical quantization of gravity (and field theories in general) and was in fact duly acknowledged in the founding papers and discussions on this approach. However, it does not seem that Weiss was directly interested in gravitation in his own research, and was more interested in the general formal structures of physical theories from a mathematical point of view—though we should note that Weiss did review several papers on general relativity (and even *quantum* gravity) for *Mathematical Reviews* and demonstrated that he had a good understanding of both areas (see, e.g., [Weiss (1946)]).

Weiss (fig. 6.1) was born in Sagan (Silesia) on April 9th, 1911, of Jewish descent. His German citizenship nonetheless made him an 'enemy alien' in the UK during the war years, resulting in his internment in Canada, along with many other notable physicists and mathematicians (including Hermann Bondi, Klaus Fuchs, Tommy Gold, Walter Heitler, and Beniamino Segrè).[21] He had been a pupil of Max Born's in Göttingen (in 1929–1930 and then 1931–1933)—from September 1930 to April 1931 he spent a 'year out' as an assistant teacher in the German public school system, at the Freie Schul und Werkgemeinschaft Letzlingen (between 1931 and 1933 he also spent some time studying in Paris and Zürich). Born clearly thought enough of him to have him to do a PhD with him at Cambridge—of course, it is highly likely that Hitler's rise to power had as much to do with Weiss's decision to leave for England since it seems that his immediate family (certainly his sister and mother) moved to Oxford at around the same time (as did Born himself). Weiss arrived in Cambridge (Downing College) to work with Born in October 1933 and lived close to Born (at Born's suggestion, writing from Bolzano in

---

[19] Letter to Walter Adams, then general secretary of The Society for the Protection of Science and Learning and later director of the LSE, dated February 4th, 1937: MS. SPSL 286/2.

[20] Indeed, the only reference I have seen to him in the historical literature (as opposed to citations to his work, that is) is in a recent article [Nossum (2012)] in which the author provides a brief account of various scientists (Weiss included) interned during the war, and the work of the Society for the Protection of Science and Learning [SPSL] in securing their release (on which more below). There is, however, a very brief appreciation of Weiss in [Gsponer and Hurni (2004)]. Keith Roberts returned to Weiss's theory ("undeservedly neglected for many years"—p. 126) as a way of setting up an interaction representation that is manifestly covariant.

[21] This was despite his well-documented and well-attested willingness to fight and work in the service of the British military. I return to Weiss's internment below.

Fig. 6.1 *Paul Weiss (1911–1991). The only photograph I have been able to find of Weiss, from his obituary [image source:* SIAM News *24(3): 2 and 6; note that this is the original photograph, supplied to* SIAM News *by Prof. Leon Brown—my thanks to Mike Cowan for making this available to me].*

1933: [BORN 1/3/2/5]), with Born's address 246 Hills Road and Weiss's address 23 Hills Road.[22]

Weiss received his PhD in 1936 with a thesis entitled *The Notion of Conjugate Variables in the Calculus of Variations for Multiple Integrals and its Application to the Quantization of Field Physics* (a rather large thesis, containing a thorough literature review, but of which an abbreviated version was published as [Weiss (1936)]). He stayed on in Cambridge after receiving his degree for a further 2 years, lecturing on QED at the Cavendish Laboratory during the academic year 1937–8.

Born expressed his puzzlement over where Weiss was getting his funds from to sustain himself after he had completed his PhD. Weiss seems to have been from a very wealthy, industrialist family. There was a connection between the Weiss and the Tugendhat

---

[22] Though in a letter to the SPSL, dated October 11th, 1939, he states that he arrived in England at the end of August. He was awarded an unconditional permit on July 26th, 1939, so this is entirely possible. Recall that Hitler was elected chancellor of the coalition at the end of January, 1933.

family.[23] Hans Weiss, Paul Weiss's brother, had been married to Grete Löw-Beer, who later married Fritz Tugendhat. There remained a connection between the families since the first marriage resulted in a daughter, Hanna. Both the Tugendhat's and the Löw-Beers were famous Brno industrialist families.

Weiss secured a position at Queens University, Belfast, at 300 pounds per annum for a period of two terms (January 1938–September 1939). Paul Ewald was in charge of the department on Weiss's arrival, and it seems that Weiss was put in charge of the department while he was there:

> Well, the nice arrangement they had in Belfast was to leave Mass's assistant, S. F. Boys, and a German refugee, Paul Weiss, in charge of the department after having appointed me, so that I could just be in the department and see how they handled it and what was going on and prepare. This was really a very wise and good idea. Paul Weiss is now here in the United States, I think with G.E. or anyway in Syracuse or thereabouts. He was a very good mathematician and physicist. http://www.aip.org/history/ohilist/4596_2.html

According to a letter Ewald sent to the SPSL in 1940 (August 17th, [MS. SPSL 213]), Weiss lectured on mathematical mechanics for at least one term (for six hours each week). He also wrote a large paper on quaternions [Weiss (1941)] while he held this position, presenting this work at the Irish Academy. This paper involves a conversion of tensor equations of special relativity into quaternion equations. By using angular variables (proper time and retarded distance), he is able to derive a quaternionic version of the classical equations of motion for a radiating charge (done using his beloved variational principle techniques).

It was just a very short time after this post in Belfast expired that Weiss was interned as an enemy alien, after his return to Cambridge. Weiss's passport was due to expire November 1939, though he received unconditional permission to stay in the UK on July 1939. In a letter dated September 5th, 1939, Weiss had already expressed a desire (to the SPSL) to do national service. He did not apply for naturalization there and then since he was waiting to find permanent work.

Like many other German academics, Paul Weiss's life during the Second World War was made considerably less painful by the work of The Society for the Protection of Science and Learning. This society began, initially envisaged as a temporary venture, as The Academic Assistance Council [AAC], in May 1933. In the words of Rutherford, then president of the AAC, its aim was "to assist scholars and scientists who, on grounds of religion, race, or opinion, were unable to continue their work in their own country" [Rutherford (1936), p. 607]. The assistance came not in the form of welfare but as a kind of clearing house for finding employment for the displaced scholars so that they might continue to contribute to the deeper "common cause of scholarship" (ibid.). In 1936 the

---

[23] The same Tugendhat family that commissioned Mies van der Rohe to build the Villa Tugendhat in Brno, and after which the famous "Tugendhat" chair was designed.

AAC was established on a permanent basis, with the change of title to SPSL[24] and an expansion to include research fellowships for exceptional refugee scholars. The society was funded via subscriptions and donations.

Despite being a 'category C' refugee (a genuine refugee, not to be viewed as an enemy or interned in the event of war), Weiss *was* interned, with other class Cs. It is possible that there was public pressure, no doubt exacerbated by hysterical news items (see [Kellerman (2007), p. 79]). There were several initial camps in the UK that were used, including the Isle of Man.

Weiss was interned on May 12th, 1940 while he was visiting Cambridge. He was released later the same year, in December, and returned to England soon after, taking up a temporary position (that would be made permanent) at Westfield College in the University of London, in February 1941. He finally became naturalized as a British citizen in June 1947.

The very process of internment overseas was not without its dangers. On July 2nd, 1940 the *SS Arandora Star* was torpedoed by the Germans, while carrying 700 internees bound for Australia (and while it flew a POW flag that the Allies naively believed would protect it from destruction). Weiss was aboard the *SS Ettrick*, the ship that took Hermann Bondi, Klaus Fuchs, E. Walter Kellermann, and Max Perutz across.[25] They arrived on July 13th in Quebec, were they were apparently greeted as enemies, due to inadequate political education. Tommy Gold recalled that the conditions on the ship were dire:

> Densely packed like sardines there, three layers, on the floor and on the tables, hammocks up above, and in the whole place, I don't know, there were maybe ten toilets for 800 people, and they all had dysentery. (http://www.aip.org/history/ohilist/4627.html)

During his internment Weiss spent his time in two camps: L and N. In camp L (Cove Fields, Quebec, overlooking the St Lawrence) it seems that the living standards were reasonably comfortable (there were "even showers" he writes: October 20th, 1940 [SPSL]). The internees in Camp L also established a kind of university (led by Max Perutz and begun while they were kept initially at a barracks at Bury St Edmunds). Concerts were also given. Weiss (ibid.) noted that there were a range of improving activities: vocational training courses, popular lectures, evening concerts and variety shows, and an art exhibition.

Hermann Bondi, Walter Heitler, and Klaus Fuchs were also held here and gave lectures (to classes of around 20 people, according to Gold). In a letter to a Mr. Skemp written after the war (November 18th, 1945) Weiss mentions that he taught a fair amount of mathematics during his internment and helped to organize this camp university.

---

[24] [Zimmerman (2006)] offers an excellent treatment tracing this transition, from AAC to SPSL. Since 1997 it has been known as CARA: The Council for Assisting Refugee Academics—though in 2014 the acronym was reinterpreted to mean "Council for At-Risk Academics".

[25] Klaus Fuchs' internment experience is briefly discussed in the first chapter of [Laucht (2012)]. For Kellermann's detailed account see chapters 7 and 8 of [Kellerman (2007)]. Note that Fuchs was also a student of Max Born's, and had written a paper [Fuchs (1939)] highlighting the general invariance of Weiss's method.

Bondi (taken in 1940) also found the internment experience to be fairly easy going.[26] He claims that his 'noteless' teaching style was forged there. On his release Bondi worked with Hoyle and Gold on radar.[27] Gold writes of this: "We didn't have any books, and so teaching from memory and puzzles was the kind of thing we did" (http://www.aip.org/history/ohilist/4627.html).

Camp N, at Sherbrooke, seems to have been a far unhappier affair. Fuchs and Bondi too went to Camp N. It was primarily Jews, and others that were on friendly terms with Jews, that were sent here. Camp N was a disused railway repair shed. Weiss claimed that they spent much of their time making repairs, building the necessary amenities, and queuing. He also noted bitterly that they had been told to do this maintenance and building work "to prove our loyalty to the British Commonwealth and that by doing this we shall prevent bad things from happening to our coreligionists in Canada" (October 20th, 1940 [SPSL]; fig. 6.2).

Nothing was done by August 1940 (despite several appeals involving high profile figures), so Paul Weiss's sister, Helene, stepped in and tried to gather some new support to petition the home office. One of those who wrote on Weiss's behalf was Paul Ewald, at Queens University, Belfast. Ewald pointed to Weiss's repeated efforts to register for National Service. He wrote: "[p]ersonally I would put full trust in Weiss's character both in private matters and with regard to the British political outlook" (August 17th, 1940 - to SPSL: MS. SPSL). Ewald suggested that Weiss be classified as 'category 8', relating to scientists, research workers, and those with academic distinction who might carry work of national importance—Bondi would be reclassified as such.[28] Weiss's old tutor from Downing College, Eric Holmes, also wrote on Weiss's behalf (pointing to his teaching potential as much as his research) as did Ralph Fowler.

There were further testimonials (in addition to earlier ones) from Born and Dirac again:

---

[26] Bondi describes his internment experience in his autobiography *Science, Churchill, and Me* [Bondi (1990)], even writing that he found the initial phase of his internment "a bit of a lark" (p. 28). He established a friendship with his long time collaborator Tommy Gold while housed in these camps. Gold gives a brief account of his internment in an interview with Spencer Weart: http://www.aip.org/history/ohilist/4627.html. It seems that Cambridge was the wrong place to be since local police forces were given orders to use their own discretion in the matter of enemy aliens. Bondi, Gold, and Weiss were all picked up in Cambridge, with Gold and Weiss unluckily just visiting at the time. As mentioned, E. Walter Kellerman, a fellow physicist internee, has also described his internment experience, in the same camps, and the events surrounding it in his autobiography: *A Physicist's Labour in War and Peace: Memoirs 1933–1999* [Kellerman (2007)]. He is less dismissive of the experience than Bondi (see, especially, pp. 83–4).

[27] Gold points out the curious twist here, writing that "the whole thing was pretty absurd, to be in internment at one time and then a few months later to go to the most secret defence establishment" (http://www.aip.org/history/ohilist/4627.html). Likewise Bondi: "there was a very short time from my being behind barbed wire because I was so 'dangerous', to my being behind barbed wire because the work I did was so secret" (http://www.aip.org/history/ohilist/4519.html).

[28] The minutes of the debates carried out in the House of Lords over the status of the internees (and the categories) can be found on the Hansard webpages: http://hansard.millbanksystems.com/commons/1940/aug/22/internees-1. In this report Sir Acland points out that even a prominent anti-Nazi, Sebastian Haffner, who had written a famous psychological study of Nazidom, was interned. Also Sigmund Freud's son. The report makes it abundantly clear that the decision to intern the vast majority of people in the Canadian (and Australian) camps was a terrible mistake.

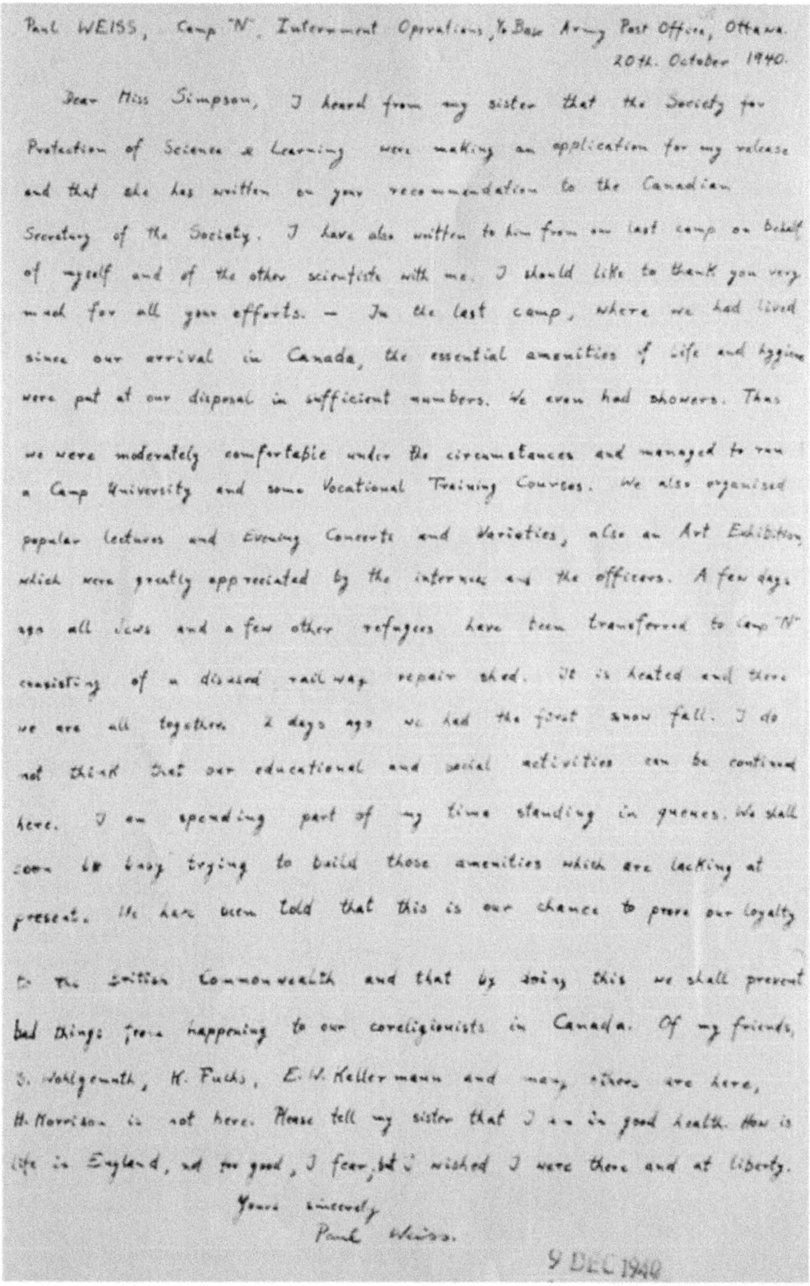
Paul WEISS, Camp "N", Internment Operations, % Base Army Post Office, Ottawa.
20th. October 1940.

Dear Miss Simpson, I heard from my sister that the Society for Protection of Science & Learning were making an application for my release and that she has written on your recommendation to the Canadian Secretary of the Society. I have also written to him from our last camp on behalf of myself and of the other scientists with me. I should like to thank you very much for all your efforts. – In the last camp, where we had lived since our arrival in Canada, the essential amenities of life and hygiene were put at our disposal in sufficient numbers. We even had showers. Thus we were moderately comfortable under the circumstances and managed to run a Camp University and some Vocational Training Courses. We also organised popular lectures and Evening Concerts and Varieties, also an Art Exhibition which were greatly appreciated by the internees and the officers. A few days ago all Jews and a few other refugees have been transferred to Camp "N" consisting of a disused railway repair shed. It is heated and there we are all together. 2 days ago we had the first snow fall. I do not think that our educational and social activities can be continued here. I am spending part of my time standing in queues. We shall soon be busy trying to build those amenities which are lacking at present. We have been told that this is our chance to prove our loyalty to the British Commonwealth and that by doing this we shall prevent bad things from happening to our coreligionists in Canada. Of my friends, S. Wohlgemuth, K. Fuchs, E. W. Kellermann and many others are here, H. Herrison is not here. Please tell my sister that I am in good health. How is life in England, not too good, I fear, but I wished I were there and at liberty.

Yours sincerely,
Paul Weiss.

9 DEC 1940

Fig. 6.2 *Paul Weiss (1911–1991). Letter written from within the internment camp. Image source: Archive of the Society for the Protection of Science and Learning, 1933–87, S.P.S.L. 286/2.*

- Born: "he is a victim of Nazi persecution and hopes for the victory of Great Britain" (letter to Miss Simpson of SPSL, dated August 5th, 1940).

- Dirac: "Mr Weiss ... is strongly anti-Nazi and thoroughly reliable. His ability could be of great service in the National Cause, and it would be a great pity not to make use of it" (August 13th, 1940).

The instructions for his release reached his sister on December 18th, 1940, thanks to persistent pressure at high levels, including the Secretary of State. Helene Weiss wrote to the secretary Miss Esther (Tess) Simpson[29] of SPSL on December 24th, from their mother's home in Oxford noting that she was very happy to be able to pass the news on to her. He was officially released in January 1941. Their mother (Babette Rosenbacher) died in 1943.

Soon after his release, in February 1941, there was some suggestion (relayed through G.H. Hardy) that Weiss might take up a post at Liverpool. It happened that he managed to find a position in applied mathematics at Westfield College, in London University, for the remainder of the war, again with a salary of 300 pounds—he was initially going to replace a Mr. Jackson who was seconded to the Admiralty (letter from Mary Stocks, Principal of Westfield College, dated August 21st, 1946: MS. SPSL 239). This position lasted from 1941 to 1950.

Weiss married and had children while at Westfield. His wife was Marliese Oppá, and they had two children together, and also a third child they adopted, a refugee from Belsen, related to his wife. His daughter, Ruth Weiss, worked for Bell labs on some of the first programing languages in the 1960s. She was also a pioneer in using programing languages to generate computer graphics of mathematical curves and surfaces.

Weiss was clearly grateful for this position, and to the SPSL, and quickly volunteered his services during the University vacation period writing to Miss Simpson (on July 15th, 1941) that he wanted to do "real war work". Miss Simpson pointed out that getting such work as a refugee really was done through personal contact, and that he should inform "the British mathematicians" that he is seeking some such work. It is very interesting to see the political power the mathematicians and scientists had at this time, behind the scenes. Weiss was in fact placed in a central register of "aliens with special qualifications," though nothing seems to have come of it.

The SPSL retained an interest in Weiss's career at least until Weiss discovered his position at Westfield had been made permanent, in 1946. Weiss also applied for a naturalization certificate through the SPSL office (now via a new secretary, Mrs. Ursell) in August 1946. It also seems that Helene and Paul Weiss applied in tandem, both through the offices of SPSL. By this time Helene Weiss had joined her brother on the staff of Westfield (some time in 1945).[30]

---

[29] Miss Simpson later wrote her memoirs, detailing her experience with the SPSL: [Cooper (1992)] (*cf.* [Kohn (2011), pp. 279–80], for more information).

[30] Helene Weiss was a philosopher and classicist of some merit, specializing in ancient philosophy, Aristotle in particular. She had been a student of Martin Heidegger between 1920 and 1934 and achieved some fame as a Heidegger scholar. Another of Weiss's sisters, Gertrud, was married to one of the founders of social and organizational psychology, Kurt Lewin.

Weiss did not completely disappear from physics after the events of the Second World War were at an end. In 1950 Weiss wrote to Born that he had received an invitation to the Institute for Advanced Study [IAS] at Princeton.[31] We see in the IAS records[32] that Weiss was indeed a visitor at the IAS during 1950–1, with his home address then given as 635 Elm Street, Syracuse 10, NY—perhaps coincident with his IAS visit in 1950, he moved to Syracuse to work for General Electric [GE], until 1957. Weiss was employed as an applied mathematician for GE. One of the projects he worked on, in 1952, was the application of operations research to the solution of business problems. While there he also designed and patented (together with Charles Johnson) a "form recognition system" based on the identification of certain invariant properties (US Patent No. US2968789).[33] He left GE in 1958, briefly taking up a position at AVCO (Aviation Corporation) until 1960, and finally settling at Wayne State University, in the mathematics department, until his death in 1991—curiously, Peter Bergmann was in Syracuse, of course, and Suraj Gupta was at Wayne State, yet there seem to have been no interactions between them, despite the fact that Bergmann was invoking Weiss's work in a fundamental way at the time in his work on quantum gravity.[34]

That part of Weiss's work that turned out to be crucially important to the program of a canonical quantization of the gravitational field (as well as being relevant to the covariant approach, as we saw in the previous chapter), was carried out in the mid- to late 1930s and involved a generalization of Hamilton–Jacobi theory, from the simple integrals, which occur in the context of point particle dynamics, to multiple integrals, in order to apply the results of the former to field theories.[35] For point mechanics, canonically conjugate variables can be defined in the following manner, using variational calculus. Given the usual action integral[36]

$$I = \int_{t_0}^{t_1} L(x^a, \dot{x}^a, t)\, dt \tag{6.1}$$

Weiss, following Cartan, performed a variation in which not just the trajectories $x(t)$ are varied (which leads to the usual Euler–Lagrange equations of motion), but also the values at the endpoint, $x(t_0)$ and $x(t_1)$, and the endpoints t_0 and t_1 themselves. The variation of the action then contains additional boundary terms:

---

[31] It turns out that this had been at Born's recommendation (Letter dated May 30th, 1950, writing from London [Born archive 1/3/2/27]). I note that by 1950 the letters between Born and Weiss had shifted from their native German to English.

[32] The relevant pages are 414–5 of the *Community of Scholars* booklet.

[33] GE was not considered a bad option for a research scientist. Both Bethe and Feynman did periods there in the summer of 1946.

[34] Paul Weiss's son, Thomas, recalls emigrating from England in November 1952, with his mother, sister, younger brother, and maternal grandmother, on the French liner Ile de France (private communication).

[35] As acknowledged by Weiss, the idea of extending the Hamilton–Jacobi theory to higher integrals came from Georg Prange, a student of Hilbert's at Göttingen, who had studied double integrals in his thesis *Die Hamilton–Jacobische Theorie für Doppelintegrale* of 1915. This work had, of course, been entirely independent of all quantum field theoretical considerations, given that quantum mechanics itself was still 10 years away.

[36] Anticipating the generalization to four space-time coordinates, Weiss chose to denote time by x (and the canonical coordinates by z^a). I have opted for the more standard notation used also by Cartan.

$$\delta I \supset \left(H(t_1)\delta_t + p^a(t_1)\delta x_1^a \right) - \left(H(t_0)\delta_t + p^a(t_0)\delta x_0^a \right) \tag{6.2}$$

That is, in the boundary terms, the Hamiltonian shows up as the function multiplying the fixed variation of the endpoints δt, while the canonical momenta p^a appear as the functions multiplying the variation of the endpoint value of the corresponding canonical coordinate, δx_1^a and δx_0^a, respectively. In fact, this can be used to define the Hamiltonian and the canonical momenta. The essential advantage of this is that the canonical momenta are now defined via the the integration boundaries appearing in the action integral. This is of no consequence in point mechanics, where there is no freedom in choosing those integration boundaries: They are always simply two points in time. For a field theory, however, defined on a multi-dimensional spacetime, one can choose arbitrarily curved hypersurfaces as integration boundaries and can thus, as Weiss outlined, arrive at a generalized notion of Hamiltonian and canonical momenta that does not single out the time coordinate, but rather defines these quantities with respect to some arbitrary hypersurface.

This brings us back into the well-trodden territory of the Heisenberg–Pauli theory to which Weiss was responding. As Weiss pointed out, "[t]he artificial separation of space and time, inherent in this quantization method, has the disadvantage that the proof of the relativistic invariance of the result becomes very complicated" [Weiss (1936), p. 193]. Of course, we have seen this problem already, since Rosenfeld was charged with figuring out a better way. Weiss's method, in referring to an arbitrary 3-surface (embedded in spacetime), secured the Lorentz invariance from the start. The orthogonal components to these surfaces (the normals to the surface) could be defined independently of the spacetime metric. And the field coordinates and canonical field momenta appearing in the canonical commutation relations were now not functions of a point in space, but rather of a 3-parameter ('surface variables') u that defined a point in that hypersurface—this constitutes the basis of the 'parameter formalism.'

Initially, Weiss did not even consider his hypersurfaces to be limited to space-like ones. Rather, he was thinking in terms of an arbitrary closed hypersurface, acting as integration boundary for the action. The canonical commutation relations were obtained by demanding that the commutation relations of Heisenberg and Pauli were recovered, when the hypersurface is given a spatial interpretation, so that $x^4 = 0$ (i.e. $t = $ const). In this way, Weiss not only made contact with the (more or less) established quantum field theory of the day, he also provided a much simpler proof of the relativistic invariance of the Heisenberg–Pauli approach from within his more general framework.

In his second paper [Weiss (1938a)], Weiss moved on to construct (already at the classical level) generalized Poisson brackets, so that the analogy with canonical particle mechanics becomes complete. This meant that he had to restrict his hypersurface, i.e., the integration boundary of the action, to two (initial and final) space-like hypersurfaces. The advantage was, however, that in this manner the quantization could be carried out in the same manner as for quantum mechanics, by turning Poisson brackets into canonical commutators. Explicit reference to the limiting case of Heisenberg–Pauli commutation relations then only served as a consistency check, but was no longer needed as a constructive principle. It was in this form that Weiss's formalism was then later taken up in the canonical quantization of the gravitational field.[37]

Weiss was not the only one to be dissatisfied with the non-manifest covariance of Heisenberg and Pauli's canonical quantization, as we saw in the previous chapter. Another approach developed in the course of the 1930s. It has its origins in the work of Jordan and Pauli in 1928, and its goal was to obtain the commutation relations between two field quantities at two arbitrary space-time points, which did not have to have the same time coordinate or even lie on the same space-like hypersurface. In this approach, the commutation relations were not derived from the classical Poisson brackets and it consequently made no reference to canonically conjugate variables. Instead, the commutators between the fields at two arbitrary points were obtained as singular solutions of the field equations. This became intractable for interacting fields, where the equations of motion were no longer linear and the singular solutions were hard or impossible to come by. Consequently, Pauli had turned to equal-time commutators in his work with Heisenberg. But a way out was pointed by Dirac's interaction picture of 1932, where the field operators obeyed the free equations of motion, so that the covariant commutation relations of the Pauli–Jordan type could still be employed in an interacting theory.[38]

It should be noted again that also the covariant Tomonaga–Schwinger quantum field theory could not be done without a foliation of spacetime into arbitrary space-like surfaces. As opposed to the attempts at a canonical quantization of gravity, discussed later, the introduction of space-like hypersurfaces by Schwinger and Tomonaga appears prima facie to have been independent of Weiss (though both can be traced to the influence of Dirac's paper on Lagrangian quantum mechanics [Dirac (1933a)]). And indeed, in the work of Schwinger and Tomonaga the hypersurfaces appeared in a quite different fashion, only on the level of the Schrödinger (or rather Tomonaga–Schwinger) equation, i.e., the equation that describes the dynamical evolution of the quantum state in the interaction picture.[39] The question of the (relativistic analog of) the Schrödinger equation was not addressed by Weiss, as he appears to have been thinking in terms of a quantum field theory formulated in the Heisenberg picture. When Schwinger, around

---

[37] In a third paper [Weiss (1938b)], Weiss extended his formalism further to deal also with the redundancies introduced into the canonical quantization procedure by the gauge invariance of Maxwell electrodynamics (or generalizations of that theory). However, rather than anything like a constraint apparatus, Weiss developed an invariant formalism using objects analogous to holonomies—John Stachel [Stachel (1982), p. 1289] identifies Weiss as the first to introduce $\oint A_\mu dx^\mu$ as the fundamental ontology of electromagnetism.

[38] As mentioned in the previous chapter, this approach further developed into the fully covariant quantum field theory of Tomonaga and Schwinger. The utilization of these approaches in a gravitational context can perhaps be considered the starting point of the covariant approach to the quantization of gravity. Alternatively, one can see this period as having no real split, instead with two aspects of one and the same basic formalism in operation. Weiss's proposal provided the starting point for a canonical quantization of gravity at a time when the covariant quantization approach might otherwise have seemed without alternative after its great success when applied to the renormalization of quantum electrodynamics.

[39] Silvan Schweber claims that Schwinger's "chief innovation" was the introduction of a state vector as a functional on a 3-dimensional surface in spacetime "which represented the state of the system as determined by the specification of a set of commuting field variables on the surface" [Schweber (1994), p. 319].

1950, shifted from the interaction picture to the Heisenberg picture in the development of his 'quantum action principle,' his commutation relations concerned field values at two points on a space-like hypersurface. The problem is then to figure out the transformation function between surfaces of this Weissian kind, in which the surfaces are the 'bearers' of data (*cf.* [Milton (2007)], pp. 86–7). So, while Schwinger's second approach to QFT is generally viewed (and was presented by Schwinger) as an elaboration of Dirac's proposals for a 'Lagrangian quantum mechanics',[40] it certainly also contains elements of Weiss's approach, indirectly attributed to Dirac.

To summarize, the basic motivation for Weiss's work on field theory was a desire to obtain a scheme for quantizing field theories that remains clearly Lorentz invariant at every step. The problem of lacking covariance had led Dirac away from the Hamiltonian method in the early 1930s and to develop both his interaction picture[41] and the Lagrangian approach. It is highly likely that Weiss's method brought Dirac around to the Hamiltonian method as a reasonable one after all for quantizing field theories in a manifestly covariant form. Dirac's focus, at least initially, lay firmly on Lorentz invariance.

6.4 Dirac's Vancouver Lectures and the First Canonical Quantizations

Dirac's return to a Hamiltonian approach is marked by a series of lectures he gave in Vancouver at the 1949 Summer Seminar of the Canadian Mathematical Congress, published as [Dirac (1950b), Dirac (1950a)]. In August and September, 1949, Dirac (fig. 6.3) presented a general Hamiltonian scheme for dealing with systems of constraints, that could be employed in a theory described by an action functional. This generalized form of Hamiltonian dynamics was, in other words, applicable when the momenta are not independent functions of the velocities, yet which can still be used for quantization purposes and which is well-suited for a relativistic description of dynamical processes, as Rosenfeld had intended in his 1930 work on the subject. We already described some of the basic ideas of his scheme in §6.1.

Felix Pirani and Alfred Schild were in attendance, and noticed that, without much ado, Dirac's new methods, if combined with Weiss's parameter formalism, could provide a framework for a manifestly covariant, canonical quantization of gravity.[42] Interestingly, T. S. Chang, a student of Dirac who had been spreading the word of Weiss's theory

---

[40] This is not to be confused with his interaction picture approach.

[41] Dirac's papers on quantum electrodynamics in the interaction picture are mainly known for his 'many-time theory.' This is, however, logically independent of the use of the interaction picture (which allows making the commutation relations invariant). Its purpose is to also make the dynamical equation for the quantum state (the Schrödinger equation) manifestly covariant. It was later generalized, as discussed above, by Tomonaga and Schwinger to the case of a pure field theory (where also the electrons are described by a fermionic field, rather than by many-particle quantum mechanics), by having the wave function defined on a space-like hypersurface.

[42] Here again, we see the complexity of the story that forbids a simple grouping into covariant and canonical at this stage. Even in what is considered to be a founding paper in canonical quantum gravity, the focus is on retaining manifest covariance along the lines of the earlier responses to the Heisenberg–Pauli theory.

Fig. 6.3 *Paul Dirac in Vancouver, in 1949. Image source: Special Collections & Archives, Florida State University Libraries, Tallahassee, Florida.*

for several years, was at the Carnegie Institute of Technology when Schild and Pirani were there, and had only just written a paper showing how Tomonaga's formalism for relativistic quantum electrodynamics could be directly deduced from Weiss formalism: [Chang (1949)]. This paper appeared March 1949, just a few months before the Dirac lectures in Vancouver, in August. Chang discussed how, in his work on localizable systems, "Dirac introduced the study of the Schroedinger wave functionals on any arbitrary space-like surface" introducing "certain deformation operators to describe the deformation of the surfaces and studied their commutation laws." It is entirely possible that some initial discussion of Weiss formalism occurred before Pirani and Schild attended the Dirac lectures, thus priming them all the more.

In their first paper [Pirani and Schild (1950)] Pirani and Schild consolidate the view of Weiss's contribution, as the originator of the paramterized arbitrary surface, in their quantization of gravity.[43] The steps to quantum gravity then seem to be as

---

[43] Jauch and Rohrlich correctly attributed this feature to Weiss in their text on QED: "A theory involving more general space-like surfaces was developed for the first time by P. Weiss (1938)" [Jauch and Rohrlich (1955), p. 9].

follows: Weiss developed a method of field quantization for theories characterized by 1st order Lagrangians and showed how it can be carried out in a manifestly invariant manner by making use of an arbitrary space-like surface. They combine Weiss's method with Dirac's prescription for dealing with singular Hamiltonian systems (which had originally been devised for Lorentz invariant field theories), in order to transform the (1st order) gravitational Lagrangian[44] into Hamiltonian form (as a function of momenta and velocities), and then using the notion of 'quantization in a metrically amorphous space' they complete the picture. The picture is then one that modern day canonical relativists will readily understand: Einstein's equations describe the evolution of the gravitational field variables g_{ab} (or rather the 3-*geometries*) defined relative to a differentiable manifold (via differential equations), and these variables are then quantized using Weiss's parameter method (with the parameters describing a family of three-dimensional surfaces) which enables the formulation of a canonical formalism with Poisson brackets. The quantization is of course purely formal, and in particular no geometrical meaning is attached to the variables initially in their case.

Bergmann and Brunings had a different motivation in their early work, in which a direct physical meaning is to be attached to the variables.[45] They wanted a canonical quantum gravity, but the reasons are connected to a desire to utilize the non-linearities in the gravitational field equations to resolve difficulties in field theory more generally (i.e. the divergences[46]). While Pirani and Schild's approach roughly amounts to a direct combination of Dirac and Weiss, Bergmann and Brunings's approach amounts to a kind of combination of Einstein–Infeld–Hoffmann and Weiss:

> Our parameter formalism has formal similarity with one developed by P. Weiss. It differs in that Weiss uses his analogue primarily to vary the domain of integration of the variational principle. His surfaces t = constant are necessarily closed, while ours are not only open, but, in the presence of particles, multiply connected. Also, Weiss has apparently taken the emergence of the x^μ as dynamical variables not as seriously as we do.
> [Bergmann and Brunings (1949), p. 487]

The last remark about treating the x^μ as the dynamical variables of the theory is, of course, one of the core advances of the modern canonical approach and con-stitutes a major conceptual shift strongly distinguishing itself from linearized, weak

---

[44] The first order form is generated by subtracting a divergence term, giving a first order Lagrangian in place of the usual (generally covariant) second-order one.

[45] Peter Bergmann left Prague for America in 1936, to work at the Institute for Advanced Study, where he would become Einstein's research assistant and collaborator. He shifted to Syracuse University in 1947, still steeped in unified field theories after working with Einstein, but quickly began to think about canonical relativity via his approach to the EIH theory. His paper on a "Unified Field Theory With Fifteen Field Variables" was the last of this work, but by the time he wrote this he was already far from this kind of theory, and presented it (an older work that he had conducted with Einstein), as a potentially plausible account of a recently published model of Jordan's. In conclusion, he wrote: "It is very difficult to judge whether any given 'unified field theory' contains a grain of truth which will contribute to that ultimate theory which must eventually replace both current field theories and current quantum theory." [Bergmann (1948), p. 263].

[46] We are still in this sense dealing with footnotes to the Heisenberg–Pauli theory.

field approaches (cf. [Bergmann (1992), p. 293]). It would soon radically distinguish Bergmann's approach from that of Dirac's at the time. Dirac expressed a superficially similar view in his Vancouver lectures:

> The concept of an instant of time is rather artificial from the relativistic point of view. It is to be pictured as a flat three-dimensional "surface" in four-dimensional space-time, with the direction of its normal lying within the light-cone. It would be more natural in a relativistic theory to replace the flat surface by an arbitrary curved one, subject to the restriction that it is everywhere space-like, i.e. the normal at every point of it lies within the light-cone. One would then work with dynamical variables referring to physical conditions on such a curved surface, as was done by Weiss.
>
> ([Dirac (1950a)], p. 2)

Again the advance, as regards relevance for canonical gravity, was in recognizing that the introduction of surface variables implied that the usual space-time coordinates could then be treated as functions of the surface parameters "on the same footing as dynamical coordinates". Bergmann then proposes to treat the coordinates themselves as dynamical variables to be quantized. As can be seen from the quotation given above, Bergmann, still tied up in the idea of a divergence-free theory based on the EIH theory, even hoped that the now dynamical space-time coordinates would be interpretable as the trajectories of singular point particles, moving in the gravitational field (hence the claims of multiply connected spaces). Yet Dirac, as mentioned, was interested in Lorentz invariance, not general covariance. This generates a very different problem in terms of establishing the physically relevant degrees of freedom required for a Hamiltonian formulation for in general relativity frames of reference and gauge transformations are entangled in such a way that one has to resolve both together. One can't, then, simply set up the canonical variables as local variables on a surface. The observables must be gauge-invariant, and this is bound up with the spacetime frame leading to a kind of non-locality: the observables must be functionals of the field variables that are invariant under spacetime mappings. Hence, Bergmann's group was bogged down early with thorny conceptual issues, while Pirani and Schild, eschewing a geometrical interpretation of the field variables, were able to bypass these.

In his discussion of the early years of the canonical quantization of gravity, Joshua Goldberg writes that "the work in Syracuse began without knowledge of the Dirac lectures or that Pirani and Schild existed" [Goldberg (2005), p. 360]. If this were true, it would constitute a remarkable case of pre-established harmony—however, we have noted the existence of a common cause through the parameter formalism, and indeed Goldberg draws attention to the importance of introducing the idea of an arbitrary hypersurface, though attributing the general parameter formalism to Dirac:

> In the second paper, Dirac introduces a field theory in Minkowski space. He is not thinking about general covariance. But, he is concerned with Hamiltonian theories with constraints and with maintaining the four-dimensional symmetry of Lorentz invariance, while at the same time introducing an arbitrary space-like surface in terms of which

to define the canonical formalism. To accomplish this, he introduces a parametric description of the Minkowski space coordinates...In order to assure that the surface $t = 0$ is space-like, Dirac introduces the time-like unit normal l^μ with the properties ∂^μ. Then he proceeds to define quantities on the surface which are covariant under a change in parameters, which leaves the surface $t = 0$ unchanged. For tensors, these are the normal components to the hypersurface.

[Goldberg (2005), p. 366]

The story Goldberg then tells about the Pirani and Schild work in relation to this development is as follows:

It was Alfred Schild who saw that, once one introduced an arbitrary space-like surface, the Dirac formalism could be applied to general relativity. In a straightforward application of the Dirac approach with some clever mathematical manipulations, together with Felix Pirani he constructed the first explicit expression for the Hamiltonian. However, they did not complete the decomposition of the metric or of the field variables for the Maxwell field with respect to l^μ, the normal. More important, they did not examine the propagation of the constraints until later.

[Goldberg (2005), p. 363]

We return to the constraints briefly in a moment. However, it is clear from what we have seen that the commonalities between Bergmann's and Pirani and Schild's approaches is due to the sharing of the arbitrary surface idea of Weiss, which they both directly refer to. The first attempts at a Hamiltonian formulation, by Pirani and Schild, and by Bergmann, then, utilized the parameter method, which was used to try and maintain manifest covariance (along the lines of Weiss's original intention).

However, Pirani and Schild already knew Bergmann and Brunings's paper, which appeared in July 1949, and which they refer to in their abstract. Dirac's lectures were in August. They had already mentioned Weiss in that paper, and the connection with quantum gravity was raised, in connection with Weiss. In seeing Dirac mention Weiss too, in connection with theories with constraints, it seems like a plausible scenario that they simply put the two together. like pieces of a jigsaw. If Bergmann and Brunings employed Weiss's method, and Dirac made an advance over that method, in relation to symmetries, then it makes sense to try to apply the improved method too. Both Pirani and Schild and Bergmann and his students were engaged in a simultaneous effort to formulate the Hamiltonian for general relativity, which required the identification of the canonical variables—Pirani and Schild sought the Hamiltonian for the combined gravitational and Dirac fields. This was not done in complete isolation, and the various teams knew of the others' work.[47] While Pirani and Schild got the Hamiltonian first, using

---

[47] For example, Alfred Schild writes to Bryce DeWitt (then Seligman, though Schild mistakenly named him "Bruce Seligman") on May 17th, 1950 that he and Pirani have been attacking the combined Hamiltonian problem for some months [CDWA]. Likewise, Bergmann and Schild were in close contact on the matter of their progress, with Bergmann noting his receipt of Schild and Pirani's Hamiltonian on November 16th, 1950 (see fig. 6.4) [Alfred Schild Papers, Box 86–27/2]. Bergmann had sent copies of his group's paper earlier that year, on January 20th. This was at the classical level, but Bergmann writes also that he knows of Schild and Pirani's plans to work on the quantization too.

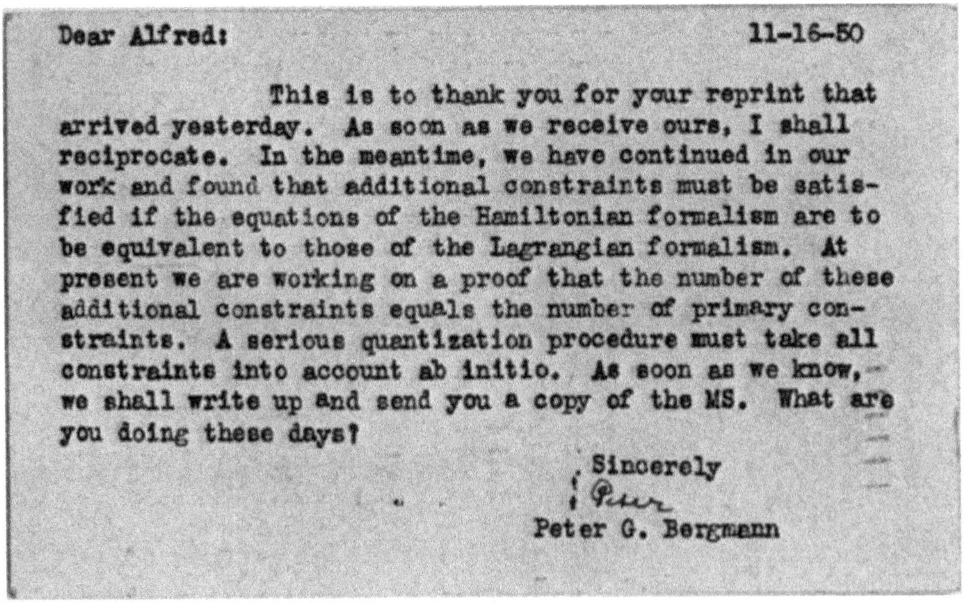

Dear Alfred: 11-16-50

This is to thank you for your reprint that arrived yesterday. As soon as we receive ours, I shall reciprocate. In the meantime, we have continued in our work and found that additional constraints must be satisfied if the equations of the Hamiltonian formalism are to be equivalent to those of the Lagrangian formalism. At present we are working on a proof that the number of these additional constraints equals the number of primary constraints. A serious quantization procedure must take all constraints into account ab initio. As soon as we know, we shall write up and send you a copy of the MS. What are you doing these days?

Sincerely
Peter
Peter G. Bergmann

Fig. 6.4 *Note from Peter Bergmann to Alfred Schild revealing the close race to find the Hamiltonian. Image source: Alfred Schild Papers, Box 86–27/2, Dolph Briscoe Center for American History, The University of Texas at Austin.*

Dirac's and Weiss's methods, they did not consider in detail the problem of constraints (and their propagation, called, by Bergmann's group, the secondary constraints[48]). As mentioned, Bergmann's group viewed the resolution of the problem of constraints as essential in pinning down the correct class of observables that would be then quantized: they would be precisely those quantities that commute with the constraints. A large part of Peter Bergmann's approach, then, consisted of getting the classical theory of gravitation into shape, before the quantization problem could be tackled. As he puts it: "the problem of quantization leads back to the nonquantum problem of an invariant description of physical situations" [Bergmann (1955), p. 493]. This involved putting the theory into canonical form—a somewhat similar motivation to that driving Rosenfeld's work on constrained systems. The key problem is finding the correct quantities that will be promoted to operators:

48 It was in fact Peter Higgs [Higgs (1958)] who revealed the link between three of the secondary constraints and transformations of the coordinates of a spatial slice—work carried out at the Chapel Hill, Institute for Field Physics, run by Bryce and Cecilé DeWitt after being prompted to work on the problem by Felix Pirani himself (as explained in a letter to Cecilé DeWitt, dated August 27th, 2009: CDWA).

> How can we identify (that is, describe unambiguously) a total physical situation in general relativity independently of the (accidental) choice of a particular coordinate system and independently of any a priori assumed identifiability of space-time points? Once we have answered this question, we have presumably found those variables that express the substance of a physical situation. Quantization should be applied to these quantities, rather than the usual field variables, whose values depend both on the physical situation and on our accidental method of description.
>
> > [Bergmann (1955), p. 492]

This brings center-stage the issue of the symmetries and constraints that Rosenfeld had already battled with, thus signifying a switch from Rosenfeld's other motivation (divergences), which previously occupied Bergmann. Since the constraints encode the non-uniqueness of the Hamiltonian, which is a by-product of the original symmetry o the Lagrangian, Bergmann's focus was specifically on the treatment and interpretation of general covariance. Given his direct interpretation of the coordinates as dynamical field variables, this took on special significance. The focus on the invariant physical description of the gravitational field would soon, therefore, eclipse the EIH foundation and the problem of divergences. This brings in the challenge already mentioned by Goldberg above. How to make sense of the gauge degrees of freedom. DeWitt expresses the challenge as follows:

> Bergmann immediately ran into major difficulties (some of which had already been foreseen by Rosenfeld) in the first stages of his program. These are referred to as "the problem of constraints," and are manifested in the following ways: Some of the field variables possess no conjugate momenta; the momenta conjugate to the remaining field variables are not all dynamically independent; the field equations themselves are not linearly independent, and some of them involve no second time derivatives, thus complicating the Cauchy problem. These difficulties are all related and arise from the existence of the general coordinate-transformation group as an invariance group for the theory.
>
> > [DeWitt (1967), p. 1114]

Bergmann's group developed their own specific way of dealing with these problems based around phase space quantization, with the focus firmly on the observables.[49] Given his link to coordinates, the observables had to be coordinate independent quantities. One approach to these was to depart from the canonical formalism of Dirac, and search for field invariants, e.g. the four (scalar) eigenvalues of the Riemann tensor which define an 'intrinsic coordinate system'—see [Bergmann and Komar (1962)].

---

[49] As Bergmann later wrote about this, "[u]nder this new symmetry group... the formulation of statements about physical reality requires much greater sophistication and care than is required when space and time are assumed to have a built-in rigid structure not subject to the dynamical equations of a physical field [Bergmann and Komar (1980), p. 227].

Dirac expressed a related problem related to this scheme in his Vancouver lecture as follows:

> To get a dynamical theory which satisfies restricted relativity, we must set up a scheme of equations which applies equally to observers with all velocities. If we work with instants, we must include instants with respect to all observers. An instant is then any flat three-dimensional surface in space-time having a normal in a direction within the light-cone. A general instant needs four parameters to describe it, three to fix the direction of the normal, or the velocity of the observer, and the fourth to distinguish different instants for the same observer. A relativistic dynamics that involves instants must enable one, given the state at any of these instants, to calculate the state at any other. We must have equations of motion showing how the dynamical variables vary as the instant varies. We can allow the instant to vary arbitrarily, with a translational motion in space-time as well as the direction of its normal varying, and the equations of motion must always apply. Thus *we need four first class ϕ's to give rise to the four freedoms of motion of the instant.*
>
> [Dirac (1950b), p. 144]

Again, however, Dirac is concerned with special relativity only. The four freedoms of motion do not refer to the physical conditions of spacetime geometry. However, these four freedoms, relating to four constraints, would serve to define the physical structure of Bergmann's approach, and indeed the Hamiltonian of general relativity turns out to be a linear combination of the constraints associated with these freedoms. However, getting to grips with the nature of the constraints from a physical point of view in the context of quantum theory proved to be an enormously complex task, not because because of operator-ordering problems, which is still a matter of controversy today. The quantization of the constraints leads, of course, to the infamous Wheeler–DeWitt equation, a wave equation from which the external time parameter t has disappeared.

6.5   Parameters in Later Work

In June 1951, Ray Skinner began work on this same problem, after moving from Toronto to Carnegie Tech with Schild.[50] Working on the secondary constraints, they dropped the parameter formalism. Interestingly, though we don't know the reasons, Schild stopped all work on quantum gravity after his work with Pirani and Skinner (and that work appears to have been primarily that of his students in any case). By December 1956, for example, in responding to Cécile DeWitt's placing of him in a session (for the 1957 Chapel Hill conference) on "Techniques of quantization of general relativity" Schild writes that he is not sure how much he has to say on the topic "if anything" (letter to C. DeWitt, December 17th, 1956: CDWA).[51] Bergmann and his group continued to

---

[50] Schild took another student, George Habetler, along with him too, though Habelter was working on classical GR problems (including a generalization of Kaluza's five-dimensional theory).

[51] One apparent return to the problem of quantization occured in 1963 when Schild applied for travel authorization to visit Jerzy Plebanski in Mexico "to discuss quantization of the model theory of gravitation"

make the analysis of the observables the central task of their program, again dropping the parameter formalism and focusing directly on the coordinates themselves.

The parameter formalism, and the focus on manifest covariance, slowly gave way to a more direct focus on the problem of quantum gravity independently of the issues relating to general field theory. The formalism was also very cumbersome to use when it came to computing things in the theory. As Goldberg notes:

> One soon found that the parameters were an unnecessary complication and after the fall of 1950 they were abandoned. As a result, the issue of secondary constraints was not examined in the parameter formalism.
>
> ([Goldberg (2005)], p. 366)

But that they were eventually believed to be unnecessary[52] should not detract from the important role they played in the early days of canonical quantization: the step was crucial in the development of the theory. Moreover, it seems that Bryce DeWitt, and other important figures, had not given up on the parameter approach, and believed that it was important for several reasons. Expanding slightly beyond our 1956 cutoff for a moment, in a report from an exploratory workshop on quantum gravity[53] held in Copenhagen June–July, 1957, Bryce DeWitt too was still firmly convinced that Weiss's parameter formalism (mentioned and named such by DeWitt) had much to offer canonical quantization, as well as other approaches and more general questions of principle.

DeWitt saw amongst the advantages (see fig. 6.5):

1. "The parameters serve to label to points in the underlying space-time manifold once and for all. They are not subject to transformation. The so-called 'co-ordinates' have then a much greater functional similarity to the metric components than in the ordinary formalism and, like the metric components, can be imposed in an arbitrary way on the underlying manifold (subject only to secondary constraints)."

2. "The change of viewpoint achieved by the use of parameters permits one to perform operations which would otherwise be impossible. In particular, the extremely important but yet to be completed task of transforming the secondary constraints, like the primary constraints, into pure momenta, can be accomplished by much simpler canonical transformations than in the ordinary formalism." In a later remark, DeWitt notes that, in particular, Weiss's formalism allowed for the problems of general covariance to be prised apart from the problem of constraints.

(request form, dated September 4th, 1963). It isn't clear what, if anything came from this, and it seems that much of Schild's work (and that of his center) was occupied with the conformal structure of general relativity.

[52] The first construction of the Hamiltonian *without* the use of parameters was carried out (in published form) by Robert Penfield [Penfield (1951)], which was also the topic of his PhD thesis.

[53] Following the 1957 Chapel Hill conference, since known as GR1. This later, closed workshop was attended by DeWitt, Stanley Deser, Oskar Klein, Bertel Laurent, Charles Misner, and Christian Møller. The hand-written draft was provided by Cécile DeWitt, from her private archive.

3. "The necessary transformations are clearly seen to have *functional* character as well as a property of *mixing up* the coordinates with the metric components in an intimate way. The subsequent extraction of the 'true observables' leaves no doubt that the resulting theory has an essentially nonlocal character which is unique among field theories. The use of parameters allows this nonlocality to be studied in a thoroughly explicit fashion."

DeWitt goes on to note that there was unanimous agreement amongst the group that there was no real winner between the Feynman quantization program and the

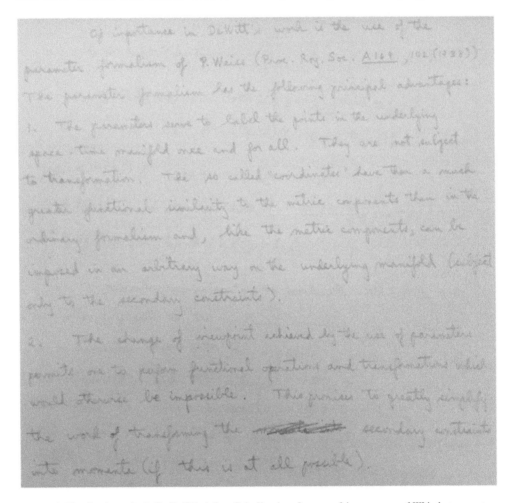

Fig. 6.5 *Handwritten draft (in DeWitt's hand) indicating the central importance of Weiss's parameter approach in the treatment of constraints and in making sense of deep conceptual problems of the theory. Taken from a draft report of an exploratory research session on quantum gravity, held in June 1957 in Copenhagen. Image source: CDWA.*

canonical program, with adherents of the former (presumably Laurent, Misner, and Deser) apparently very impressed by "how far it actually turned out to be possible to push the canonical formalism and also being shown explicitly how easy it is to pass from one to the other". Again, notes DeWitt, the parameter formalism "facilitated the demonstrations". Hence, the formalism of Weiss provided a transparency in the canonical transformations, permitting difficult manipulations having to do with the constraints. The transparency extended to communication between the participants, providing "an aid in understanding one another."

This discussion I hope clearly solidifies the pivotal role played by the parameter formalism in the early stages of the canonical approach to quantum gravity. Though in a convoluted way, it essentially allowed for the jettisoning of historical links to the Heisenberg–Pauli theory, which infused the initial phases of research, paving the way for the study of canonical quantum gravity in its own right.

6.6 Conclusion

Although it commenced relatively early, the canonical approach was slow in its subsequent development. This had two sources: (1) it required the introduction of tools and concepts from *outside* of quantum gravity proper (namely, the constraint machinery and the parameter formalism); (2) by its very nature, it is highly rigorous in a conceptual sense, demanding lots of groundwork to be established, in terms of the structure of physical observables, before the actual issue of quantization can even be considered. Work was further complicated by the fact that these two sources of difficulty happened to be entangled. For example, along the latter lines, to set up the theory in Hamiltonian form, enabling quantization, one must have nailed down the structure of the phase space, the inner product on that space, and the notion of an observable (all complicated by the constraints). This essentially had the effect of pausing the approach until the very end of the 1940s. However, it would be a mistake to view this as an absence of progress during those slow years. Ironically, the elements that needed to be filled in in the meantime turned out to be associated with what we would now think of as *manifestly covariant* approaches, especially relating to advances made by Paul Weiss, in particular the parameter formalism. Modern *canonical* quantum gravity was in no small way, then, a byproduct of the search for a manifestly covariant field theory.

. . . .

REFERENCES

Bergmann, P. G. (1992) Quantization of the Gravitational Field, 1930–1988. In J. Eisenstaedt and A. J. Kox (eds.), *Studies in the History of General Relativity* (pp. 364–6). Boston: Birkhaüser.
Bergmann, P. G. (1955) Fifty Years of Relativity. *Science* **123**: 486–94.

Bergmann, P. G. (1948) Unified Field Theory with Fifteen Field Variables. *Annals of Mathematics* **49**(1): 255–64.

Bergmann, P. G., and J.H.M. Brunings (1949) Non-Linear Field Theories II. Canonical Equations and Quantization. *Reviews of Modern Physics* **21**: 480–7.

Bergmann, P. G. and A. Komar (1980) The Phase Space Formulation of General Relativity and Approaches Toward its Canonical Quantization. In A. Held (ed.), *General Relativity and Gravitation: One Hundred Years After the Birth of Albert Einstein, vol. 1* (pp. 227–54). New York: Plenum Press.

Bergmann, P. G. and A. Komar (1962) Status Report on the Quantization of the Gravitational Field. In *Recent Developments in General Relativity* (pp. 31–46). Oxford: Pergamon Press.

Bondi, H. (1990) *Science, Churchill and Me*. Oxford: Pergamon.

Born, M. (1975) *My Life: Recollections of a Nobel Laureate*. New York: Charles Scribner's Sons.

Chang, T. S. (1949) Relativistic Field Theories. *Physical Review* **75**: 967–71.

Cooper, R. M. ed. (1992) *Refugee Scholars: Conversations with Tess Simpson*. Nottingham: Moorland Publishing.

DeWitt, B. S. (1967) Quantum Theory of Gravity: I. The Canonical Theory. *Phys. Rev.* **160**: 1113–48.

Dirac, P. (1958a) The Theory of Gravitation in Hamiltonian Form. *Proceedings of the Royal Society of London* **A246**: 333–43.

Dirac, P. (1958b) Fixation of Coordinates in the Hamiltonian Theory of Gravitation. *Physical Review* **114**: 924–30.

Dirac, P. A. M. (1950a) The Hamiltonian Form of Field Dynamics. *Canadian Journal of Mathematics* **3**: 1–23.

Dirac, P. A. M. (1950b) Generalized Hamiltonian Dynamics. *Canadian Journal of Mathematics* **3**: 129–48.

Dirac, P. A. M. (1933) The Lagrangian in Quantum Mechanics. *Phys.Z.Sowjetunion* **3**: 64–72.

Fuchs, K. (1939) On the Invariance of Quantized Field Equations. *Proc. Roy. Soc. Edin.* **A59**: 109–21.

Goldberg, J. (2005) Syracuse: 1949–1952. In A J. Kox and J. Eisenstaedt (eds.), *The Universe Of General Relativity* (pp. 359–73). Einstein Studies Volume 5. Boston: Birkhäuser.

Gsponer, A. and J.-P. Hurni (2004) Lanczos's Functional Theory of Electrodynamics: A Commentary on Lanczos's PhD Dissertation. In W. R. Davis et al. (eds.), *Cornelius Lanczos. Collected Published Papers With Commentaries, Vol.1* (pp. 2–23). Raleigh: North Carolina State University.

Heisenberg, W. and W. Pauli (1929a) Zur Quantendynamik Der Wellenfelder. *Zeitschrift Für Physik* **56**: 1–61.

Heisenberg, W. and W. Pauli (1929b) Zur Quantentheorie Der Wellenfelder. Band II. *Zeitschrift Für Physik* **59**: 168–90.

Higgs, P. W. (1958) Integration of Secondary Constraints in Quantized General Relativity. *Physical Review Letters* **1**: 373–4.

Jauch, J. and F. Rohrlich (1955) *The Theory of Photons and Electrons*. London: Palgrave Macmillan.

Kellerman, E. W. (2007) *A Physicist's Labour in War and Peace: Memoirs 1933–1999*. Hertford: M-Y Books.

Kohn, R. (2011) Nazi Persecution: Britain's Rescue of Academic Refugees. *European Review* 19(2), 255–83.

Laucht, C. (2012) *Elemental Germans: Klaus Fuchs, Rudolf Peierls and the Making of British Nuclear Culture 1939–59*. Boston: Addison-Wesley.

Milton, K. A. (2007) Julian Schwinger: From Nuclear Physics and Quantum Electrodynamics to Source Theory and Beyond. *Physics in Perspective* 9: 70–114.

Nossum, R. (2012) Emigration of Mathematicians from Outside German-Speaking Academia 1933–1963, Supported by the Society for the Protection of Science and Learning. *Historia Mathematica* 39: 84–104.

Penfield, R. (1951) Hamiltonians Without Parameterization. *Physical Review* 84: 737–43.

Pirani, F. A. E. and A. Schild (1950) On the Quantization of the Gravitational Field Equations. Physical Review, 79: 986–91.

Rickles, D. and A. Blum (2015) Paul Weiss and the Genesis of Canonical Quantization. *The European Physical Journal H* 40(4–5): 469–87.

Rosenfeld, L. (1932) La théorie Quantique des Champs. *Annales de l'I. H. P.* 1: 25–91.

Rosenfeld, L. (1930a) On the Quantization of Wave Fields. *The European Physical Journal H* 42(1): 63–94.

Rosenfeld, L. (1930b) Zur Quantelung der Wellenfelder. *Annalen der Physik* 397(1): 113–52.

Rutherford, E. (1936) A Society for the Protection of Science and Learning. *British Medical Journal* 1: 607.

Salisbury D. (2009) Léon Rosenfeld and the Challenge of the Vanishing Momentum in Quantum Electrodynamics. Studies in History and Philosophy of Modern Physics 40: 363–73.

Salisbury, D. C. (2008) Rosenfeld, Bergmann, Dirac and the Invention of Constrained Hamiltonian Dynamics. In H. Kleinert and R. T. Jantzen (eds.), *The Eleventh Marcel Grossmann Meeting* (pp. 2467–9). Singapore: World Scientific.

Schweber, S. (1994) *QED and the Men Who made It*. Princeton: Princeton University Press.

Stachel, J. (2014) The Hole Argument and Some Physical and Philosophical Implications. *Living Rev. Relativity* 17: http://www.livingreviews.org/lrr- 2014–1.

Stachel, J. (1992) The Cauchy Problem in General Relativity—The Early Years. In J. Eisenstaedt and A. J. Kox (eds.), *Studies in the History of General Relativity* (pp. 407–18). Boston: Birkhäuser.

Stachel, J. (1982) Globally Stationary but Locally Static Space-times: A Gravitational Analog of the Aharonov–Bohm Effect. *Physical Review D* 26(6): 1281–90.

Utiyama, R. (1948) On the Interaction of Mesons with the Gravitational Field. II. *Progress of Theoretical Physics* 3(1): 14–25.

Utiyama, R. (1947) On the Interaction of Mesons with the Gravitational Field. I. *Progress of Theoretical Physics* 2(2): 38–62.

Weiss, P. (1946) Review of Tonnelat, Marie-Antoinette "La particule de spin 2 et la loi de gravitation d'Einstein dans le cas de présence de matière" *C. R. Acad. Sci. Paris* **218**, (1944). 305–8. *Mathematical Reviews: MR0013051.*

Weiss, P. (1941) On Some Applications of Quaternions to Restricted Relativity and Classical Radiation Theory. *Proc. Roy. Irish. Acad.* **46**: 129–68.

Weiss, P. (1938a) On the Hamilton–Jacobi Theory and Quantization of a Dynamical Continuum. *Proceedings of the Royal Society of London* **A169**(936): 102–19.

Weiss, P. (1938b) On the Hamilton–Jacobi Theory and Quantization of Generalized Electrodynamics. *Proceedings of the Royal Society of London* **A169**(936): 119–33.

Weiss, P. (1936) On the Quantization of a Theory Arising from a Variational Principle for Multiple Integrals with Application to Born's Electrodynamics. *Proceedings of the Royal Society of London* **A156**(887): 192–220.

Zimmerman, D. (2006) The Society for the Protection of Science and Learning and the Politicization of British Science in the 1930s. *Minerva* **44**: 25–45.

7

Quantum Gravity as a Resource

In his review of the state of relativity in 1955, for the 50th anniversary of special relativity, Peter Bergmann wrote:

> The general theory of relativity for many years appeared to have its applications principally in cosmology and cosmogony, themselves fields as yet in a highly unsettled state. But recently general relativity is also being considered in connection with questions affecting microphysics. Its relationship to quantum theory is still highly problematical. *But the quantum theory of the atomic nucleus and of elementary particles is not in such a satisfactory state that it can afford to disregard possible assistance from whatever source.* General relativity offers us a new approach to the ultimate properties of space and time, and these may bear on the physics of the very small as much as we know they do on the physics of the very large. Many of our present efforts are still in a very early stage. The very fact that interest in general relativity has recently increased throughout the world is indicative of the fact that its implications have not yet been fully worked out and exploited for our understanding of the physical universe as an organic whole.
>
> [Bergmann (1955), p. 494; my emphasis]

Einstein was one of the first to suggest that general relativity might have something novel to say about the special problems facing the theory of elementary particles [Einstein (1919)]. He kept this attitude, of course, though not along the lines Bergmann was pursuing. Not so very long before Bergmann wrote those words, Einstein wrote, in the foreword to the former's textbook on general relativity:

> It is quite possible...that some of the results of the general theory of relativity, such as the general covariance of the laws of nature and their nonlinearity, may help overcome the difficulties encountered at present in the theory of atomic and nuclear processes.
>
> (Einstein in [Bergmann (1942)], p. iii)

Given, as we have seen, that most of the earliest work connected with quantum gravity had an eye on the more general problems facing field theory (especially quantum electrodynamics[1]), it is no surprise that a sudden increase in interest in quantum gravity

---

[1] A central theme was the existence of a variety of divergences, especially the problems generated by the self-energy of elementary particles. A good historico-philosophical treatment of the reactions to these divergences is [Rüger (1992)].

Covered with Deep Mist: The Development of Quantum Gravity (1916–1956). Dean Rickles, Oxford University Press (2020). © Dean Rickles.
DOI: 10.1093/oso/9780199602957.001.0001

from the elementary particles community occurred when it was realized, initially, it seems, by Lev Landau in 1955, that a gravitational field may provide a 'natural cutoff' for high frequency components of quantum fields, much as Einstein guessed.[2]

There are two ways (not necessarily distinct) in which gravitation (and quantum gravity in particular) was expected to play a role in the theory of elementary particles: (1) in eliminating the divergences resulting from singularities in the particle propagators— the problems with the ultraviolet divergences had particle physicists flummoxed; (2) in contributing something to the structure of the elementary particles themselves (e.g. accounting for their stability and properties). Bergmann himself adopted a position along the lines of (1) in his earliest work on quantum gravity, arguing that the Einstein– Infeld–Hoffmann equations of motion (which showed that the motions of sources of the gravitational field were entirely determined by the field equations) might offer a way of completely sidestepping the divergent quantities so rife in theories involving point particles [Bergmann (1949)]. Hence, albeit in a different way to the cutoff approach, Bergmann proposed that gravity could function as a kind of regulator, because of its unique properties (related to the non-linearities) which allowed one to determine particle motions without divergences.[3] This foundation of divergence-avoidance was, as we saw in the previous chapter, the starting point of Bergmann's canonical approach to quantum gravity, which was a bid to realize his initial regulatory aim.

Though we can find early suggestions of a fundamental length leading to a taming of divergences, e.g. in the work of Heisenberg, Flint, and others (as we will see), the notion of a *gravitationally* induced cutoff took some time to form, and had to wait for a more thorough understanding of the non-linear aspects of the theory, which had remained largely buried under the more tractable linear approximation.[4] But this work can be seen as emerging from the well-worn path involving the usage of these other cutoffs (minimum lengths) to tame the infinite behavior of field theories: if gravity suggests a minimum length, then previous work on such ideas can be transferred across from context to context, and so the previous cutoff ideas at least provide a conceptual background. Hence, many of the ideas presented in this chapter were not written with quantum gravity in mind; this was a connection only made later once the problem of quantum gravity itself had undergone several changes—not least the idea that quantum gravity involves some notion of 'quantum spacetime.'

---

[2] Recall from Chapter 5, however, that Frederik Belinfante also presented a theory (using Nathan Rosen's 'bimetric' idea) in that same year that also provided momentum cutoffs, as a result of 'holes' in the spacetime which would mean that fields around them could not probe within.

[3] We will further explore this apparently unique 'self-sufficiency' of general relativity in the next chapter, in the context of John Wheeler's work. Wheeler would attempt to escape problems with Bergmann's line of attack, namely that general relativity represented particles as structureless field singularities (i.e. infinities with no masses or charges derived from the theory), by pursuing something more along the lines of strategy (2).

[4] What the non-linear analysis eventually revealed was that gravity's divergences were more complex than those of electrodynamics, since they involved a shifting of the light cone so that light cone singularities would be 'smoothed out' by fluctuations in a quantum theory of gravity (a nice result of the quantum geometry inherent in quantization of the metric): [Klein (1955), Landau (1955), Deser (1957)]. We will discuss this further in the remainder of this section.

7.1   Gravity as a Regulator

The potential utility of introducing gravity into elementary particle physics, so as to eliminate divergencies, spurred on work on quantum gravity enormously. The most obvious strategy is to impose a cutoff. There were many suggestions that gravity might act in this way, as some kind of 'regulator.' The divergences in question were those of QED, and meson theories, which were still, pre-WWII, a somewhat mathematically murky territory (quantum field theory still is, of course). It was known, by 1955, that the pion (π-meson) interaction behaves quite differently to the electrodynamical case, so that the same renormalization techniques failed to remove infinities—presumably a component in the realization that gravity too might not be much like electromagnetism. The general problem concerned the transitions between quantum states, during which time (a very short time, determined by the uncertainty relations) energy conservation is violated. In more detail, the electron self-energy involves the electromagnetic interaction of an 'undressed' (or 'bare') charge with virtual photons of all possible momenta from 0 to infinitely large, leading to divergent integrals. A fairly obvious idea is to impose a cutoff to prevent momenta reaching arbitrarily high values, thus generating a finite electron mass. The great hope for introducing gravitation into elementary particle physics was that it would provide a physically well-motivated way to terminate the wavelengths before they have a chance to reach the problematic high-energy (ultraviolet) wavelengths. Landau [Landau (1955)] appears to be the first to have suggested this idea based on his more general desire to achieve something *beyond* quantum field theory, since Landau had a deep distrust of quantum field theory. More specifically, Landau's field theoretic investigations revealed that even given renormalization, the short-distance behavior continued to generate infinities ('Landau ghosts') on account of the approach to the infinite bare masses and couplings (past the 'screening' effect that Landau had himself discovered).[5] Pauli also makes several comments to this effect,[6] including the following remarks in a letter addressed to Abrikosov, Khalatnikov, and Pomeranchuk (in which he assigns priority of the idea to Landau):

> I was very interested in *Landau's* remarks on the possibility of a connection of the cut-off moment of quantum electrodynamics with *gravitational* interaction (his article "on quantum theory of fields" in the Bohr-festival volume). It appeals to me, that the situation regarding divergences would be fundamentally changed, as soon as the light-cone itself is not any longer a *c*-number equation. Then every given direction in space-time would have some "probability to be on the light-cone", which would be different from zero for a small but finite domain of directions. I doubt, however, that the *conventional* quantization of the $g_{\mu\nu}$-field is consistent under this circumstance.
>
> Zürich, August 15th, 1955; in [von Meyenn (2001), p. 329]

---

[5] Later work would reveal that such Landau ghosts could be dealt with in the context of renormalization group theory, but at the time it appeared as though quantum field theory was suffering from an incurable illness—for more on this, see Brown (ed.), *Renormalization: From Lorentz to Landau (and Beyond)* (Springer, 1993).

[6] See also, in von Meyenn (2001), Peierls' letter to Pauli (dated May 9th, 1955: pp. 228–9) Pauli's letter to van Hove (dated May 11th, 1955: 230–1); Heisenberg's letter to Pauli (dated May 11th, 1955: pp. 234–5).

Likewise, in a letter to Källen, dated April 24th, 1955, he wrote that it was "grati-fying to see that [Landau] hits on the same conjecture as me, concerning quantum electrodynamics—(His idea/attempt to relate the cutoff scale to gravity is amusing)" [von Meyenn (2001), pp. 207–208].

Given his obvious expertise in both general relativity and quantum field theory, one wonders why Pauli didn't do more work in the area of quantum gravity—as we saw in earlier chapters, he was not so impressed by his own important results in the field. Though as he suggests, it is clear that novel (i.e. "unconventional") approaches are probably required (a far cry from his superficially optimistic remarks in the 1929 paper with Heisenberg in which he claims that the general relativistic case would be much the same as electromagnetism)—no doubt this refers to the fact that conventional quantiza-tions require exactly the kind of causal structure put under pressure by such fluctuating light cone phenomena. He was, nonetheless, certainly preoccupied with general relativity towards the end of his life—a 'later-life' proclivity to study the unification of quantum and gravity which was shared by Eddington and Schrödinger, of course.

Developing his earlier remarks to Abrikosov et al. (in the discussion after Klein's talk at the Berne conference), Pauli mentions Landau's argument that for large cutoff momentum P, the gravitational coupling between a pair of electrons is of the same magnitude as the Coulomb forces. He notes that Landau's relation $GP^2 \sim 1$ is the same as Klein's $P \sim \hbar/l_0$ (where $l_0 = \sqrt{G\hbar c}$). He writes:

> [T]he connection... of the mathematical limitation of quantum electrodynamics with gravitation, pointed out by LANDAU and KLEIN, seems to me to hint at the indetermi-nacy in space-time of the light-cone, which is governed by probability laws in a quantized field theory, invariant with respect to the wider group of general relativity. It is possible that this new situation so different from quantized theories, invariant with respect to the LORENTZ group only, may help to overcome the divergence difficulties which are so intimately connected with a *c*-number equation for the light-cone in the latter theories (Pauli's comments after Klein's talk, in [Mercier and Kervaire (1956), p. 69]).

Pauli's thoughts were borne out in one way (the light cone structure is affected as he suspects); however, it leaves a challenge behind in dealing with new divergences. As Bryce DeWitt pointed out, in Louis Witten's important collection, *Gravitation: An Introduction to Current Research*, from 1962 (less than seven years after Pauli's remarks),

> [I]t must constantly be borne in mind that the 'bad' divergences of quantum gravidynam-ics are of an essentially different kind from those of other field theories. They are direct consequences of the fact that the light cone itself gets shifted by the non-linearities of the theory. But the light-cone shift is precisely what gives the theory its unique interest, and a special effort should be made to separate the divergences which it generates from other divergences.
>
> [DeWitt (1962), p. 374].

DeWitt sought to revisit Rosenfeld's work on the computation of gravitational self-energies. DeWitt would also revisit this idea of Landau's that gravity might act as a natural regulator [DeWitt (1964)]. Though Landau didn't explicitly mention the Planck scale (he placed the location of the cutoff much higher), Pauli clearly appeared to think that Landau had quantum gravitational effects in mind (or that he *ought* to have). It is clear that if there is a 'fundamental length,' below which ordinary quantum field theoretic processes cannot operate, then one has what Landau sought. DeWitt was able to confirm that (at lowest order of perturbation) when gravity is included, the self-energies of charged particles (and the gravitons themselves) remain finite (though often very large). Here again, as in earlier parts of this book having to do with issues of measurability, we see a link between minimum length scales and the notion of *limits* and *domains of applicability* of theory and concepts. The question of whether there is a physical cutoff naturally has theoretical links with programs concerning the existence of a fundamental length, and discrete space(time) in general.[7]

Landau notes that the energies at which the divergence difficulties manifest themselves are very large. In such circumstances the gravitational interaction can (as a result of the relation between energy and mass) be in excess of electromagnetic interactions. Given this, the closed system idealization in which gravitational interactions are turned off is no longer appropriate.

There are interesting hints of the direction that quantum gravity research would follow in the future, namely that it must involve some kind of revision of spacetime:

> We have seen in the case of electrodynamics that a point interaction can lead to the absence of any interaction, even if its intensity increases without limit. The possibility cannot be excluded that this is a general property of point interactions. In this case, the construction of meson theories is possible only by abandoning the point interaction, that is, by renouncing essentially all the methods at present existing. The great difficulties which arise in a physical "smoothing out" of particles, as opposed to a purely formal "smoothing out". [...]
>
> We emphasise that the physical "smoothing out" related to the introduction of some "fundamental length" of the order 10^{-13}cm must inevitably have some effect on electrodynamics, although at these energies no logical difficulties arise.
>
> [Landau (1955), pp. 67–8]

Klein's contributions to the problem of quantum gravity can be found in a pair of festschrift volumes, aptly those of Bohr and Einstein from 1955—the Bohr volume consisted of papers prepared for his 70th Birthday (October 7th, 1955). He is very much still in the mindset of analogies and connections between electrodynamics and gravitation. Though he relates his approach to the standard divergence problems of quantum field theories, Klein has a slightly different agenda. He speculates (p. 99) that the amelioration of the divergence problems would be a possible by-product of a "generalized quantum-relativity theory." He describes his approach as a contribution

---

[7] As DeWitt puts it: "The dimension 10^{-32}*cm* constitutes a fundamental limit on the smallness of allowable measurement domains. Below this limit it is impossible to interpret the results of measurement in terms of properties or states characterizing individual systems under observation" [DeWitt (1962), p. 373].

to "an intimate alliance of the two fundamental viewpoints of present physics, that of complementarity and that of relativity" [Klcin (1955), p. 117]. We might think of this as a 'principle theory' approach in which the two broadest and most solid characteristics of the ingredient theories are brought together:

> [W]e shall tentatively take the point of view that general relativity is fundamental for the formulation of the laws of quantum field theory and that the demand of an adequate formulation of other invariance claims, e.g. that of gauge invariance, should be regarded as an indication of the need for a natural generalization of the relativity postulate.
>
> [Klein (1955), p. 9]

Klein questions the standard folklore that general relativity and quantum theory must remain separate bedfellows on account of their very different domains of applicability:

> Now, it is very usual to regard the point of view of general relativity as insignificant in quantum theory because the direct effects of gravitation in ordinary atomic phenomena are very small. This, however, may easily be the same kind of fallacy, which it would have been to regard the electron spin as unimportant for the formulation of the laws of chemical binding, because the direct interaction between spin magnetic moments is, in general, negligible compared with chemical binding energies.
>
> [Klein (1955), p. 98]

More specifically, Klein adopts a group-theoretic approach to the structure of the two ingredient theories, and so quantum gravity, exploring the idea that isolating the appropriate invariances could unlock the problem of their unification:

> The operators to be used in quantum field theory should have a simple connection to a transformation group (so far insufficiently known) which contains the general coordinate transformations in spacetime as a subgroup. The quantum conditions ought to characterize the group in question. In trying to develop a theory according to such a programme it should be kept in mind that the direct quantization according to the ordinary scheme of quantum mechanics of the Einstein equation meets with difficulties of the same type, but very much enhanced, as those met with in the quantization of the Maxwell equations. Also from this point of view it would seem preferable to start with the quantum conditions expressing group properties instead of starting with a Lagrangian density. This would probably make the theory still more symbolic and remote from direct observation than ordinary quantum field theory.
>
> [Klein (1955), pp. 98–99]

In a paper written just a little later, Klein is fully on board with the Landau idea:

> It is perhaps not unreasonable that the rigorous consideration of gravitational and perhaps other similar non-linear effects would do away with the remaining divergencies of electron theory.
>
> [Klein (1956), p. 61]

This viewpoint stands in marked contrast to the standard perturbative way of conceiving of quantum field theories in which the non-linearities are discarded at the outset. Now the non-linearities are absolutely vital.

As already mentioned, but related to importance of the non-linearities of Einstein's equations, however, was Peter Bergmann's method of utilising the fact that the gravitational field equations determined particle trajectories free of any notions of divergences. He believed this would follow from the analysis of Einstein, Hoffmann, and Infeld, according to which the assumption of geodesy for a free particle's motion was redundant, since it already could be seen to follow (by a method of successive approximation) from the field equations alone. The idea was that the vacuum field equations determine the motions of such particles, with interactions between them determined by non-linearities in the equations; because these motions are determined without incurring self-energy issues, it was argued that there might be some relevance to the problem of quantum gravity.[8]

7.2 Minimal Length Scenarios

Developing the cutoff idea, and the idea that there might be a minimal (fundamental) length, leads one quite naturally into the idea that space and time might not be continuous, but better modeled instead by a discrete lattice or similar structure. In the early days the cutoff was implemented in the kinematical structure, rather than having it emerge dynamically—whether the cutoff (discreteness) is fundamental or not is a different issue. This was suggested by several people, usually independently of matters gravitational, but the ideas nonetheless have a relevance to quantum gravity. Ambarzumian and Iwanenko [Ambarzumian and Iwanenko (1930)] argued for the introduction of a spatial lattice structure for physical space as a way of eliminating the infinite divergences from the self-energy of the electron. The basic idea was that the existence of a minimal length would imply a maximal frequency (ibid., p. 567). Alfred Schild investigated the properties

---

[8] I might also note here that ultimately string theory emerged from the divergence problems facing quantum field theories of fields other than the electromagnetic field (particularly the strong interaction). In particular, since the perturbative approach breaks down when the coupling constant determining interactions strengths is high (as in strong interaction physics), alternative approaches were sought in the late 1950s and throughout the 1960s. One of the more popular of these approaches combined Heisenberg's S-matrix theory with dispersion theory. The S-matrix is a tool to encode all possible collision processes. Heisenberg suggested that one take this to embody what was relevant about the physics of collision processes. In particular, all that was observable were the inputs and outputs of collision processes, observed when the particles are far enough apart in spacetime to be non-interacting, or free. This black box approach to physics was very much inspired by the Copenhagen philosophy. The dispersion relation approach to physics tried to construct physical theories on the basis of a few central physical axioms, such as unitarity (conservation of probabilities), Lorentz invariance, and causality (effects can't precede causes). These two approaches were combined, by Geoffrey Chew amongst others, so that the focus was on the analytic properties of the S-matrix. One model for the S-matrix, incorporating some other principles thought to be involved in strong interaction physics, was the Veneziano model. This used the Euler beta function to encode the various desirable properties of the S-matrix. The model was found to be generated by a dynamical theory of strings. (See [Cushing (2005)] for a detailed historico-philosophical account of the early development of string theory, or [Rickles (2014)] for a more recent account.)

of such a discrete lattice in order to see if it would break essential symmetries. In particular, he was responding to the objection that discrete theories would violate Lorentz invariance, which could manifest experimentally resulting in inconsistencies with known results.[9] He wasn't able to devise a model to preserve all such symmetries, but enough to provide a plausible candidate for a background for a physical theory. Here again we find constraints operating on the various approaches to provide some sort of mechanism for the rejection and selection of theories or approaches—in this case the Lorentz symmetry of the classical theory.

Another discrete approach, of David van Dantzig ([van Dantzig (1938), van Dantzig (1955)]), was motivated by a combination of general covariance (as expressed in Einstein's 'point-coincidence' argument) and the definition of observability in such a theory. He argued that in a generally covariant theory the observable things will be coincidences: events (not shuffled by diffeomorphisms). Van Dantzig argues that in order to not introduce unmeasurable structure into the interpretation or formulation of one's theory, one should dispense with the existence of a four-dimensional continuum, in favour of a discrete manifold of events. Peter Bergmann describes one such approach as one of "constructing 'spaces' that have certain topological properties similar to those of point spaces in the large but do not possess 'points' as elementary constituents" (Bergmann, following a talk of Wigner's: [Wigner (1955), p. 226]. The general approach lives on in several of the current approaches, including causal set theory and dynamical triangulations—though the conceptual basis (especially *observability through invariance* is absent from the latter case). Bergmann's comments also draw attention to the

---

[9] This same objection to discrete models surfaces again in present-day discussions of discrete space in quantum gravity (a fairly generic prediction of several approaches), especially in the context of loop quantum gravity which directly predicts (at least at the kinematic level) geometrical operators with a discrete spectrum. Given that there is supposed to be a fundamental length (namely the Planck length, and corresponding fundamental times and masses) in these approaches, it makes sense to ask if observers in relative motion will agree on this length: why no Lorentz–FitzGerald contraction for boosted observers, rendering the notion of a minimum length incoherent? Why is a length measurement for the minimum length case not subject to the usual frame dependence? According to Carlo Rovelli (one of the primary architects of loop quantum gravity that itself appears to face the problem) and Simone Speziale, quantum mechanics is the key to avoiding this 'discreteness/invariance' conflict: "the minimal length (more precisely, minimal area) does not appear as a fixed property of geometry, but rather as the minimal (nonzero) eigenvalue of a quantum observable [so that the] boosted observer can see the same observable spectrum, with the same minimal area. What changes continuously in the boost transformation is not the value of the minimal length: it is the probability distribution of seeing one or the other of the discrete eigenvalues of the area" [Rovelli and Speziale (2003), p. 064019]. They elaborate as follows, linking directly with issues of quantum spacetime: "The geometry of space comes from a quantum field, the quantum gravitational field. Therefore the observable properties of the geometry, such as, in particular, a length, or an area, are observable properties of a *quantum physical system*. A measurement of a length is therefore a measurement in the quantum mechanical sense. Generically, quantum theory does not predict an observable value: it predicts a probability distribution of possible observable values. Given a surface moving in spacetime, the two measurements of its area performed by two observers O and O' boosted with respect to one another are two entirely distinct quantum measurements. Correspondingly, in the theory there are two distinct operators A and A', associated to these two measurements. Now, our main point is the technical observation that A and A' do not commute: $[A, A'] \neq 0$. This is because A and A' depend on the gravitational field on two distinct 2d surfaces in spacetime...and a field operator does not commute with itself at different times". [Hagar (2014), §8.4.4] contains a useful, detailed discussion of this problem—though we must point out that strictly speaking since all of this discussion is at the level of the kinematical (non-physical) Hilbert space, much of this is moot (a point not addressed by Hagar).

'emergence' of continuous spacetime from a discrete structure (a problem at the root of causal set theory, though one in which progress has been made: see, e.g., [Major, Rideout, and Surya (2007)]).

Focusing in on the key idea in this section, that a discretization of space might go some way towards resolving the issues of short-distance problems issuing from field theories, is of course a rather natural and almost obvious idea. The initial developments were, however, not linked to gravitational physics, but many of the results originally couched in non-gravitational work were carried over into the former. It was eventually realized, for example, that gravitation itself might be able to provide a *physical foundation* for discrete space and that given the dual nature of the metric field, quantum gravity should lead one to expect a discrete spacetime. Given this, the various results pursued independently of the quantum gravity problem (violation of Lorentz invariance and so on), become of direct relevance.

There are three motivations underlying the notion of discrete space(time) in the early work:[10]

- An *ad hoc* discretization using a lattice structure—often used as an approximation, for which the continuum limit would be taken later on.
- An *operational* discretization using fundamental measurement limitations imposed by the uncertainty relations.
- A discretization using a *physical cutoff* imposed (e.g. by gravity).

The first steps for which a field theory over a discrete space—along the first motivation (in the context of field theory)—were, as mentioned, taken by Ambarzumian and Iwanenko (ibid.) in 1930. This paper also includes a discussion of whether time would need to be quantized, along with space, as a corollary. The argument is simple: a minimum length implies a maximum frequency which implies a minimum time interval $\Delta t = \frac{1}{c}\Delta x$.[11] They are concerned solely with the infinite self-energies that arise from the point-like nature of electrons. As they note, there seem to be two broad ways out of the predicament: give the electrons a finite size, or else restrict the spatial resolution to which one can probe (placing a limit on the validity of the theory—motivation two above). Since the former was thought to be not possible in quantum mechanics, they

---

[10] As Rüger makes clear [Rüger (1992), p. 317], too, there was earlier on, prior to the 1930s, a sense that the infinities were simply a hangover from the classical theory that if cured first (classically) would not reassert themselves at the quantum level. This was not the case, and it became clear that there existed specifically quantum divergences. As we saw in Chapter 5, the early study of quantum gravity was a key part of this project.

[11] In another paper from the following year, Iwanenko reiterates that the value $\lambda \sim \frac{h}{mc}$ also determines a 'chronon': "Dieser Wert hat schon als kleinste definierbare Entfernung zu gelten und nicht der Elektronen-radins. Mit der kleinsten Entfernung hängt die kleinste Zeitspanne zusammen" [Iwanenko (1931), p. 623]. As Kragh and Carazza note, there were earlier speculations, with similar results, about time atoms from Pokrowski and Fürth [Kragh and Carazza (1994), pp. 457–8]. Indeed, they show that the 20s and 30s were positively teeming with discrete space, time, and spacetime proposals. However, many of them are detached from the central problems of field theories that concern us here.

opt for the latter strategy. They resolve this "problem of space" by introducing a cubic lattice with grid points separated by some constant factor, a, to be determined (such that ordinary quantum theory is recovered as $a \to 0$). Differential equations are then replaced by discrete, difference equations.

This was followed by Heisenberg,[12] Ruark, March, and several others, including, in England, Henry Flint. Flint was an interesting case, since he had his eye on the problem of unification of relativity and quantum mechanics in his work on fundamental length (via "ultimate measurements"—again, corresponding to the second motivation).[13] The Ambarzumian and Iwanenko paper was also directly cited by Schild, in his paper on discrete spacetime.

There were some other interesting attempts at "quantizing space" in the 30s. The most interesting is perhaps John Von Neumann's (unpublished) proposal from 1937.[14] Von Neumann distinguishes two kinds of singularity: the point-particle singularity and the infinite degree of freedom singularity (resulting from the infinite number of parameters needed to describe a field). In a letter to Rudolf Ortvay from 1938 he describes his model for discrete spacetime as follows:

(1) The x, y, z coordinates and the t are *non-commuting* operators.

(2) The order of magnitude of commutators is $\frac{h}{mc}$. (That is to say, this is the uncertainty associated with a simultaneous measurement of coordinates.)

(3) The whole structure has the Lorentz symmetry.

(4) Each of the x, y, z coordinates has a discrete spectrum: $\pm 1/2, \pm 3/2, \ldots$

(5) The spectrum of the time t is continuous, from $-\infty$ to $+\infty$.

(6) When (4) and (5) are combined with (3) this comes out:
Given four real numbers $\alpha, \beta, \gamma, \delta$, the spectrum of the operator $\alpha x + \beta y + \gamma z + \delta t$ is as follows:

   (a) If $\alpha^2 + \beta^2 + \gamma^2 - \delta^2 > 0$ then it is discrete: $\pm \epsilon/2, \pm 3\epsilon/2, \ldots$, where $\epsilon = \sqrt{\alpha^2 + \beta^2 + \gamma^2 - \delta^2}$.

   (b) If $\alpha^2 + \beta^2 + \gamma^2 - \delta^2 < 0$ (indeed even when $= 0$) then it is continuous from $-\infty$ to $+\infty$.

So this a "discrete" crystalline space with "continuous" time, which has not only spherical symmetry even though it is a "crystal"!), but is even invariant with respect to changes

---

[12] In fact, in the acknowledgements to their paper, Ambarzumian and Iwanenko refer to analogous work of Heisenberg that they had only just become aware of at the time of publication.

[13] In his notebook (from 1950) there is a section on "The Theory of Relativity and the Quantum Theory" in which he nails down his project: "The underlying theme of this work is the union which exists between the theory of relativity and the quantum theory and the purpose is to portray it by means of geometry and a theory of measurement" (Henry Flint Archives, B53: p. 1).

[14] The manuscript is entitled "Quantum Mechanics of Infinite Systems" (see [Redei (2005), pp. 21–2]).

of the reference system given by Lorentz transformations, and so shows the proper Lorentz–FitzGerald contraction phenomenon. (This is made possible, of course, by the non-commuting nature of the coordinates.) (von Neumann, letter to Rudolf Ortvay: March 17th, 1938 [Redei (2005), p. 22])

In a letter to Dirac (dated January 27th, 1934) he writes:

It should be perhaps desirable, to have operators X, Y, Z which gave discrete (point) spectra, in order to avoid the difficulties connected with the point electron (in electrodynamics).

([Redei (2005)], p. 21)

Dirac replied the following month (February 28th, 1934) pointing out that the model was not invariant under displacement of the origin of the reference frame defined by the X, Y, Z.

Quantization here is viewed, then, as a cutoff to prevent the ability to resolve to point-like distances. The problem with such accounts is that they are physically *ad hoc* (corresponding motivation one from above). Von Neumann did not pursue the idea further for this very reason: "because [he] considered it very artificial and arbitrary" [Redei (2005), p. 22].

Heisenberg was inspired primarily by the second motivation, though it mixed with the first, in order to tame the infinite self-energy of electrons.[15] His first thoughts about discretization can be found in a long letter to Bohr from March 1930 (translated into English in [Carazza and Kragh (1994)], along with a reconstruction of the logic of the argument it contains)—one wonders whether he was aware of Ambarzumian's and Iwanenko's work, which is remarkably similar (as mentioned, Ambarzumian and Iwanenko note, at the proofs stage of their paper, that they *were* aware of Heisenberg's attempt, though it is hard to discern whether their work was initially written without knowledge of this). The idea is also to divide space up into a cubic lattice, where the cells have volume $r_0^3 = (h/Mc)^3$. The length $\sqrt[3]{r_0}$ (the electron radius) was then the 'elementary length.' He called the world described by this theory "gitterwelt" ("lattice world"). The self-energy of an electron would be rendered finite in the gitterwelt—a point Heisenberg returned to in his paper "Die Selbstenergie des Elektrons" (submitted in August of that year). As Heisenberg also notes, in the given scheme differential equations would have to be replaced with difference equations.[16] A central problem, as Heisenberg saw it (and as would deter others from the discrete space idea) was that

---

[15] Interestingly, Heisenberg had already briefly considered the idea of letting spatial coordinates be non-commuting in 1930 in order to generate a minimum length from uncertainty relations. He put this idea to Rudolf Peierls asking for any suggestions, including any input from Pauli. Julius Wess [Wess (2001), p. 1] claims that Heisenberg relayed it to Peierls (his student), who relayed the idea to Pauli who relayed it to Oppenheimer (his student), who relayed it to Hartland Snyder (his student: see §7.3 below)! This occurs over a period of 15–16 years.

[16] Carazza and Kragh [Carazza and Kragh (1994)] argue that Heisenberg did not really endorse a discrete space at this stage, but rather used discreteness only at the level of derivatives with respect to spatial coordinates (which are indeed replaced by discrete, finite differences).

relativistic invariance was spoiled by any scheme that introduced a fundamental length—this assumption was progressively taken apart in papers from the late 1930s onwards.[17] (Heisenberg also pointed to difficulties in making the space isotropic; as well as with energy, momentum, and charge conservation: for these reasons he asked Bohr whether he thought the idea "completely mad"!) But beyond this breakdown of Lorentz invariance, the other target of Heisenberg's 1930 paper was to show that there are wider problems with field theory that go beyond the problem of infinite self-energy—this became part of a general program of getting clearer on the distinct kinds of divergences in physical theories.

In his 1938 paper "Über die in der Theorie der Elementarteilchen auftretende universelle Länge", Heisenberg explicitly ignores gravitational interactions "which hardly play a role in nuclear physics," focusing on $\hbar$ and c alone. This is part of a general to and fro with respect to the role of gravitation in elementary particle physics. However, when discussing the 'universal length' he does briefly return to the issue, though again to dismiss gravity's role in the fundamental length. As is standard, he considers the electromagnetic analogy, comparing the gravitational interaction of photons with the electrical interaction of electrons. But he notes a crucial disanalogy: introducing the gravitational constant (Heisenberg uses γ rather than G) together with $\hbar$ and c can be combined to generate the (Planck) length: $l = \sqrt{\hbar\gamma/c^3}$ (which Heisenberg computes to be 4×10^{-33} cm). However, given the vast distances separating these domains, Heisenberg points out that the problems associated with his r_0 (the electron radius) ought to be resolved first, as the most urgent task. In other words, there is a practical argument here for the *neglect* of issues having to do with quantum gravity.

The reason for this urgency was the difficulties faced by Fermi's theory of β decay, based on Pauli's neutrino hypothesis, which was found to suffer from divergences of an extreme (i.e. unrenormalizable in modern parlance) kind—involving the divergence of (Born approximation) cross sections as the energy of the incident particles went to infinity—so that the perturbation technique for treating interactions didn't give sensible answers.[18] Of course, we know that this problem was pointing to a limit with the then current quantum field theory. But Heisenberg, viewing Fermi's theory as a fundamental (and unified, in terms of weak and strong forces, with a single coupling constant) theory, took it to point to another source in which one could only resolve distances to certain

---

[17] Rosenfeld had referred to this lack of relativistic invariance as "a fatal flaw" of gitterwelt models [Rosenfeld (1932), p. 78]. Interestingly, Rosenfeld links this trouble with gravitation, noting that the same kind of difficulties arise "independently of any material field," pointing to his own computation of the self-energy of the field of gravitation generated by a photon, in which there are also infinite terms "due to a point singularity" (ibid, pp. 78–9).

[18] The is the famous four-fermion coupling G_F which was not properly understood until the electroweak theory was developed, and the machinery of gauge theory was applied, along with Yukawa's idea of mediation by a new kind of boson (the "U-quantum" or mesotron) which replaced the four-fermion term. The evidence for mesons came in 1937, when they were isolated in cosmic rays. It should perhaps also be said that the realization that all would not be plain sailing with respect to the other forces of nature shifted the focus onto the peculiarities of gravitation. See Chapters 3 and 4 of [Brown and Rechenberg (1996)] for a historical study of the Fermi-field theory.

distances, again close to his r_0.[19] In this case Heisenberg drew attention to the particle multiplicity ("explosionen") in cosmic ray showers in which many particles are created: the particle production would limit the resolution (so that r_0 represents a fundamental limit in this sense: physics becomes 'turbulent' at shorter lengths as the coupling blows up). Of course, there was a limit, but the limit was theoretical rather than practical: there was a layer of particle physics below that captured by Fermi's theory. The short-distance, strong interaction physics that followed this was a major impetus to quantum gravity physics since gravitational and strong interactions had similar non-linearities (due to the self-interacting nature of the forces)—though, of course, gravity is universal (couples to all sources of energy equally). Hence, a new analogy between these forces, and less so with electrodynamics, took hold.[20]

We have seen that Bronstein had already written on related issues in 1936, pointing out that there exist quantum measurement restrictions beyond the commutation relations in the case of gravitational measurements, since there cannot be bodies of arbitrarily large mass density (cf. [Gorelik (1994), p. 106]). Bronstein thought this called for a revision of spacetime concepts (as did Heisenberg, though for different reasons). Unfortunately, his untimely death means that we don't know how, or if, he intended to pursue this revision.

The third motivation was discussed, mostly informally, by Pauli, Klein, and Landau, as discussed earlier, but the idea was not fully developed in published form beyond mere suggestions. More rigorous work came in the period after 1957, with Stanley Deser probing the idea that general relativity could be a universal regulator of the divergence problems of quantum field theory at the Chapel Hill conference, with the work published as a paper: [Deser (1957)].[21] Bryce DeWitt also investigated the issue of whether gravity could function as a universal regulator, in 1964, showing that gravitational self-energies and electromagnetic self-energies can be impacted by gravity's regulation abilities.

7.3   Quantized Spacetime

Hartland Snyder's is probably the best known early work on discrete spacetime—the paper is truly a citation classic, with 1714 citations at the time of writing.[22] In this case the

[19] Now the length involves the mesotron mass $\mu, \hbar/\mu c$, derived from Yukawa's theory. Note that Heisenberg's persistent belief in a fundamental, universal length can be seen as more reasonable given that there is a remarkable coincidence between the electron radius and this meson mass (and so the range of the nuclear forces).

[20] As Brown and Rechenberg make clear, the existence of cosmic ray phenomena was pivotal precisely in that it served to delineate the borders of the known physical theories, pointing out exactly when they would break down (see, [Brown and Rechenberg (1996)], p. 72). Heisenberg was, of course, wrong in thinking that Fermi's theory was fundamental: there was new physics to which Heisenberg was not then privy to.

[21] Though he claimed to have become interested in the subject after the Berne conference, through his encounters with Oskar Klein, in 1955 (see [Deser (1995), p. 3])—Deser married Klein's daughter, Elsbeth, shortly after.

[22] Of course, this is the same Snyder that had worked with Robert Oppenheimer, in 1939, on the fate of very massive collapsing stars (approximated by an homogeneous, zero pressure ball of dust), showing that a one-way membrane (an 'event horizon' in modern parlance) would emerge from the process and that a

resulting spacetime is explicitly presented as *quantized*, with the spacetime coordinates themselves represented by Hermitian operators with discrete spectra. We have already seen this basic idea, of course, with Heisenberg and von Neumann. The innovation is to properly formalize the idea and establish that the discrete space idea need not violate Lorentz invariance.[23]

Snyder returned to the subject the following year, with a paper applying the quantized spacetime concept to the electromagnetic field: [Snyder (1947)]. There is then a trail leading from Snyder to Schild, in which the mathematical properties of quantized spacetime are uncovered.[24]

On January 7th, 1947, Bohr received a letter from Pauli [Kalckar (1985), p. 450–1]:

> I am looking as critical as you on this idea of a so-called "universal length". If this length– let us call it l_0 – is understood to be of geometrical nature, such theories or models will always lead to strange consequences for large momenta of the order of $\hbar/l_0$ in a field of purely classical experiments where the quantum of action should not play any role. Recently, we discussed here in Zürich a mathematically "ingenious" proposal of Snyder, which, however, seems to be a failure for reasons of physics of the type just mentioned.

Here we see that Bohr and Pauli had clearly discussed this issue of minimal length theories earlier. Such models were being increasingly suggested as a way out of the divergence problem. Pauli would later come to accept the idea of a physically imposed cutoff, thanks to Landau's discussion at the Bohr fest in 1955. What is interesting about this remark is that way that empirical constraints are at work in a field of research that is widely understood to suffer from a complete detachment in this respect.

C. N. Yang tackled a serious issue with Snyder's model, namely that it violates translation invariance whenever the coordinates are not a continuum. However, a

---

final singularity would also result—Landau had earlier noted the existence of a critical mass in 1932, and Chandrasekhar had shown in 1931 that the electron degeneracy pressure could not withstand further collapse for stars greater than 1.3 solar masses. It is rather odd that Snyder never made any link between these two streams of his work, continued collapse to a singularity and discrete space, since the former involves the reduction of a system's dimensions to values small enough (perhaps indefinitely small) to be relevant for the latter. (Oppenheimer and Snyder even wrote, "Physically such a singularity would mean that the expression used for the energy-momentum tensor does not take account of some essential physical fact which would really smooth the singularity out", [Oppenheimer and Snyder (1939), p. 456]. Later, John Wheeler would bring the two together via the Planck length: two areas where the "dynamics of geometry" fails to lend itself to classical analysis [Wheeler (1968), pp. 253–4]. This work would lead, ultimately, to Wheeler's notion of 'spacetime foam' (see below, and next chapter).

[23] Pauli had been looking at Snyder's proposal in Zürich, and notes that he found the idea "ingenious" at a mathematical level, but physically problematic since it would lead to unphysical consequences for large momenta of order $\frac{\hbar}{l_0}$ (where l_0 is the universal length resulting from the discretization of Snyder's proposal) where $\hbar$ should not be manifesting itself (Pauli to Bohr, January 28th, 1947, [von Meyenn (1993), p. 414]).

[24] Bergmann and Brunings briefly refer to Snyder, if only to distance their quantized metric variables from his: their coordinates, as they say, "commute with each other, but not with the energy-momentum densities". They continue: "The dynamical character of any particle coordinates follows automatically, but probably does not exhaust the physical significance of the coordinate commutation relations" (see [Bergmann and Brunings (1949)]). This highlights the continuity, at least, between Snyder's (and the other related) work on quantized/discrete spaces and quantum gravity research.

continuum clashes with the fundamental (i.e. non-epistemic) minimum length of the model.[25] Yang resolves the translation issue, but a problem of scale persists, namely in the form of a curvature of the universe at odds with what we observe (curvatures are of the order of the Planck scale rather than the Hubble radius).

It is rather interesting that Alfred Schild published his work on discrete spacetime around the same time that he transitioned into research on the canonical quantization of general relativity, following Dirac's influential 1949 lectures at the International Mathematical Congress in Canada, which Schild attended with his Masters student Felix Pirani.[26] Yet there is no mention of gravity in his paper on discrete spacetime, despite the fact that together with Pirani, Schild explicitly quantized the spacetime metric. This clearly reveals (perhaps rather surprisingly) that the project of 'quantization of gravity' had not yet been linked to what we now call 'quantum spacetime.' The focus is instead on the construction of a discrete model of spacetime that is as close as possible to Lorentz invariant, and the context is the problematic divergences of standard quantum field theory. Schild's basic object is a hypercubic lattice, with a time coordinate included amongst the spatial coordinates. He deals with ($c = 1$) Lorentz transformations that map a 3-lattice onto itself (where the 3-lattice takes on integer coordinates).

While Snyder's approach was indeed Lorentz invariant, it made use of the rather awkward idea that spacetime coordinates were non-commuting operators (so that spacetime functions become Hilbert space operators) and was not translation-invariant. Schild uses coordinates that are integer multiples of a fundamental length (rather than having eigenvalues that are integer multiples as with Synder), and so more along the lines of the proposals of Ambarzumian and Iwanenko *et al.*. Schild's goal is likewise to show that a common objection against discrete approaches to eliminating the divergences— that they violate Lorentz invariance due to the frame dependence of the 'minimum' cell size—is only partially correct since one can construct models that are invariant under a large subgroup (the discrete subgroup) of the Lorentz group. These, he suggests are in fact physically viable (unlike Snyder's and Yang's), and cast in a model closer to ordinary spacetime, thus undermining a host of common objections and making discrete models in principle a genuine possibility for fundamental physical theory—though, as he admits, his own model suffers from physical inconsistencies to do with a radically oversized minimal velocity.[27]

---

[25] We saw above, in footnote 9, how Rovelli and Speziale managed to sidestep the problem by introducing probabilities for measurement outcomes.

[26] Indeed, Schild's paper appears in the very same journal as Dirac's paper, in the issue directly preceding that containing the paper that would inspire Schild's work on the quantization of the gravitational field.

[27] This shortcoming was partially eliminated by E. L. Hill in 1955 by restricting the values of spacetime variables to rational numbers—partially, because the resulting space does not quite live up to the 'discrete' moniker. As Hill notes in a footnote in this paper, his Master's student, C. N. Kelber was working on this same problem of Lorentz invariance violation at the same time as Schild. There is some correspondence between Schild and Kelber, where the latter explains that he has a model that involves non-homogeneous Lorentz transformations so that the origin is not fixed for all observers (Kelber, letter to Schild, June 21st, 1948—Alfred Schild Papers [Box 86–27/2]: Briscoe Centre for American History).

Nathan Rosen introduces statistical considerations into the treatment of a discrete space: his elementary volumes are related to position measurement uncertainties (that is, to *practical* limitations: no infinitesimal measurement rods, therefore no physical point-like measurements). More specifically, the measurement of spatial coordinates of elementary particles (electrons) introduces inaccuracy into the measurement results such that repeated measurements will generate values sitting around the mean of a Gaussian distribution. His aim is, as with other proposals we've considered, to eliminate singularities (relating to the second motivation again). The resulting picture is not so very different from the Synderian one of a non-commutative space. However, the discreteness here is epistemological, coming from the difficulties involved in pinning down a spacetime point.

There is a very (later) Eddingtonian quality to this, especially the splitting of the abstract space from the observable space, which corresponds to Eddington's geometrical and physical frames.[28] Volumes in the observable space correspond to points in the abstract space. Lorentz invariance is preserved in this scheme only in the abstract space; yet Rosen suggests that a kind of translation manual could be established between transformations in this space and real physical transformations in the observable space.

There is also an interesting parallel to some of the issues over the "reality of spacetime points" (e.g. in the context of the hole argument in general relativity) here. Rosen argues that the value of a physical quantity at a point is not directly observable, so that physical laws should not be based on such quantities. What is not clear is whether, according to Rosen, the world (ontology) tracks epistemology so that our laws must be written this way because the world is that way so that only the mean values of quantities over volumes have any physical meaning at all.

Rosen reviewed a closely related paper in Russian by V. Averbah and B. Medvedev [Averbah and Medvedev (1949)] in 1949.[29] He also later returned to a similar idea, writing with Asher Peres, in 1960, though this time explicitly linking to measurement of the gravitational field. By this stage they viewed the existence of quantum uncertainties in these measurements (in the mean values of the Christoffel symbols) as pointing to the necessity of quantizing the gravitational field. Though we don't see any explicit discussion of the 'discrete space-gravitation' connection, the work presented here nonetheless contains crucial evolutionary steps. The recognition that playing around with the structure of space(time) might offer up cures for some of the difficulties of quantum field theory was an early one; linking this up with the way in which general relativity includes the geometrical structure of spacetime as one of the dynamical variables took somewhat longer.

---

[28] Rosen adds a statement to this effect, though pointing out that he was unaware of Eddington's work at the time of writing. Interestingly, this information was relayed to Rosen by M. F. M. Osborne, more famous now perhaps as an early econophysicist, but who also did early work on measurement restrictions, and minimal length, in quantum general relativity, as we saw in Chapter 4.

[29] For *Mathematical Reviews* (MR0029337). The paper was "On the theory of quantized space-time," in which the authors give a review of Snyder's work on non-commutative spaces.

In 1956 Takao Tati [Tati (1956)] presented an early example of what we would now call 'emergent spacetime,' though still falling within the cutoff camp. The idea here is simply to make do without the concept of spacetime in the formulation of the fundamental laws of elementary particles, with the concept of spacetime then arising statistically ("expressed as a mean value of some sort, of more fundamental quantities"). As a result of this proposal, field equations (including those of general relativity, though the discussion is only couched in terms of $\hbar$ and c) would have to be derived from the theory in the style of thermodynamics. Without spacetime coordinates, the theory becomes one of relations between the fundamental variables. This scheme also implies that any wave aspects of particles are also statistical. In this example, then, there is no quantum spacetime at all, which is instead retained as a purely classical concept. However, the emergence of spacetime is not the crucial aspect of this proposal, but rather the supposed *finiteness* of the theory which Tati supposes results from the removal of the continuum. Of course, this is employed, as with the other proposals in this chapter, to treat the divergence problem without the necessity of renormalization.

7.4 Gravitational Collapse and Singularities

A fairly standard response to the question of whether we need a quantum theory of gravity, given the absence of a direct empirical need, is that the *classical* theory of GR contains, as Peter Bergmann famously put it, "the seeds of its own destruction." A better way of putting it is that there are matter configurations (stress-energy tensors) that push GR to its limits of applicability, the limit beyond which the theory is no longer able to make predictions or provide physical descriptions. Most of the important work on the understanding of collapse and singularities in general relativity took place in the 1960s, but there were important and relevant early forays—some of which we already touched upon in footnote 22, but we will briefly say a little more here since it at least rubs shoulders with quantum gravity, though in this case we find that our previous focus on general relativity and gravitation as a 'resource' for solving problems in quantum theory is reversed.

Eddington posed the basic problem of the final stage of the collapse of stars as follows: "I do not see how a star which has once got into this compressed state is ever going to go out of it" [Eddington (1926)]. Eddington thought that the star must be in "an awkward predicament" once the store of subatomic fuel has expired. In response to this problem, the analysis of electron degeneracy pressure (utilizing Fermi–Dirac statistics), as a force battling against gravitational collapse, was given by Ralph Fowler in 1926 [Fowler (1926)]. Of course, Chandrasekhar provided an analysis, famously rejected by Eddington as "absurd,"[30] showing that for stars with a mass exceeding $6.6\mu^{-2}M_\odot$, the matter will not become degenerate, so that Fermi–Dirac statistics cannot save the star from

---

[30] See [Bonolis (2017)] for an excellent account of the early history of stellar structure and collapse, of which we skim only the surface.

collapse to a singularity [Chandrasekhar (1932)]. Chandrasekhar ended his paper with a puzzle, that would eventually form one of the key motivations for studying quantum gravity: "what happens if we go on compressing the material indefinitely?" (ibid., p. 327).

Though it didn't involve general relativity, another early quantum mechanical analysis of gravitational collapse (using Newtonian gravitation) was carried out, along the lines suggested by Chandrasekhar, by Lev Landau in 1932. Landau considered a star modeled by a cluster of fermions (a fermi-gas) interacting under the influence of gravitation alone (with the system at zero temperature so that it is not generating energy). He discovered the existence of a critical mass (about 1.5 solar masses) beyond which no equilibrium states could be found. Instead, the star undergoes gravitational contraction. Landau didn't see any alternative to the conclusion that such a star would collapse down to a singular point of infinite density. Landau's own suggested resolution of the (unacceptable) singularity was to suppose that quantum mechanics was violated beyond the critical mass. Of course, Landau's zero energy approximation ignores the thermonuclear properties of stars which do indeed fight against collapse, but only while the reserves of thermonuclear energy survive: Landau's catastrophe remains.

A series of related papers by Oppenheimer and his students, from 1938 and 1939, amount to a kind of 'quantum gravity near miss.'[31] Oppenheimer was engaged in the forces acting in neutron stars, and initially (with Robert Serber) focused on the battle between gravitational and nuclear forces in such stars. This led to a paper on the stability of stars, with George Volkoff. And, finally, and most relevant, a paper on continued gravitational contraction (collapse), co-authored with Hartland Snyder. This latter paper proves (albeit from some heavy assumptions: a spherically symmetric collapse of pressure-free dust) that collapse will continue to occur *indefinitely* when thermonuclear sources of energy have run out.[32] Such a situation, as we now know, leads to a black hole, which they themselves couch in terms of light being able to escape "over a progressively narrower range of angles" as it becomes progressively red shifted. More, they say, an "observer comoving with the [infalling] matter would not be able to send a light signal from the star; the cone within which a signal can escape has closed entirely" [Oppenheimer and Snyder (1939), p. 459]. However, neither the mathematical nor conceptual machinery were present at this stage, nor was a solid set of problems pertaining to the 'final state' of collapsed stars. There was certainly no sense that such objects might be *observable* entities at this stage—even high energy quasars would have to wait until the 1960s.[33] The missed opportunity most likely occurred because the focus was not on gravitation (and spacetime) in its own right, but simply stellar structure itself. De-idealizing the Oppenheimer–Snyder results, from a simple model of collapsing dust,

---

[31] There are certainly some sociological factors lurking behind the scenes: the day of publication of his "On Continued Gravitational Contraction" paper with Hartland Snyder was September 1st, 1939, the same day as the German invasion of Poland! Note that of the 1723 citations this paper has received, only 5 of those were ≤1956.

[32] The rather severe assumptions were not relaxed until the 1950s when John Wheeler, together with his students, relaxed them to include nuclear forces, though still in the context of spherically symmetric solutions.

[33] Walter Baade and Fritz Zwicky had postulated the existence of neutron stars in 1934.

would lead to greater appreciation of the physical nature of the collapsed stars. John Wheeler was one of those to tackle this problem, as mentioned earlier. This led him to accept the black hole concept as an astrophysically realistic object:

> The most significant piece of work in relativity, as I see it, was establishing the inescapability of gravitational crunch in a star. No equation of state could prevent matter from being crunched. [This is] in contrast to the work of Oppenheimer and Snyder of 1939, which had taken the escapability of resistances for granted and analyzed the time development of crunch. They took for granted that one could arrive at circumstances where resistance of matter to being crunched was negligible and could be forgotten about, whereas the analysis of the equation of state that I had gone through made it clear that no equation of state compatible with special relativity would prevent crunch. That was therefore an intellectual doorway-opener to the black hole.[34]

The inevitability of a crunch left the nature of the end state (the completely gravitationally collapsed object) as a mystery. It seems fairly obvious to us now to consider such end states, black hole formation in general, and singularities, in the context of the microstructure of spacetime, since one can envisage approaches to the Planck scale and, hence, domains in which quantum gravitational effects can no longer be ignored. It was Wheeler who linked such phenomena to quantum gravity considerations, initially fighting the notion that a spacetime singularity would be inevitable in gravitational collapse. However, even when John Wheeler considered the issue of collapsing stars in his 1954 Richtmeyer Memorial Lecture, "Fields and Particles," (John Archibald Wheeler Papers, American Philosophical Society, Philadelphia, Box 182), his focus was not on the ramifications for general relativity and spacetime structure, but rather on meson decay (gravitation still in the service of elementary particle theory). We turn to Wheeler's work in the next chapter.[35]

7.5 Conclusion

This chapter focused on the central motivation for much of what can be labeled 'quantum gravity' in the earliest phases of research, namely that it provides a potentially abundant resource for curing problems in quantum field theory. While it was rare to have fully worked out examples along these lines, it provided a much needed impetus to the study

---

[34] Interview of John Wheeler by Kenneth W. Ford on March 15th, 1994, Niels Bohr Library & Archives, American Institute of Physics: https://www.aip.org/history-programs/niels-bohr-library/oral-histories/5908-10.

[35] Strictly speaking, Wheeler's broader project fits into the theme of the present chapter, since it was concerned with deriving elementary particle physics from the geometry and topology of spacetime (including from quantum gravitationally induced topological entities, such as wormholes, which Wheeler associated with a Feynman path-integral treatment of gravity). However, there are many facets to Wheeler's work into quantum gravity, that have yet to be studied by historians, so we devote a chapter to it—however, Aaron Wright has a forthcoming book, *More Than Nothing: A History of the Vacuum in Theoretical Physics, 1925–1980*, that will include Wheeler's geometrodynamical research.

of quantum gravity at a time when there were few other reasons to bother with it. The primary problem was the ubiquitous divergences, which proved extremely stubborn and worrying to field theorists. Not all of the approaches were looked at involved gravitation directly, however, and focused more on ways of generating a discrete structure (with a minimal length or maximum energy) that would provide a physical cutoff, thus grounding a finite theory. These filtered through into gravitational research only later than our timeframe, in a variety of ways, including the small scales necessarily reached in gravitational collapse.

··

REFERENCES

Ambarzumian, V. and D. Iwanenko (1930) Zur Frage nach Vermeidung der unendlichen Selbstrückwirkung des Elektrons. *Zeitschrift für Physik* **64**: 563–7.

Averbah, V. L. and B. V. Medvedev (1949) On the theory of quantized space-time. *Doklady Akad. Nauk SSSR (N. S.)* **64**: 41–4.

Bergmann, P. G. (1955) Fifty Years of Relativity. *Science* **123**(3195): 486–94.

Bergmann, P. G. (1949) Non-Linear Field Theories. *Physical Review* **2**(75): 680–5.

Bergmann, P. G. (1942) *Introduction to the Theory of Relativity*. New York: Dover Publications.

Bergmann, P. G. and J. H. M. Brunings (1949) Non-Linear Field Theories II. Canonical Equations and Quantization. *Rev. Mod. Phys.* **21**: 480–7.

Bonolis, L. (2017) Stellar Structure and Compact Objects before 1940: Towards Relativistic Astrophysics. *The European Physical Journal H* **42**(2): 311–93.

Brown, L. and H. Rechenberg (1996) *The Origin of the Concept of Nuclear Forces*. Bristol: Institute of Physics Publishing.

Carazza, B. and H. Kragh (1994) Heisenberg's Lattice World: The 1930s Theory Sketch. *American Journal of Physics* **63**(7): 595–605.

Chandrasekhar, S. (1932) Some Remarks on the State of Matter in the Interior of Stars. *Zeitschrift für Astrophysik* **5**: 321–7.

Cushing, J. (2005) *Theory Construction and Selection in Modern Physics. The S Matrix*. Cambridge: Cambridge University Press.

Deser, S. (1995) *Oskar Klein: From his Life and Physics*. CERN-TH/95–9.

Deser, S. (1957) General Relativity and the Divergence Problem in Quantum Field Theory. *Reviews of Modern Physics* **29**: 417–23.

DeWitt, B. S. (1964) Gravity: A Universal Regulator? *Physical Review Letters* **13**(3): 114–18.

DeWitt, B. S. (1962) The Quantization of Geometry. In L. Witten (ed.), *Gravitation: An Introduction to Current Research* (pp. 266–381). New York: John Wiley and Sons.

Eddington, A. S. (1926) *The Internal Constitution of the Stars*. Cambridge: Cambridge University Press.

Einstein, A. (1919) Spielen Gravitationsfelder im Aufbau der materiellen Elementarteilchen eine wesentliche Rolle? *Sitzungsberichte der Königlich Preußischen Akademie der Wissenschaften Berlin* **1**: 349–56.

Fowler, R. H. (1926) On Dense Matter. *Monthly Notices of the Royal Astronomical Society* 87(2): 114–22.

Gorelik, G. (1994) *Matvei Petrovich Bronstein: and Soviet Theoretical Physics in the Thirties.* Basel: Birkhäuser.

Hagar, A. (2014) *Discrete or Continuous? The Quest for Fundamental Length in Modern Physics.* Cambridge: Cambridge University Press.

Iwanenko, D. (1931) Die Beobachtbarkeit in der Diracschen Theorie. *Zeitschrift für Physik* 72(9–10): 621–4.

Kalckar, J. ed. (1985) *Foundations of Quantum Physics I (1926–1932), Volume 6 (Niels Bohr – Collected Works).* Amsterdam: North Holland.

Kragh, H. and B. Carazza (1994) From Time Atoms to Space-Time Quantization: The Idea of Discrete Time, ca 1925–1936. *Studies in History and Philosophy of Science* 25(3):437–62.

Klein, O. (1956) Generalisations of Einstein's Theory of Gravitation Considered from the Point of View of Quantum Field Theory. In A. Mercier and M. Kevaire (eds.), *Fünfzig Jahre Relativitätstheorie* (pp. 58–71). Helvetica Physica Acta. Supplementum 4. Basel: Birkhaüser.

Klein, O. (1955) Quantum Theory and Relativity. In W. Pauli, L. Rosenfeld, and V. Weisskopf (eds.), *Niels Bohr and the Development of Physics* (pp. 96–117). Oxford: Pergamon Press.

Landau, L. (1955) On the Quantum Theory of Fields. In W. Pauli, L. Rosenfeld, and W. Weisskopf (eds.), *Niels Bohr and the Development of Physics* (pp. 52–69). Oxford: Pergamon Press.

Major, S., D. Rideout, and S. Surya (2007) On Recovering Continuum Topology from a Causal Set. *Journal of Mathematical Physics* 48(3): 032501.

Mercier, A. and M. Kervaire (eds.) (1956). *Fünfzig Jahre Relativitätstheorie.* Helvetica Physica Acta. Supplementum 4. Basel: Birkhaüser.

Oppenheimer, J. R. and H. Snyder (1939) On Continued Gravitational Contraction. *Physical Review* 56: 455–9.

Redei, M. (ed.) (2005) *John Von Neumann: Selected Letters.* Washington, DC: American Mathematical Society.

Rickles, D. (2014) *A Brief History of String Theory: From Dual Models to M-Theory.* Berlin-Heidelberg: Springer-Verlag.

Rosenfeld, L. (1932) La théorie Quantique des Champs. *Annales de l'I. H. P.* no1: 25–91.

Rovelli, C. and S. Speziale (2003) Reconcile Planck-scale Discreteness and the Lorentz-Fitzgerald Contraction. *Physical Review D* 67: 064019.

Rüger, A. (1992) Attitudes towards Infinities: Responses to Anomalies in Quantum Electrodynamics, 1927–1947. *Historical Studies in the Physical and Biological Sciences* 22(2): 309–37.

Snyder, H. (1947) The Electromagnetic Field in Quantized Space-Time. *Physical Review D* 72: 68.

Tàti, T. (1956) An Attempt in the Theory of Elementary Particles. *Il Nuovo Cimento* 4(1): 75–7.

van Dantzig, D. (1955) "On the Relation Between Geometry and Physics and the Concept of Space-Time." In A. Mercier and M. Kervaire (eds.), *Fünfzig Jahre Relativitätstheorie, Bern, July 11–16, 1955* (pp. 48–53). Basel: Birkhaüser Verlag.

van Dantzig, D. (1938) "Some Possibilities for the Future Development of the Notions of Space and Time." *Erkenntnis* 7: 142–6.

von Meyenn, K. (2001) *Wissenschaftlicher Briefwechsel mit Bohr, Einstein, Heisenberg u.a. Band IV, Teil III: 1955–1956 | Scientific Correspondence with Bohr, Einstein, Heisenberg, a.o. Volume IV, Part III: 1955–1956*. Berlin-Heidelberg: Springer-Verlag.

von Meyenn, K. (1993) *Wissenschaftlicher Briefwechsel mit Bohr, Einstein, Heisenberg u.a. Band III: 1940–1949 | Scientific Correspondence with Bohr, Einstein, Heisenberg, a.o. Volume IVII: 1940–1949*. Springer.

Wess, J. (2001) Non-Abelian Gauge Theories on Non-commutative Spaces. *Communications in Mathematical Physics* **219**(1): 247–57.

Wheeler, J. A. (1968) Superspace and Quantum Geometrodynamics. In C. M. DeWitt and J. A. Wheeler (eds.), *Battelle Rencontres*. New York: W. A. Benjamin Inc.

Wigner, E. (1955) Relativistic Invariance of the Quantum Mechanical Equation. In A. Mercier and M. Kervaire (eds.), *Fünfzig Jahre Relativitätstheorie, Bern, July 11–16, 1955* (pp. 210–226). Bäsel: Birkhaüser.

8

Geon Wheeler

[H]ow does one know what to invent, or even that any invention is necessary, until one has explored further the rich consequences of what one already has?

John Archibald Wheeler[1]

Bryce DeWitt recalls that Richard Feynman teased John Wheeler by calling him "Geon Wheeler" at the 1957 Chapel Hill conference on the Role of Gravitation in Physics.[2] 'Geon' was John Wheeler's term for a source-free solution to the combined Einstein–Maxwell equations. That he described as a 'gravitational-electromagnetic entity' that is held together by gravitational force—that is, an electromagnetic (or gravitational) wave which remains highly localized for a long time in a small region of space due to the gravitational attraction of its own field energy; a configuration somewhat like a latter-day vortex-atom of Lord Kelvin.[3] It is, as Wheeler himself puts it in the paper introducing the concept (and his first paper devoted to general relativity), "a model of something particle-like and yet not a particle: a geon, a gravitational wave going around in a circle held in orbit by the mass-energy of that wave itself" [Wheeler (1955)].[4]

It is quite clear from this description of a geon that the problem of the nature of elementary particles was, at this stage, firmly in Wheeler's mind, as it was with many

---

[1] John Wheeler, letter to John von Neumann, October 12th, 1954.

[2] Interview of Bryce DeWitt and Cecile DeWitt-Morette by Kenneth W. Ford on February 28th, Niels Bohr Library & Archives, American Institute of Physics, College Park, MD USA, www.aip.org/history-programs/niels-bohr-library/oral-histories/23199. This conference was pivotal in the transformation of gravitational physics (including quantum gravity) from backwater into a respectable discipline—John Wheeler himself played an outsized role in the founding of the Institute (The Institute for Field Physics) of which this conference was the inaugural event. The report from the conference is reproduced in [DeWitt and Rickles (2011)].

[3] Kelvin was struck by the permanence and apparent indivisibility of smoke rings in a demonstration of Peter Guthrie Tait's. Kelvin's own account can be found in [Thompson (1867)]. See [Alkemade (1994)] and [Silliman (1963)] for historical discussions. Interestingly, a similar debate to the 'particle versus field' controversy discussed here was present in Kelvin's day, where there was a tension over corpuscular versus fluid (i.e. ether) theories of matter, with Kelvin attempting to show, as Wheeler would (following Einstein's general idea), that corpuscular features could emerge from a continuous, fluid theory (as knotted and linked vortices): 'corpuscles without corpuscles'! See [Kragh (2002)] for more on this debate.

[4] The paper was submitted September 8th, 1954, shortly after he began teaching a course on general relativity at Princeton.

Covered with Deep Mist: The Development of Quantum Gravity (1916–1956). Dean Rickles, Oxford University Press (2020). © Dean Rickles. DOI: 10.1093/oso/9780199602957.001.0001

other physicists who became interested in gravitation around this period.[5] 'Elementary' particles ('carriers' of mass and charge) in this case would not be elementary, but would be derived from purely classical (zero rest mass, zero charge density) fields (albeit multiply-connected manifolds, or 'Misner–Wheeler' wormholes). Unlike Einstein, Wheeler was not trying to simply extract quantum behavior from classical theory, but rather wished to extract *objects*, with their multifarious qualities. Hence, the geon project is intimately tied to the project of *geometrodynamics* (understood here in Wheeler's stronger sense of getting *everything* from curved space, putting physical flesh on William Kingdom Clifford's "space theory of matter" [Clifford (1982)][6]). However, as we will see (and saw briefly in the previous chapter), the issue of gravitational collapse was a further motivation guiding his earliest general relativistic speculations. Wheeler's later quantum gravitational interests (e.g. summing over spacetime geometries) also point back to the more basic idea of getting charged, massive particles from zero-mass fields.

Wheeler believed that the existence of geons[7] in general relativity would show that the theory was uniquely comprehensive, accounting for both fields generated by bodies, as well as the motions of those bodies (thanks to the Einstein–Infeld–Hoffmann [EIH] result: [Einstein, Infeld, and Hoffmann (1938)]), but also accounting for the very existence and stability of the bodies themselves (absent from the EIH result, in which they had *assumed* existence and stability)—though as mentioned above and in footnote 7, qua bodies geons are hardly eternal and the kinds of 'charge without charge' wormhole geons are simply not possible (classically at least). In many ways, this scheme corresponds to Einstein's own later 'unified field theory' approaches,[8] and (though less so) also to Peter Bergmann's initial foray into quantum gravity in 1949 which also considered the problem of how to get particles out of a classical field theory (though in this case in a bid

---

[5] Indeed, Wheeler had several years earlier delivered a review lecture to the American Philosophical Society entitled "Problems and Prospects in Elementary Particle Research," in 1945, in which his "8th problem" concerned "what modifications are required in the field concept to describe gravitation, electromagnetism, the creation of electron-positron pairs and the forces between elementary particles in a unified manner, unambiguously and without divergences" ([Wheeler (1946)], p. 36). The geon program can be viewed as one possible way of responding to this, and it is interesting to see that though we often think of Wheeler as steering clear from all things gravitational until the 1950s, he was mulling over it relatively early on. We see in the next section that he had another preliminary skirmish, also relevant to his 8th problem, in the form of an action-at-a-distance approach to gravitational forces. Charles Misner recalls that "Wheeler said at some point to me – not during my graduate student years I don't think, but not [much later]) that he always was interested in gravity only never put a lot of time into it until he taught a course in it, which was his way of learning it." (Interview of Charles Misner by Christopher Smeenk on May 22nd, 2001, Niels Bohr Library & Archives, American Institute of Physics, College Park, MD USA, www.aip.org/history-programs/niels-bohr-library/oral-histories/33697).

[6] Adolf Grünbaum has studied the relationship between Wheeler's and Clifford's approaches (which he labels 'monistic,' with geometry as the 'ultimate substance') in [Grünbaum (1973)].

[7] In fact, the existence of geon solutions (qua globally stationary, stable ('solitonic') configurations solving the full set of equations of motion) was not known to Wheeler. A 1975 theorem of Dennis Gannon [Gannon (1975)] shows that any solution of the vacuum Einstein equations evolving from a non-simply connected Cauchy surface will develop a singularity. More physically: starting with some non-simply connected spacetime, one will end up with collapse into black holes (i.e. geodesic incompleteness). This would hide the non-trivial topological structure behind event horizons ('topological censorship'). See [Giulini (2009)] for more on this.

[8] For more details on Einstein's aims for his unified theory, see §3 of [Lehmkuhl (forthcoming)] (which includes new material from Volume 15 of Einstein's collected papers (Princeton University Press, forthcoming)).

to have a quantum theory without divergences of self-mass and similar, directly using the EIH result). To some extent, Wheeler's geon project was at least able to disprove Pauli's quip to Einstein, that "classical field theory is a completely squeezed out lemon that can't possibly produce anything new" (Pauli letter to Einstein, September 19th, 1946: Einstein archive: 19–182[9]): there was at least some juice remaining. Indeed, this is an instance in which quantum gravity-type considerations showed that the classical theory of general relativity is not quite a closed book.

Though the geon concept (certainly as a *resource* for particle theorists) fell out of favour fairly rapidly, despite some initial strong interest (from the likes of Bergmann and Bondi, amongst others[10]), it was undoubtedly Wheeler's main gateway into general relativity and quantum gravity, and set the scene not only for the kinds of very fundamental issues he tackled, but also the adventurous ('radical conservative') style he adopted that would both remain at the centre of his research until his death.[11] The initial 1955 geon paper also introduced many of Wheeler's characteristic 'free lunch' concepts: 'mass without mass', 'charge without charge', etc.—though not quite couched in this epigrammatic form yet. We also find in this and subsequent papers (<1960) a wealth of new terminology and concepts that still fill the field of quantum gravity research, including the notion of spacetime fluctuations (later 'spacetime foam') as a result of introducing a summing over all possible metrics (and topologies) in the context of quantized general relativity, along with the explicit linkage to Planck units.[12]

---

[9] Reproduced as Letter 835 in the third volume of Pauli's scientific correspondence ([Pauli (1993)], p. 383).

[10] Wheeler evidently gave a talk on geons to Bondi's group at Cambridge in July 1954, at what Bondi calls a "convention" (letter from Bondi to Wheeler, dated July 12th, 1954: John Archibald Wheeler Papers, American Philosophical Society [henceforth, JAWP], Box 5). Bondi apparently found a "detailed solution for a geon," though he appears to have never published on it and does not give any details in his letter (nor can I find the solution amongst any of Wheeler's or Bondi's papers).

[11] He didn't stay with gravitational physics until his death, however. Wheeler wrote to Paul Cohen on January 4th, 1973 that he was "winding up twenty year's involvement with gravitation physics and relativity with the conclusion that the mystery of things lies still deeper, in the quantum principle" (John Archibald Wheeler Papers, I. Princeton University Correspondence and Files, Box 7).

[12] Wheeler gives credit to Peter Bergmann and James Anderson ([Wheeler (1955)], p. 535) for initial discussions of quantum gravity, though he clearly favoured his own student's (i.e. Richard Feynman's) path-integral approach to quantization for the possibility it offered of generating topological complexity, such as multiply connected metrics that could then be used as an abundant resource for explaining the particle structures that were his primary focus at this early stage. Around the time of this paper (in the fall of 1954), Wheeler would pass on the problem of the Feynman quantization of general relativity (which he had merely sketched the very broadest outlines of) to another student, Charles Misner—Misner changed track soon after to demonstrating that general relativity is an 'already unified field theory' [see §8.3] in which both gravitation and electromagnetism are reduced to spacetime geometry. However, he discovered this had in fact been carried out earlier by George Yuri Rainich—it was Peter Bergmann who pointed out this duplication to Misner via Wheeler (see [Barrett and Byrne (2012)], p. 300; the comment can be found in a letter Bergmann wrote to Wheeler on May 29th, 1956, in which he points out that he is not entirely convinced by the already unified field theory idea—JAWP, Box 4). Wheeler and Rainich later corresponded (letter from Rainich to Wheeler, January 30th, 1957: JAWP, Box 22) on the already unified theory after Rainich took issue with Wheeler's view (expressed in print) that Einstein would have been impressed with the approach. It seems, however, that Einstein had expressed opposition to Rainich himself. Misner later wrote back to Bergmann (June 26th, 1956: JAWP, Box 4) offering thanks, and pointed out that he had first heard the idea that the electromagnetic field can be reconstructed from its gravitational imprint from Hugh Everett. On discovering this, Misner returned to his original PhD project of Feynman quantization of gravitation. (As Lehmkuhl [Lehmkuhl (forthcoming)] notes, the disagreement turned on whether Rainich's theory could account for electrons, which Einstein doubted—the

The geon program was pushed hard, but the conclusion of Misner, with the agreement of the rest of the participants at a 1957 exploratory research session on quantum gravity, in Copenhagen,[13] was that though 'topological tricks' (such as wormholes) could go a long way in getting out features of particle physics from spacetime, they could not be coaxed into giving a representation of sources of spinor fields from the Einstein field equations.[14] A little earlier, at the Chapel Hill Conference on the Role of Gravitation in Physics in January 1957, Wheeler had himself already stopped thinking of geons as plausible candidates for the elementary particles (but see footnote 15)—beyond the theoretical issues were also direct empirical problems to do with the instability of geons in our universe. In responding to a request from Peter Bergmann asking what the present motivation of geon research was, Wheeler notes that the geon doesn't have "the slightest to do with an elementary particle, nor with astrophysical objects" but rather is an exercise in making sense of non-linear equations ([DeWitt and Rickles (2011)], pp. 142–3). By the time of the Royaumont conference *Les Théories Relativistes de la Gravitation* in June 1959, Bergmann was able to state that Wheeler's wormholes do not represent elementary particles, but instead the "froth" of a manifold with a metric subject to quantum fluctuations ([Bergmann (1959)], p. 464)—the wormholes in this case were many orders of magnitude below what would be of relevance to the problem of elementary particles. By this point, Wheeler was immersed in 'quantum geometrodynamics' and the problem of quantum gravity and the microstructure of gravity more generally.[15]

Einstein–Rainich correspondence will appear in the forthcoming Volume 15 of the collected papers of Einstein, Princeton University Press.)

[13] Aeronautical Research Laboratory, Contract No. AF 33(616)-5367, Wright Air Development Center, Air Research and Development Command, United States Air Force, Wright-Patterson Air Force Base, Ohio. Along with Stanley Deser, Charles Misner had examined the possibility of representing scalar meson fields with these Wheelerian topological tricks and found that it is indeed possible (in both the massless case as well as for non-vanishing mass). So far as I have been able to ascertain, this was the first meeting devoted *solely* to quantum gravity—Blum and Hartz have reproduced this report, along with a useful analysis [Blum and Hartz (2017)].

[14] This sentiment was echoed by Wheeler in his 1957 paper on geometrodynamics, writing: "How can a classical theory endowed with fields of integral spin possibly give on quantization a spin 1/2 such as is required to account for the properties of the neutrino, the electron, and other particles? . . . Is there anything about the process of formulating Feynman's sum over histories, anything about two choices for the orientation of each elementary space-time volume, or any other feature, that forces the introduction of any such nonclassical two-valuedness? Unless there is, pure quantum geometrodynamics must be judged deficient as a basis for elementary particle physics" ([Wheeler (1957)], p. 613). However, the inability of the geon project to achieve "spin without spin" was possibly somewhat premature. In 1959 Misner and Finkelstein [Finkelstein and Misner (1959a)] cast some doubt on this orthodox view that one cannot engineer geons that have half-integer spin (and so that are capable of modelling particle physics) by proving that general relativity can permit the definition of double-valued wavefunctions. However, they were not able to show that the closed path which flips the wave function's sign corresponds to a 2π rotation, so that it is left open whether this is in fact representing half-integer spin—indeed, in an update (presented at Royaumont in 1959: [Finkelstein and Misner (1959b)]) they explicitly show that the sign is not changed by a spatial rotation. Still later, however, in 1980, Rafael Sorkin and John Friedman [Friedman and Sorkin (1980)] claim to have shown that it is in fact perfectly possible to get spin-1/2 from gravity alone—see [Giulini (2016)] for a discussion of their result. However, this holds only in the 'quantum gravitational' case, and involves approximations related to the fact that there is (still) no such full theory available.

[15] But we mustn't be too quick here. Both Dieter Brill and Charles Misner (doctoral students under Wheeler in the mid-50s) recall that things were not so clearcut: "BRILL: I mean Wheeler was always, well, not ambivalent really, but he somehow did protest too much, it felt like, on the question of whether he meant the geons to be

This chapter looks at the prehistory, birth, evolution, and demise of the geon research program, and Wheeler's entry into gravitational research more generally. We also look at the issue of gravitational collapse and its link to the genesis of the geon concept, as well as the introduction of spacetime fluctuations at the Planck scale ('quantum foam,' depicting the small scale structure of spacetime) along with the conceptual linkages between these important ideas, especially as they relate to quantum gravity, which they all flow into in some way or other. The progression is: 'all is particles' → 'all is fields' → geons → geometrodynamics → quantum gravity (with the latter finding some degree of coverage in both the geons and geometrodynamics stages).[16] We begin by charting a major shift in Wheeler's thinking concerning the fundamental ontology of the universe, from a world of particles (with fields as convenient fictions) to a world of fields (from which particle-like aspects are to be derived). This shift seems, at the same time, to have radically altered Wheeler's creative process.[17]

8.1 From "All is Particles" to "All is Fields"

It is worth spending some time plotting Wheeler's rather late entrance into the field of relativity—a seemingly radical shift that appears to have occurred in Spring of 1952. However, as mentioned, there was an important prelude in the form of an investigation of gravitational action-at-a-distance theory[18] (initiated to a certain extent while working

---

elementary particles. Because he would often say, 'Well there's no reason to think that they have anything to do with elementary particles.' Still, he wanted . . . –MISNER: Wanted people to think about elementary particles when they thought about geons, [laughter] whether negative or positive as long as they were there" (Interview of Dieter Brill and Charles Misner, March 16th, 2011—conducted with Don Salisbury and Dean Rickles).

[16] For completeness, we should note that in good dialectical fashion, Wheeler later advocated a kind of synthesis of his two ontological standpoints ('all is particles' and 'all is fields'), arguing for a view (based on what he called 'pregeometry') in which spacetime structure and particle structure are intertwined (see, e.g., the report in *Nature*, 1972, **240**, p. 382).

[17] He coined this process "radical conservative-ism": on which, see [Misner et al. (2009)]. As the passage from the letter to von Neumann (from the opening of this paper) makes plain, the trick in this approach is to assume the validity of current laws and push them and mine them as far as is possible, no matter how seemingly outlandish: one must not introduce 'new inventions'—there are clear historical links between this methodology and the so-called 'principle of plenitude,' according to which, roughly speaking, anything not forbidden is possible. (However, as Dennis Lehmkuhl has pointed out to me, in the area of the 'already unified theory,' in which Misner and Wheeler claim to use nothing beyond standard Einstein–Maxwell theory, the field equations they employ are of 4th order—constraining the curvature, through the so-called Rainich conditions, in a way that standard Einstein–Maxwell theory does not—and is not quite as 'conservative' as they make out).

[18] In addition to the review lecture mentioned in footnote 5, there is another earlier example that shows that Wheeler had at least been giving some thought to general relativity, in the form of a letter to Philip Frank (dated November 8th, 1941: JAWP, Box 142), in which Wheeler comments on Frank's discussion of Mach's principle (in Frank's book *Between Physics and Philosophy*), noting that "there are many who fail to recognize that Einstein's general theory in principle sets up a connection between the inertial properties of the individual particle and the distribution of matter throughout space" (something Wheeler would later encapsulate in the phrase 'inertia here arises from mass there'). Later, Wheeler points out that his interest had been whetted by Lorentz's 1927 book *Problems of Modern Physics*, dealing with both quantum theory and relativity, though not in combination (see Interview of John Wheeler by Kenneth W. Ford on February 21st, 1994, Niels Bohr Library & Archives, American Institute of Physics, College Park, MD USA, www.aip.org/history-programs/niels-bohr-library/oral-histories/5908–8).

on his second action-at-a-distance paper with Feynman [Feynman and Wheeler (1949)]) from 1949, just prior to his involvement with the Matterhorn Project (see footnote 28), as scientific director, in which he directed a bomb study group at Princeton—this was the bomb design component of a two-pronged project, the other being Lyman Spitzer's plasma/fusion project code-named (by Wheeler) 'the stellarator'. Interestingly, both Martin Kruskal and Martin Schwarzschild (son of Karl Schwarzschild, who died when Martin was four years old) participated in Project Matterhorn[19]—both would, of course, go on to produce important work in gravitational physics, especially on the structure and formation of black holes. One can certainly see evidence that a disenchantment with his more practical national security work led to a desire to pursue something more pure and fundamental in Wheeler's work. However, the seeds of his shift were already planted prior to 1952 in this earlier pre-bomb work—Wheeler himself dates the decision to work on the bomb project to early February, 1950 ([Wheeler (1998)], p. 190), a short while after he had begun thinking about gravitation.

During a leave of absence in Paris, in 1949, Wheeler was (amongst other projects) focusing his attention on removing the field concept from gravitational physics, just as he had earlier removed the electromagnetic field with his then student Richard Feynman. As Wheeler later explained in a letter to Alfred Schild, the action-at-a-distance was intended to make "particles the really fundamental elements in a description of nature" (Letter dated December 30th, 1953: JAWP, Box 24). Feynman and Wheeler had started this work before the war, but had to interrupt it, finishing the first paper in 1945 (during Wheeler's visits to Los Alamos). Speaking of his thoughts at this time, Wheeler refers to a sense that the action-at-a-distance theory "seemed more direct and more natural than field theory".[20] The 1945 paper simply tried to derive the retardation aspect of electromagnetic forces using action-at-a-distance. A second paper was written in 1949, drawing out many of the interesting features associated with this so-called 'Wheeler–Feynman theory'. Wheeler continued this work, without Feynman, during a research visit to Bohr in 1949–1950 and a little earlier with his research student John Toll in France.[21]

The ramifications of this application of Wheeler–Feynman theory to gravitation promised, of course, to be rather more spectacular since gravitation is related to spacetime, and so would amount to "sweeping away the spacetime from between particles"[22]—in his letter to Bohr, Wheeler speaks explicitly of constructing a

---

[19] A good historical account of Project Matterhorn, and cold war fusion research more generally, can be found in [Bromberg (1982)]. The final report from Project Matterhorn can be found at The National Security Archive: http://nsarchive.gwu.edu/nukevault/ebb507/docs/doc%207%2053.08.31%20PM-B-37.pdf.

[20] Interviewed by Kenneth W. Ford, Jadwin Hall, Princeton University, February 14th, 1994: https://www.aip.org/history-programs/niels-bohr-library/oral-histories/5908–7).

[21] Letter to Bohr, September 3rd, 1949 (JAWP, Box 136)—see fig. 8.1. Immediately prior to this Wheeler had established a cosmic ray laboratory at Princeton but decided to pass the responsibility onto George Reynolds so that he could divert his attention to more fundamental issues in physics (see Interview with Charles Weiner and Gloria Lubkin, Princeton University, Wednesday, April 5th, 1967, https://www.aip.org/history-programs/niels-bohr-library/oral-histories/4958).

[22] As he later put it: "In Paris my great hope was to translate the idea of action at a distance, sweeping out fields between particles, to translate that idea from electric forces to gravitational forces. Well, the field in the case of gravitation is space and time, so what this amounted to was the thought of sweeping out space and time

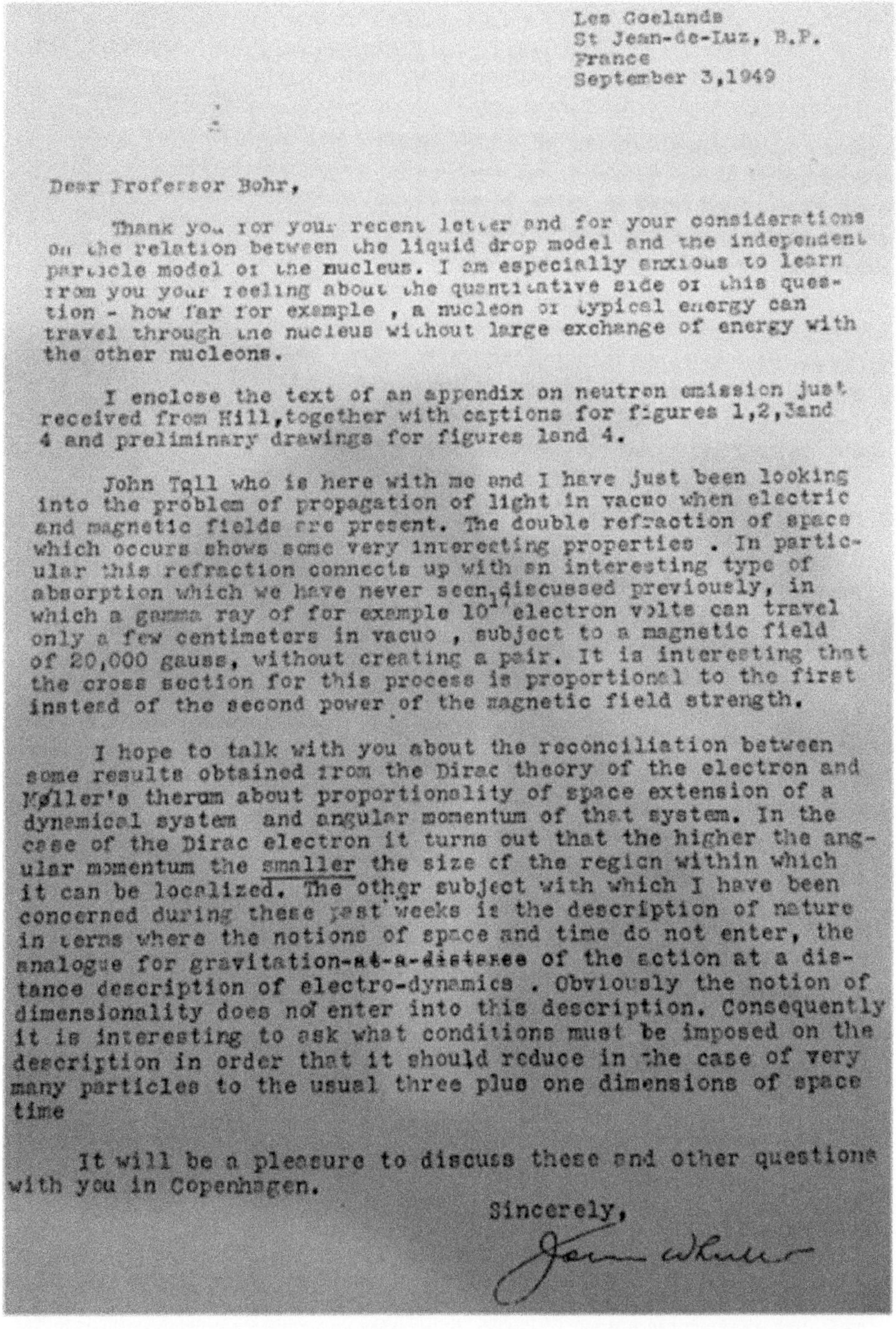

Les Goelands
St Jean-de-Luz, B.P.
France
September 3,1949

Dear Professor Bohr,

Thank you for your recent letter and for your considerations
on the relation between the liquid drop model and the independent
particle model of the nucleus. I am especially anxious to learn
from you your feeling about the quantitative side of this question - how far for example , a nucleon of typical energy can
travel through the nucleus without large exchange of energy with
the other nucleons.

I enclose the text of an appendix on neutron emission just
received from Hill, together with captions for figures 1,2,3 and
4 and preliminary drawings for figures 1 and 4.

John Toll who is here with me and I have just been looking
into the problem of propagation of light in vacuo when electric
and magnetic fields are present. The double refraction of space
which occurs shows some very interesting properties . In particular this refraction connects up with an interesting type of
absorption which we have never seen discussed previously, in
which a gamma ray of for example 10^{11} electron volts can travel
only a few centimeters in vacuo , subject to a magnetic field
of 20,000 gauss, without creating a pair. It is interesting that
the cross section for this process is proportional to the first
instead of the second power of the magnetic field strength.

I hope to talk with you about the reconciliation between
some results obtained from the Dirac theory of the electron and
Møller's theorm about proportionality of space extension of a
dynamical system and angular momentum of that system. In the
case of the Dirac electron it turns out that the higher the angular momentum the smaller the size of the region within which
it can be localized. The other subject with which I have been
concerned during these past weeks is the description of nature
in terms where the notions of space and time do not enter, the
analogue for gravitation-at-a-distance of the action at a distance description of electro-dynamics . Obviously the notion of
dimensionality does not enter into this description. Consequently
it is interesting to ask what conditions must be imposed on the
description in order that it should reduce in the case of very
many particles to the usual three plus one dimensions of space
time

It will be a pleasure to discuss these and other questions
with you in Copenhagen.

Sincerely,

Fig. 8.1 *Letter to Bohr, written during Wheeler's leave of absence in Europe, in which he describes his
wish to tackle an action-at-a-distance version of gravitation. Image source: John Archibald Wheeler
Papers [JAWP], American Philosophical Society, Box 136.*

"description of nature in terms where the notions of space and time do not enter".[23] Though Bohr did not refer in his reply (dated December 24th, 1949: JAWP, Box 136) to the gravitational project Wheeler described, Wheeler is quite persistent (though points out that the subject is "rather specialized"), returning to a slightly more detailed outline of his idea in his reply to Bohr (letter dated January 21st, 1950: JAWP, Box 136). As he describes it here, his (and Toll's) approach sounds something like what we would now call an 'emergent gravity' approach (i.e. one in which the gravitational field, or spacetime, does not appear in the fundamental description but only in some limit in the manner of effective theories):

> The idea is to construct a description of nature which - while maintaining in the limit of many particles a close correspondence with general relativity theory - nevertheless is more general in two ways: (1) it is not strictly imbeddable in a Riemannian 3+1 space-time in small dimensions - in fact, is not then even accurately describable be [sic.] the notion of dimensionality - and (2) it makes force dependent upon the number of particles in the universe.
>
> (ibid.)

During this period, Wheeler has claimed that he was working on topics that included gravitation "without benefit of any knowledge, in any deep sense, of general relativity".[24] However, Wheeler didn't seem to think this yet constituted a shift of his interests over to general relativity.

This early gravitational work was, as mentioned, a direct outgrowth of the work he had carried out with Feynman in the context of electromagnetism and the reasons are similar in both cases: fields are problematic in ways that particles are not.[25] It is possible that the initial idea to apply the theory to the gravitational field came from the first presentation of their work in a colloquium talk at Princeton. It was Feynman's first talk, and various luminaries were present, including Pauli and Einstein. Feynman recalls that Pauli was frequently discussing the talk with Einstein and at one stage Einstein raised a potential problem with making the theory compatible with gravitational interactions—

---

between the particles. I won't go into the details of the idea. I don't think I ever published anything on it. I'm still hoping someday a student will come by who would like to take a whang at it" (Interview of John Wheeler by Kenneth W. Ford on February 21st, 1994, Niels Bohr Library & Archives, American Institute of Physics: https://www.aip.org/history-programs/niels-bohr-library/oral-histories/5908–8).

[23] As Domenico Giulini has pointed out to me, this cannot really be achieved in such a scheme since spacetime and its causal relations still enter through a backdoor, via the propagators of the particle interaction.

[24] Interview of John Wheeler by Kenneth W. Ford on February 21st, 1994, Niels Bohr Library & Archives, American Institute of Physics, College Park, MD USA, www.aip.org/history-programs/niels-bohr-library/oral-histories/5908–8.

[25] As Misner recalled, "John told me at some later stage that he spent 20 years of his life trying to say everything is particles—scattering, action at a distance, etc.—and then he switched and decided to take the other tack and say everything is fields" (Interview of Charles Misner by Alan Lightman on April 3rd, 1989, Niels Bohr Library & Archives, American Institute of Physics: https://www.aip.org/history-programs/niels-bohr-library/oral-histories/33955).

though Einstein wasn't pushing his objection strongly since he admitted that "I am not absolutely sure of the correct gravitational theory" ([Mehra (1994)], pp. 95–6).

No publication seems to have come out of this action-at-a-distance formulation of gravitation, the cracks of which can already be seen in 1953. For example, in the letter to Schild, mentioned above, Wheeler writes:

> I now think it [the action-at-a-distance approach as a theory of fundamental reality] is misguided, for two reasons. First, if there were to be a fundamental particle out of which everything else was built up this particle would presumably have to be the electron. But the electron has not enough degrees of freedom in the way of spin and statistics to act as the building block for elementary particles which we now know. Therefore some other entity would have to be introduced over and above the electron—such as the neutrino. Second, the electron itself seems nowadays not to be at all a simple thing. Indeed, the distinction between the "dressed" and "undressed" electron brings out this point in an especially clear way and even leaves open the possibility that the "undressed" electron is an entity of zero rest mass. Both reasons therefore suggest that one ought not to take a field of finite rest mass at all as the structural element in a theory of the elementary particles but rather fields of zero rest mass: gravitation field, electromagnetic field and neutrino field. (letter dated December 30th, 1953—Alfred Schild Papers [Briscoe Center for American History], Box. 86–27/2).

This is the central insight leading to Wheeler's search for geons, and, later, to the geometrodynamic project. The underlying aim (prior to the actual construction of geons, and which geons would exemplify) is to do physics with nothing but zero mass fields, with no *ad hoc* external sources (the charges and masses). Reflecting later on the switch from particle theory to field theory (gravitation), Wheeler wrote:

> [W]hen I went to Copenhagen, I wanted to go on with this idea that the interaction between electrons was the great thing, and also, in the same idea, to take out the field in the space between the particles – the gravitational field – and that's quite a deep thing. And I would still like to come back to that, although I don't regard that today as the great white hope. You've heard that old phrase, "Somebody who is a reformed drunkard gets religion more strongly than anybody," and so I became more strongly field theoretic after that. So that's how come I moved into gravitation. That's a field theory that, at that time at any rate, that was the master field.[26]

Thus originated Wheeler's about-turn, from trying to do away with spacetime, to trying to make the world from nothing but spacetime.[27]

---

[26] Interview by Finn Aaserud, Princeton University, May 4th, 1988: https://www.aip.org/history-programs/niels-bohr-library/oral-histories/5063–1.

[27] One might think that Wheeler's proximity to its chief architect might have played some role in his shift to general relativity. But Wheeler seems not to have been particularly impressed, as he notes: "To see and hear Einstein at Princeton in October 1933 when he first settled in this country was interesting, but I can't say it was inspiring. I didn't get the feeling he had any great vision that one could subscribe to or develop or go on with. He seemed to be trying equations on the wholesale scale without any great physical idea to guide it." (Interviewed

8.2   Engineering Geons and Teaching Relativity

As noted above, Wheeler was once again diverted from action-at-a-distance by another war effort, this time the hydrogen bomb project, in which he led a stream, based at Princeton University, dealing with bomb design, Project Matterhorn.[28] Wheeler deferred the Guggenheim fellowship on which he had intended to work, amongst other things, on the gravitational work with his student John Toll which would, ultimately, stall his entrance to gravitational research for three years. But even during the Matterhorn work, Wheeler's focus was shifting to gravitation and general relativity proper. He recalls how, on May 6th 1952, still immersed in Matterhorn, he started his first notebook on relativity ("Relativity I"), with the first entry: "5:55pm. Learned from Shenstone 1/2 hour ago the great news that I can teach relativity next year" ([Wheeler (1998)], p. 228).[29]

1953 marked the end of Wheeler's involvement with Project Matterhorn and allowed him the freedom to focus on general relativity full time, heralded by the teaching of a (now rather famous) course on the subject at Princeton: Physics 570.[30] One can see

by Kenneth W. Ford, Location: Jadwin Hall, Princeton University, Interview date: Monday, January 3rd, 1994: https://www.aip.org/history-programs/niels-bohr-library/oral-histories/5908–3).

[28] A parallel group dealing with fusion energy was led by Lyman Spitzer (who had come up with the name Project Matterhorn)—hence, there was Matterhorn A and Matterhorn B, where project B was devoted to the bomb and project A was devoted to fusion (based on plasma physics). This group featured an impressive array of astrophysicists (not least Spitzer himself), but also, as mentioned earlier, Martin Schwarzschild and Martin Kruskal. While it was a failure in terms of its mission to develop efficient energy sources, it did lead to advances in plasma research. Interestingly, Spitzer had never received any formal training in general relativity, and claims to have gotten what he needed from Eddington's books (Interviewed by David DeVorkin, Princeton University, Princeton, New Jersey, Friday, April 8th, 1977 https://www.aip.org/history-programs/niels-bohr-library/oral-histories/4901–1). There are clear and interesting overlaps between this plasma research and astrophysics. As Spitzer himself points out: "Oh, I'm sure it had some connection. The scientific interest in the bomb research and my background in plasma astrophysics together made me receptive to the problems involved in a controlled fusion reactor. When that subject was brought to my attention by the Argentine announcement, I became very actively interested in it, and decided I would like to pursue it" (Lyman Spitzer, Interviewed by Joan Bromberg, Wednesday, March 15th, 1978, https://www.aip.org/history-programs/niels-bohr-library/oral-histories/4900). Charles Misner notes how Wheeler had, coming out of his bomb project work, developed something of a 'computational mindset' with respect to physics problems: "Wheeler would take the point of view that if you want to tackle a problem and understand it well, you should imagine that you're programming a computer, because the computer needs actual details–you can't leave choices to be waived later" (Interview of Misner by the author and Donald Salisbury, University of Maryland, March 16th, 2011). Interestingly, the numerical integration for the reduced equation for a spherical geon was performed on an IBM punch card-programmed electronic calculator, by Robert Goerss ([Wheeler (1998)], p. 524–5).

[29] As Wheeler ([Wheeler (1998)], p. 228–9) recalls his motivation for studying general relativity (with the teaching of the course being a core and crucial part of that learning process), it was the issue of gravitational collapse of stars that clinched things. In January 1952 Wheeler had studied Oppenheimer's papers on continued gravitational contraction (co-authored with Volkoff and Snyder respectively), and was concerned by the singularities that they predicted as the end state of such collapse following thermonuclear burnout. Wheeler was right away considering escape routes from this conclusion by invoking quantum gravitational considerations to modify the predictions of classical general relativity (in much the same way that others had earlier considered invoking quantum gravitational effects to tame the infinities of quantum field theory). In other words, the issue of collapse appeared to lead to a genuine problem for physics in which one of the central theories appeared to break down. While this might have been the trigger to teach the course on general relativity, it is clear from his pre-Matterhorn work that he was already leaning towards more relativistic matters.

[30] Oddly, especially since Einstein was a presence there (at the Institute for Advanced Study), this was the first time such a course had been taught in the physics department—though as David Kaiser has pointed out,

from the table of contents taken from this course that the integration of quantum and gravitation was already a central issue in Wheeler's mind (see figs. 8.2 and 8.3).

> November 1st, 1952 was the test of the Mike H bomb. I can remember flying the helicopter over the waters surrounded by Eniwetok atoll and looking down at the sharks circling underneath and feeling they were a greater source of danger than what would happen to this H bomb. But I was already eager to get back to pure physics and had asked Shenstone, by that time the Chairman of our physics department, if I could give a graduate course in relativity. I evidently already realized at that time the reason universities have students is to teach the professors. It was a great thing to give that course and learn about gravitation, and one of the outcomes of it was a model of something particle-like and yet not a particle: a geon, a gravitational wave going around in a circle held in orbit by the mass-energy of that wave itself.[31]

But Wheeler had already amassed a book full of notes during his Matterhorn days, and was well-prepared for his course by 1953. Interestingly, Wheeler begins these lectures with the concepts of particle, field, and action-at-a-distance (see figs. 8.2 and 8.3 for course contents—albeit of what might be a slightly different course: see footnote 30). The geon concept appears early on. The basic principle behind the geon is simply the same as that behind the deflection of light by strong gravitational fields of the kind demonstrated by Eddington in 1919: given enough mass, a beam of passing radiation (e.g. light) can be bent into a circular orbit around the mass. By increasing the radiation intensity to match the gravitational field induced by the original mass causing the bending, one can have a self-subsistent quasi-object consisting of the light itself held together by its own gravity. What a geon does is use the energy of the radiation itself, creating an entity that seems to behave like any other massive object. In this case, then, mass can become a soliton-like feature of sufficiently curved space rather than a property of a 'carrier': what he would later call "mass without mass"—indeed, it was later realized that one could have purely gravitational geons.

There is no doubt that initially at least geons were intended to provide a model for elementary particles, despite the fact that Wheeler's work on geons begins with his disenchantment with particles. As he points out:

> It had become clear through the work of Feynman and others in the field of elementary particle physics – quantum electrodynamics – that the picture of the electron as a

---

it had been offered in the mathematics department as 'Mathematics 570' ([Kaiser (2000)], pp. 567–8). As Kaiser also notes (ibid., p. 569, footnote 6), though Wheeler states in his memoirs that he first gave the course in the fall of 1953, it does not appear in the Princeton list of courses until 1954. In fact, it appears as though many relativity components were shoehorned in as part of a course on 'advanced quantum mechanics' (Physics 561–see JAWP, Box 81) in 1953–5 (see figs. 8.2 and 8.3—note, however, that the dates on the title page have been scribbled out and replaced with 1954–1955; I shall be using the course notes from this offering since they overlap in dates and, one can presume, content). Charles Misner recalls that Wheeler had taken his first class (the first time the course ran) to see Einstein, and that was the year before he joined Wheeler as a graduate student (Interview of Misner by the author and Donald Salisbury, University of Maryland, March 16th, 2011).

[31] Interviewed by Kenneth W. Ford. Location: Jadwin Hall, Princeton University. Interview date: Monday, February 14th, 1994: https://www.aip.org/history-programs/niels-bohr-library/oral-histories/5908–7.

Advanced Quantum Mechanics

1953-1954

DETAILED LIST OF TOPICS, FIRST TERM

The star refers to the second lecture given each Monday. This lecture comes at 4.30 pm except on days of faculty meetings when it is moved forward to 4.00 pm (double star).

1   Sept. 27   Survey of the problem of fields and particles: historical review; electromagnetism; the idea of the quantum and the neutrino.

2*   Sept. 27   Survey of the problem of fields and particles concluded. Developments of electron theory from the simple Dirac starting point to hole theory to a field theory; conclusion that the electron is not a simple particle; the geon.

3   Sept. 30   Resumé of relevant principle of relativity, general and special.

4.   Oct. 4   Resumé of classical electrodynamics and related boundary value issues.

5*   Oct. 4   The geon, an entity held together by gravitational forces, as a self-consistent, non-singular completion of the scheme of classical physics.

6   Oct. 7   The quantum idea. The relation of the Feynman and the Schroedinger pictures.

7   Oct. 7   The quantum idea. Boundary conditions; contact transformations; and indeterminism relations, seen from the Feynman point of view.

8*   Oct. 11   The quantum idea. Alternative forms of the action principle. Relativistic treatment of one particle and difficulties thereof.

9   Oct. 14   Classical relativistic source-free field theory; action principles. Lagrange and Hamiltonian descriptions.

10   Oct. 18   Classical relativistic source-free field theory; contact transformations; Hamilton-Jacobi functions; functionals and functional derivatives.

11* Oct. 18   Feynman method of field quantization. Definition of state; correspondence to classical theory; question of equivalence of alternative action principles.

12   Oct. 21   Schwinger form of action principle for field in Schroedinger representation.

Fig. 8.2 *Table of contents from the first term of John Wheeler's course on advanced quantum mechanics, 1953–4. Source: JAWP, Box 81.*

13	Oct. 23	Schwinger form of action principle for field in Heisenberg representation.
14[*]	Oct. 25	Open for a topic not in chronological sequence.
15	Oct. 28	Covariant commutation laws.
16	Nov. 1	Emission and absorption operators; spontaneous emission.
17[**]	Nov. 1	Bohr-Rosenfeld analysis of field fluctuations.
18	Nov. 4	Details of the Bohr-Rosenfeld analysis.
19	Nov. 8	Interaction representation.
20[*]	Nov. 8	Open for a topic not in chronological sequence.
21	Nov. 11	Interaction representation continued.
22	Nov. 15	First applications of the interaction representation.
23[*]	Nov. 15	Two photon emission processes.
24	Nov. 18	Line width.
25	Nov. 22	Further details on line width; line shift.
26[*]	Nov. 22	First analysis of mass renormalization.
27	Nov. 29	Summary of quantum electrodynamics to this point.
28[*]	Nov. 29	Consequences of quantum for general relativity.
29	Dec. 2	Geometrical interpretation of spinors.
30	Dec. 6	The algebra of the spin matrices; Lorentz transformations.
31[**]	Dec. 6	Survey of neutrino physics.
32	Dec. 9	Dirac wave equation and general relativity; specialization to neutrinos
33	Dec. 13	Solution in radial coordinates; stress-energy tensor.
34[*]	Dec. 13	Neutrino geon; comparable to what extent to particles?
35	Dec. 16	Dirac theory of hydrogen atom.
36	Jan. 10	Photo effect and single quantum annihilation for large Z.
37[*]	Jan. 10	Critique of Dirac electron theory.
38	Jan. 13	Nuclear scattering of electrons before radiative corrections.

Fig. 8.3 *Table of contents (page 2) from the first term of John Wheeler's course on advanced quantum mechanics, 1953–4. Image source: JAWP, American Philosophical Society, Box 81.*

primordial entity – a point-like object, an independently existing thing, and everything built out of electrons or things like electrons, even if they do go backward and forward in time – is not a good way to do physics, because in the neighborhood of what you think is an electron – experimental electron – there are in effect these things going back and forth in time; there's a whole atmosphere of virtual electrons. So it's much more a field theoretic thing than the particle picture admits, or realizes.[32]

In the earliest phase of geon research the term 'kugelblitz' (ball lightning) was used instead, most likely revealing some connection to the plasma research. Wheeler's shift in nomenclature, from 'kugelblitz' to 'geon,' is indicative of the way he conceptualized his new solutions over time.[33] During the kugelblitz period, Wheeler's concern was with gravitational collapse and its independence, in the context of general relativity, from matter. As he puts it, "[y]ou could have empty space, radiation, collapsing too". This itself signals the possibility of using empty space as a resource, which leads to the geon period (certainly established by the beginning of 1954), in which the '-on' suffix signifies an ontological treatment and a direct link to elementary particles.

The first lecture Wheeler gave on the geon concept (the fifth lecture in his course given on October 4th, 1953—or, likely, 1954) appears to have been based on a draft of the manuscript of his 1955 Geons paper. There are, then, no lecture notes. However, there are three slides and a brief summary in the course notebook (see figs. 8.4, 8.5, 8.6). Also noted is that fact that two questions were posed (but not answered):

1. "What are small geons if they are not elementary particles?"

2. "How can one know what new physical concepts to invent, or even if invention is necessary, if one does not explore the rich consequences of the concepts one already has?"

The second question is, of course, repeated verbatim in his letter to John von Neumann.[34] With regard to the first question, Wheeler still holds up the geon as "an impression

---

[32] Interview of John Wheeler by Finn Aaserud on May 23rd, 1988, Niels Bohr Library & Archives, American Institute of Physics: www.aip.org/history-programs/niels-bohr-library/oral-histories/5063–2.

[33] Wheeler knew the significance of nomenclature, having served on an international committee on symbols, units and nomenclature while vice-president of the International Union of Pure and Applied Physics (1951–1954)—see [Wheeler (1998)], p. 172.

[34] Wheeler was seeking reassurance from a variety of sources about how to proceed in general relativity in such a way as to stick to the maxim expressed in his second question. For example, on November 3rd, 1953 (JAWP, Box 15), he wrote to Behram Kursunoglu (calling himself just an "innocent student" of the subject) to ask if a "conservative physicist" (that is one "unwilling to introduce new ideas, new concepts, and particularly unwilling to introduce any quantity with the character of a fundamental length except as called for by inescapable evidence") has to use the standard action in general relativity (combined with electromagnetism), namely: $\int\int\int\int \left\{ \frac{c^3}{16\pi G}R - \frac{1}{16\pi G}F_{\alpha\beta}F^{\alpha\beta} \right\} \sqrt{-g}dx^1 \cdots dx^4$ (the quadruple integral is Kursunoglu's own choice, which should perhaps be replaced by a single one). Kursunoglu had been developing a unified field theory based on just such a fundamental length, and Wheeler points out that it reduces to the old action when the fundamental length parameter goes to zero, and this is the action, insists Wheeler, that any good conservative physicist must adopt—he is only unhappy, says he, that "no one has quantized it and investigated its consequences".

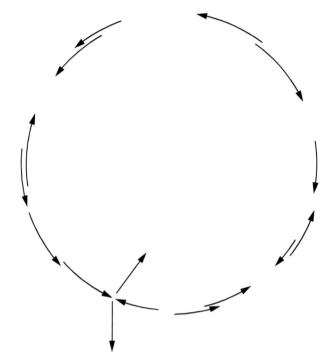

Fig. 8.4 *Slide from Wheeler's first lecture on the geon, as part of his general relativity course at Princeton. This is a representation of a system of photons or geons held together by mutual gravitational attraction, with escape due to collision of two quanta—as the arrows indicate, the energy flows in both directions. Source: JAWP, Box 81.*

of a conceivable theory of particles" (Lecture 32) in a context in which there is no respectable alternative model. In his letter to John von Neumann (October 12th, 1954: JAWP, Box 31) he writes that he is "still trying to understand elementary particles in terms of existing concepts" (that is geons and multiply connected spacetime). However, in the 1955 paper on geons he notes that there is a critical value of $(m^2 c^3/e\hbar)$ (with m the mass, e the elementary charge, c the speed of light, and $\hbar$ the reduced Planck constant) above which classical physics cannot be applied,[35] and electron–positron pairs will be produced from the vacuum. The implication of this is that classical geons must be a minimum of 10^{39} grams (with no maximum mass): a classical geon with a physically reasonable field strength value would be staggeringly enormous—see fig. 8.6. What it nonetheless reveals, however, is that even at this stage, Wheeler's eye was on particle

---

[35] This critical quantity possesses the physical dimension of an electric field strength (it is, as Wheeler notes, the critical field of the theory of electron pairs) and gives the upper bound above which the linear displacement parallel to the electric field and amount $\bar{\lambda}$ (the reduced Compton length $= \hbar/mc$) involves an energy larger than mc^2. This limit is today known as the Schwinger electric field strength or Schwinger limit (with unit dimensions $[mass][length][time]^{-3}[current]^{-1}$).

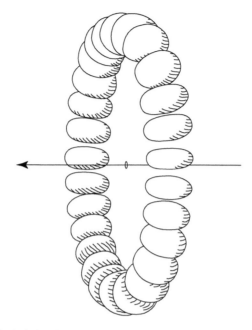

Fig. 8.5 *Slide from Wheeler's first lecture on the geon, as part of his general relativity course at Princeton. This is a representation of the distribution of the electric field energy in a toroidal geon. Source: JAWP, Box 81.*

physics, with gravitation amounting to an explanatory resource—despite the fact that classical, static gravitational fields themselves are too unstable to be useful in elementary particle problems.

This course features several brief forays into quantum gravitational territory.[36] Including the following remark that reads as a battle cry:

> The mathematics of the quantization of general relativity theory has not been thoroughly investigated because of general fear that it is non-linear in just the sense that will throw out [the propagation kernel mapping between field configurations–DR] and leave no prescription for the calculation of a substitute. Obviously the non-linearity of a well founded physical theory is no adequate reason for shutting one's eyes to the consequences of that theory. Hydrodynamics is not left unexploited because it is non-linear. Similarly one cannot expect the rich ground of general relativity to lie fallow indefinitely.

He also considers the Dirac equation in curvilinear coordinates, and the complications (or, rather, addition) that spinors make to Einstein's general theory, noting that they

---

[36] A paper submitted to the 1957 Gravity Research Foundation essay competition, entitled "Gravity Can Glue Together Energy to Make Matter," explicitly linked small geons to quantum mechanics and quantum gravity.

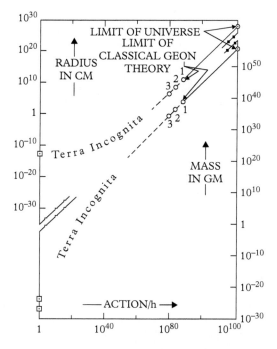

Fig. 8.6 *Slide from Wheeler's first lecture on the geon, as part of his general relativity course at Princeton. This diagram indicates the range of masses and radii for which geons can be given a classical treatment (along with the comparison magnitudes for elementary particles). Source: JAWP, Box 81.*

provide freedom to describe at a point "directions – and rotations of directions – that have nothing whatever to do with changes in the Einstein coordinate system" (Lecture 33).[37] Much of the groundwork for his later pioneering work on gravitation was established in this course, despite the fact that it was ultimately directed at explaining the existence of elementary particles. He also cultivated in this period a handful of students that would, to a large extent, explore his second question (albeit initially in ways directly pushed by Wheeler), and impress upon them a similar mindset in how to do physics.

8.3   Geometrodynamics and the Already Unified Theory

The geon project, in which geons are understood as completing classical physics by grounding the notion of body, in zero rest mass and zero charge fields ('zero fields': electromagnetic, neutrino, and gravitational), leads quite naturally to geometrodynamics. According to the geometrodynamic program, spacetime does not contain real examples

---

[37] There also appears an appendix, written by Charles Misner, based on a set of lectures Wheeler had given at a seminar devoted to the study of the Dirac equation in general relativity.

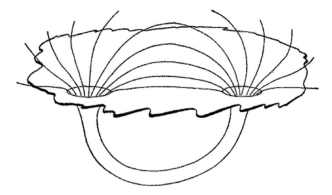

Fig. 8.7 *Lines of force threading through a 'wormhole' (then simply referred to by it's topological name: 'doubly connected space'. The idea (due originally to Weyl), is that the mouths behave as equal and opposite charges. Image source: [Wheeler (1955)], p. 535.*

of mass and charge (i.e. they are not really 'elementary'): these are constructs from spacetime geometry and topology (in particular, from geons and multiply connected regions of space: see fig. 8.7[38]). The example of charge coming from topology is in fact drawn from Hermann Weyl who argued that in the case of a multiply connected space, one cannot say "here is charge, but only that this closed surface, which is located in the field, contains charge" ([Weyl (1924)], p. 57).[39] There is a further precedent in the form of the Einstein–Rosen bridge of 1935. In their paper "The Particle Problem in the General Theory of Relativity" (Einstein and Rosen (1935)), they had attempted to derive an atomistic theory of matter and charge from the field variables of general relativity (i.e. the gravitational field $g_{\mu\nu}$) and Maxwell's theory (the electromagnetic field, ψ_ν). They draw attention to the possibility of deriving quantum phenomena, though do not tackle that directly.[40]

---

[38] In modern topological terminology "n-fold connectedness" means that the universal cover is n-sheeted, or, equivalently, that the order of the fundamental group is n. In this case, Wheeler's wormhole is *infinitely* rather than doubly connected (irrespectively of whether we consider the picture two- or three-dimensional)— my thanks to a referee for pointing this out.

[39] As Wheeler puts it: "One key point, one key inspiration, for what we [i.e. Wheeler and Misner–DR] were doing was a remark by Hermann Weyl back in the 1920s, 'An electric charge is not a point out of which something showers, but it's a region of space out of which electric lines of force diverge.' One could envisage a multiply connected spacetime in which every positive charge is at one mouth of the wormhole in space and negative charges at the other mouth" (Interview of John Wheeler by Kenneth W. Ford on March 4th, 1994, Niels Bohr Library & Archives: https://www.aip.org/history-programs/niels-bohr-library/oral-histories/5908–9.). A result by Misner, in 1956, that would have been crucial for Wheeler's geometrodynamic program, had it persisted, was the proof of charge conservation in wormhole worlds—i.e. that the flux lines of force through the wormholes is constant over time (given fixed topology, but not fixed metric)—the result is indicated in a summary of Misner's work, dated "as of 30th March, 1956," an earlier mention of the result also appears in a letter to James Fisk (December 8th, 1955), in which Wheeler notes that Misner's flux invariance result naturally suggests an association with charge: JAWP, Box 18.

[40] John Synge also claimed to have developed the idea of charges appearing as "U-tubes" in a multiply connected space in 1947 (letter from Synge to W. S. Ament, May 26th, 1957: JAWP, Box 29), though left the work unpublished, speaking on the subject twice only (at a colloquium at Carnegie Tech, and in Dublin, both in 1947).

But Wheeler's vision in geometrodynamics was more all-encompassing, including all aspects of the bodies and the Maxwell equations too, all from curvature:

> So the electromagnetic field makes footprints on spacetime so characteristic that from those footprints and the curvature of spacetime one can read back to almost all of the electromagnetic field. In this sense the electromagnetic field could be regarded as just a feature of geometry. Actually it's not quite as simple as that. There is a feature of the electromagnetic field which is left undetermined by the geometry—so-called complexion. Anyway, we thought this is an open invitation to ask to what extent the whole show—particles and fields of force—can be understood to be geometrical in origin.[41]

The already unified theory played an important role in Wheeler's shift to geometrody-namics:[42] it suggested the possibility that geometry (and, by extension, topology) could be used as a vast explanatory resource capable of describing the whole of classical physics (more or less in line with Clifford's vision expressed in [Clifford (1982)]). It is clear that both Wheeler and Misner viewed geometrodynamics as belonging firmly within the 'unified field theory' camp—though unified field theory *without any new inventions in physical theory*. Indeed, Misner, reacting to the fact that Rainich had already done much of what he and Wheeler had discovered, is explicit about this (see fig. 8.8). The geons and the threaded wormholes were to work in tandem to provide an overall package that realized the unified field theorist's dreams.[43]

As we saw in the previous chapter, the potential value of gravitation as a resource had already been probed in the context of providing a method of eliminating divergences. Wheeler speculated that it could do a whole lot more.[44] However, rather than having *a priori* a multiply connected spacetime capable of allowing electric lines of force to be threaded, to create charge without charge, Wheeler wanted to *derive* such features. It was for precisely this reason that he was led to consider the Feynman quantization of general relativity—indeed, Wheeler speaks of being "forced to consider such topologies when one adjoins Feynman's way of quantization, á la sum over field histories, to the formalism of general relativity" (Letter to John von Neumann, October 12th, 1954: JAWP, Box 31).

---

[41] Interview of John Wheeler by Kenneth W. Ford on March 4th, 1994, Niels Bohr Library & Archives: https://www.aip.org/history-programs/niels-bohr-library/oral-histories/5908−9.

[42] We can distinguish between 'strong' and 'weak' geometrodynamics here. According to the latter, it simply refers to the fact that general relativity is about the dynamics of geometry. According to the former there is a kind of metaphysical element too: the dynamics of geometry constitutes *all there is!* I have in mind the stronger sense here. See [Lehmkuhl (2018)] for a philosophical analysis of this stronger form (which is discussed there under the label 'super-substantivalism').

[43] There is a recent resurgence of the importance of the wormhole concept in the context of the AdS/CFT duality. Wormholes are viewed, in this framework, as 'dual' to entanglement so that, as the slogan says, 'ER=EPR' (that is, the Einstein—Rosen bridge, or wormhole is equivalent to a pair of entangled particles). This has been expanded to the more all-encompassing 'GR=QM,' so that general relativity and quantum theory are equivalent in terms of physics.

[44] In a later paper, Misner and Finkelstein relate the machinery of geons to the old 'space theory of matter' of Riemann and Clifford. They in fact propose using Wheeler's 'geon' terminology to apply more broadly, as "pidgin-Greek," for any geometrical entity in the context of the space theory of matter ([Finkelstein and Misner (1959a)], p. 241).

As I mentioned to you as you were leaving, the unified field theory I have been looking into had been published by G. Y. Rainich in 1924. The fact that it has been ignored makes it no less valuable, and your geon together with my work filling in the details of the idea of charge you suggested show that in content it acheives many of the hopes that have motivated all attempts at unified field theories. Thus I feel that we are bound to publish a paper whose aim would be to revive this unified field theory. But more than this, the acheivements of this theory should be so persuasively presented as to eliminate the motivation for inventing novel unified theories, and summarized in a check list openly presented as a "put up or shut up" challenge to inventors of new theories.

11 Aug 1956 from Misner to me at St. Gilgen

For this purpose a variation principle is quite important since it is a criteria that most unified field theorists seem to value, and they will of course only be deterred from further invention by a theory which satisfies their own sensibilities. The null case is not so important since it can be argued that it is a fiction, a limiting case which never exactly occurs in nature; so even if the Einstein-Maxwell equations allow it and the Rainich unified theory does not, Rainich's theory is not in contradiction with experiment and shows a much closer relationship to classical electromagnetism than has ever been demonstrated for any other unified theory. Still, the question of whether or how the null case can be included in Rainich's theory is an answerable question, so one feels guilty if he doesn't take the trouble to answer it.

Fig. 8.8 *Letter from Charles Misner to John Wheeler, August 11th, 1956: Box 18.*

Charles Misner had, by January 1956, already made progress on this problem and was able to state that if the Feynman path integral is used "as a starting point for a quantum theory of gravity, then the requirement of general covariance specifies uniquely the relative weights which must be assigned to different field histories" (report dated January 1st, 1956: JAWP, Box 18)—essentially, Misner uses the Haar measure to do the sum. It seems, from a letter of Misner's (March 10th, 1957: JAWP, Box 18), that there was an expectation that all sorts of quantum numbers would come out of the Feynman quantization (strangeness, isotopic spin, etc.).[45]

8.4   Spacetime Fluctuations and The Planck Scale

Already in his initial teaching of general relativity in 1953, Wheeler is convinced that the principles of quantum theory might provide insights into various "deeper questions" of general relativity and gravitation, in particular the questions of why spacetime is four-dimensional with signature $(+++-)$, which must still be inserted *a priori* in general relativity. Indeed, he already has the basic insight that applying Feynman's sum-over-histories approach to general relativity leads to fluctuations of spacetime that include distinct topologies, in addition to the already-voiced idea that the geometry might fluctuate:

> Feynman's formulation of quantum mechanics expresses the wavefunction for a wave field as the sum-over all field histories consistent with the preassigned initial and final conditions–of an expression that contains the classical integral, $\mathcal{J}_s$, for the field history:

$$\Psi \sim \sum_{\text{Histories}} \exp(i\,\mathcal{J}/\hbar) \tag{8.1}$$

> Here the sum goes over all conceivable histories of the field, not merely those that are consistent with the variational principle of the classical field equations, $\delta\mathcal{J} = 0$. Consequently we have to sum in the case of general relativity not only over configurations of the metric field with signature $(+++-)$, but over all signatures; not merely over singly connected topologies, but over doubly and multiply connected topologies of the space time field. Such an approach is not a mere matter of choice; it is demanded by the very spirit and method of quantum theory. (Wheeler, Lecture Notes on Advanced Quantum Mechanics, 1953–4, JAWP: Box 81.)

Thus he was led inexorably into quantum gravitational considerations via his work on multiply connected spaces as a resource for physics: if the original main motivation of the geon project was to get 'mass without mass' and objects for free, the topological complexities of the 'charge without charge' idea were more fruitful for Wheeler's later work, pushing him to a means of generating those complexities (the charge of a bare particle becomes a fluctuation phenomenon). He even contemplates using this quantum gravitational scheme to get out an explanation for the four-dimensionality of spacetime,

---

[45] Misner concludes this letter by referencing a paper 'in preparation' with the title "On Quantum Numbers Characteristic of the Gravitational Field." It never appeared, but it was clearly an attempt to provide a more all-encompassing version of a unified field theory that avoided the usual objections about Einstein's approaches, that they ignored recent discoveries in particle physics.

pointing out that two and three dimensions are "trivial," and that five and higher dimensions might receive low weighting in the path integral.[46]

Wheeler visited Leiden, where he was also giving a set of Lorentz lectures, in 1956. He was once again back in touch with Bohr, again keen to discuss a range of topics. Amongst these was the issue of how several fields can combine to give a vacuum with zero mass density. However, he was also still focused on his effort to recover elementary particles from zero rest mass fields, of which, as he puts it, he has not "been able to discover yet any sign that this what I call conservative and invention-free approach is crazy" (letter to Bohr, dated April 24th, 1956: JAWP, Box 5). He took three graduate students along with him: Charles Misner, Peter Putnam,[47] and Joe Weber. With the latter two, Wheeler discussed geons and gravitational radiation. With Misner he was discussing the problem of quantizing gravity and, in particular, the status of spacetime concepts as one approaches ever smaller distances. It is here that the notion of the Planck length is once again placed centre-stage as of pivotal importance in the limits of general relativity and quantum theory, with gravitational fluctuations leading to new physics at distances $(\hbar G/c^3)^{1/2}$.[48] Writing to Bohr[49] about this latter seemingly quantum gravitational issue, he writes:

---

[46] This was only schematic, and many difficult technical problems remained, not least making sense of the measure for path integration. He mentioned this in his letter to von Neumann, noting that a new formalism would be required to extend Feynman's sum to "an omnidimensional or continuous space;" continuing: "What one uses for action function is one question on this score. But a much bigger and prior question to me, with my ignorance of geometry, is this – what sort of formalism does one use in the first place to deal with such queer spaces?" (October 12th, 1954: JAWP, Box 31). It was Charles Misner who would be set to work on this problem. Wheeler would be led via his work on quantum geometrodynamics to the canonical approach to quantum gravity, and a similar problem arose there. In the first of the papers from his 1962 trilogy on quantum gravity, Bryce DeWitt noted how a "fundamental question" would recur in his discussions of (canonical) quantum gravity with Wheeler: "*What is the structure of the domain manifold for the quantum-mechanical state functional?*" ([DeWitt (1967)], p. 1115). This led, of course, to Wheeler's notion of 'superspace,' or the space of all possible 3-geometries over which the quantum state functionals of quantum gravity are defined—Wheeler's hope was to get topological transitions to generate his 'particle friendly' multiply connected spacetimes (though as a referee has pointed out, only later work by Mikhail Gromov allowed one to define a 'Superspace' of geometries without fixing the topology, so that topology change could be described).

[47] Putnam had also traveled to Europe almost a decade earlier with Wheeler (in November 1947), during which they had visited Bohr together in Copenhagen—see letter to Bohr, dated November 11th, 1947: JAWP, Box 15. Wheeler was focused on the action-at-a-distance work with Feynman, and also the idea of taking existing theories down to smaller and smaller distances (along with the self-energy problems), with a view to better understand elementary particles. One wonders if there was a continuity in his thinking between this line of reasoning and the conclusion that one is forced to consider spacetime fluctuations at 10^{-33} cm. This would certainly line up with his early focus (in terms of his gravitational interests) on gravitational collapse, which indicated the necessity to look at very small distances.

[48] Wheeler has remarked in several interviews that he coined the terminology of 'Planck units' in this paper. However, it is nowhere to be found. What we do find is the *expressions* for what we now call the 'Planck length' and 'Planck energy,' along with the physical interpretation that such lengths correspond to the dimensions of the wormholes connected to fluctuations in the gravitational field (what would later be called 'spacetime foam'—indeed, Wheeler describes the character of the fluctuating space as 'foamlike' in the aforementioned letter to Bohr).

[49] Interestingly, Wheeler also makes references to a draft of Everett's thesis, though at this stage (April 24th, 1956) the thesis title is "Wave Mechanics Without Probability," which faces the common objection to Everettian interpretations (that they don't make sense of the quantum probabilities) rather head on!

[W]e have to take seriously fluctuations in the gravitational field down to dimensions of the order of $L_g = (\hbar G/c^3)^{1/2} \, 10^{-33}$ cm. Then changes in the metric, of the order $g_{ik}L_g/L$, are great enough in comparison with unity to force consideration of multiply connected topologies of the space-time continuum, such as are necessary for the explanation of charge in a theory that does not start with the idea of charge. The flux through a "handle" or "wormhole" represents the charge. ...I see no escape from saying that the flux coming out of the end of a tunnel mouth is to be identified, not with the charge of an elementary particle, but with the charge of the "bare particle" as it is envisaged in quantum electrodynamics. In other words, the charge of the bare particle is regarded as of fluctuation theory origin. (Letter to Bohr, dated April 24th, 1956: JAWP, Box 5)

Thus, while Wheeler's focus is still on explaining properties of elementary particles here, he is led to considerations of the fluctuations of the spacetime. He views such quantum gravitational predictions as *forced* by the joint consideration of then-known physics (and *only*, he insists, the then-known physics, or gravitation, electromagnetism, neutrino physics, and quantum theory).

Just before Leiden, in October 1955, Wheeler was having discussions with Joe Weber and Charles Misner on the quantization of gravitation in combination with electromagnetism.[50] Central to these discussions was the need to develop machinery to deal with the quantum mechanical sum-over-histories in the context of general relativity. Again, the expectation was that from a physical point of view, topology change would be a natural part of the sum-over-histories, rather than just summing over all metrics compatible with some one topological structure (see fig. 8.9).

The discussion also focused on Wheeler's earlier idea of producing a kind of explanation of four-dimensionality by virtue of the weighting of the 'special properties' of $3+1$ more heavily in the sum, though this was still left as more of a hope than a proven feature. A further direction, of interest from the point of view of the *split* between approaches to quantum gravity, involves the blending of aspects associated more with

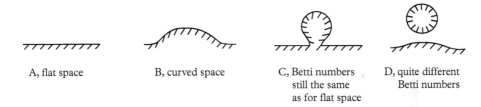

| A, flat space | B, curved space | C, Betti numbers still the same as for flat space | D, quite different Betti numbers |

Fig. 8.9 *Notes from discussions between Wheeler and Joe Weber and Charles Misner, October 1955. A version of these drawings would later appear in Wheeler's first paper on quantum geometrodynamics: [Wheeler (1957), p. 607].*

[50] Weber got his start of gravitational waves (a kind of geon-type solution) during this visit. His initial problem had been the quantization of electromagnetism in multiply connected space (see fig. 8.10). However, as part of this general discussion group, he began to think about the polarizability of geons as an analog to the polarizability of elementary particles.

NOTES ON QUANTIZATION OF GRAVITATION AND ELECTROMAGNETISM (Outgrowth of conversations with Charles Misner and Joseph Weber; motivation partly to summarize present state of our thinking, partly to lead up to Joseph Weber's proposal to investigate quantization of electromagnetism in a multiply connected continuum. -- J.A.W. -- 15 October 1955)

Selection of a particular framework of ideas from among several possibilities

Possibility (1) -- To forget temporarily the electromagnetic field and look at how to do the sum over histories for the metric field. Under investigation by Charles Misner. Expect that appropriate volume element for history integration will be defined by the action principle of general relativity itself, as in the more elementary examples discussed by C. Morette, Phys. Rev. 81, 848 (1951). Consequently anticipate definability of volume element by means of techniques of differential geometry. The issue is not so much the do-ability of the definition, as the do-ability of the integration. The non-linearity of gravitation theory makes one fear the worst. Contrast this non-linear theory with electromagnetism. There the integration goes like a dream. The exponent is quadratic in the field variables. The answer is the classical action for the classical history that leads from the initial to the final field configuration. Hopes for such a simple outcome in the non-linear gravitation problem may be low, but should absolutely not be abandoned. The action principle there is founded on the most simple geometrical notions. Consequently the answer to the integration ought also to be simple. After all, what can it depend on? Only on simple invariants. By exploiting this line of reasoning one might be able to get the answer by no calculation at all. As a preliminary way to test this possible approach, one can try it on electromagnetism itself, where one knows the answer.

Possibility (2) To make the machinery of the discussion still more abstract, but to keep the general plan and scope the same as in possibility (1). By "abstract" is here meant a mathematical machinery that bears the same relation to the many alternative coordinate systems of general relativity that the abstract theory of groups bears to the theory of group representations. An October 12

Fig. 8.10 *Notes from discussions between Wheeler and Joe Weber and Charles Misner.*

Bergmann's 'canonical quantization' school with the Feynman quantization approach. The idea was to utilize Arthur Komar's invariant coordinates (using the invariants of the Riemann tensor) to nail down the physical content that would go into the sum-over-histories. This appears to have been invoked, later, in the gauge-fixed path-integral formalism of Fradkin and Vilkowisky, in the context of non-Abelian Yang–Mills theories and also to general relativity itself (albeit with difficulties in doing the integral).

What is important to note in this, as with Wheeler's earlier work in gravitational physics, is that even here gravitation and general relativity are being used as a resource or aid in the study of elementary particle physics. Ironically, Wheeler's journey from particle physicist to general relativist took far longer than is often supposed. Even in 1963, in his lectures for Les Houches (Relativity, Groups, and Topology), Wheeler, in considering the issue of the final state (of gravitational collapse), writes that "It cannot be supposed... that even a purely gravitational configuration, nor pure gravitational radiation, are entirely divorced from issues of elementary particle physics" ([Wheeler (1964)], p. 326). Indeed, he views the issue of collapse (and matter destruction and creation as a result of such) as "a decisive area of elementary particle physics" (ibid., p. 504). Wheeler goes on to repeat his ideas relating to particles, but also a host of other structures from the theory of elementary particles, to topological features of space (see, e.g., his Table XX in ibid., p. 512): he remains convinced that the physics of the vacuum at small distances is dominated by the multiply connected nature of space.[51] However, the modern day picture in which gravitational collapse is linked with new physics at the Planck scale is firmly established at this stage, and its relevance as a motivating factor for quantum gravitational research solidified. As Wheeler poetically puts it in the close of his 1963 Les Houches lectures:

> Whatever the outcome, one feels that he has at last in gravitational collapse a phenomenon where general relativity dramatically comes into its own, and where its fiery marriage with quantum physics will be consummated.
>
> ([Wheeler (1964)], p. 518)

8.5 Conclusion

Although Wheeler's work on geons and geometrodynamics points to many concepts and results that would be of importance to quantum gravity research, these projects, initially, were rather old fashioned, harking back to the classical 'unified field theory' work of Einstein.[52]

---

[51] Likewise, in his 1967 Battelle Rencontres lectures, quantum geometrodynamics is still viewed as offering a comprehensive account of physics (in which geometry is the "ultimate dynamical object"), from which particle physics emerges—in this case, however, a particle is now viewed as "a quantum state of a collective excitation of the entire geometrical continuum" ([Wheeler (1968)], p. 268).

[52] Perhaps this is why Wheeler and his geometrodynamical ideas (and his general style) were so mercilessly lampooned at the Royaumont conference (I quote at length for full effect, cited in [Schucking (1990)], pp. 481–2):

Moreover, this work that we now tend to think of as foundational in quantum gravity—e.g., we often think of 'quantum geometrodynamics' as just another phrase for 'quantum gravity'—had its roots firmly embedded in the quest for understanding the elementary particles. It wasn't until after 1957 that Wheeler began to look seriously at general

"This paper is entitled 'Space-time without space-time.' It forms Part 23 of a critique of classical field theory. Parts 24 to 30 have been published already. Parts 1 to 22 will appear during the next few decades.

I shall speak about the already-unified field theory. I should like to begin with a quotation from Sir Isaac Newton, who has reminded us so often and with such importance that 'force is proportional to acceleration.'

In the already-unified field theory, we have a set of field equations, some of the 2nd differential order, some to the 4th order, in the metric tensor. The already-unified theory provides simultaneously a description of the gravitational and electromagnetic fields.

Recently, a young student at Cambridge has proposed a new description of the theory which seems to me to represent a major contribution to the field. One may understand it in this way:

In order to reduce the order of the equations from 4 to 2 so that one may apply the well-known techniques of the theory of hyperbolic differential equations, and also formulate a satisfactory variational principle, one introduces an auxiliary anti-symmetric tensor field F_{ik}, which we propose to name the electromagnetic field tensor, since by its use one may in a very elegant and beautiful way separate in the already-unified theory the gravitational field. One might describe the form of this new theory as 'unification without unification.' It turns out that the new electromagnetic field tensor satisfies two sets of equations of the first order, which are very simple and which one may hope to solve explicitly in some cases. At the same time, the electromagnetic field, through a symmetric tensor, which we intend to refer to as the Maxwell energy tensor, after the student who proposed this formalism. A very promising student, I might say: the best since Misner. The Maxwell energy tensor has the same canonical structure as the energy-momentum tensors in Lorentz-invariant field theories, a remarkable fact of which one has not yet a complete understanding. One hopes that the exploration of these ideas will make it possible to develop the great richness of general relativity up to the hilt, in the exciting arena of geometrodynamics. And the chorus sang:

It's a long way to quantization,
It's a long way to go.
It's a long way to quantization
As Bergmann ought to know.
Good-bye, Palatini,
So-long Inverse Square
It's a long way to quantization,
But my heart's right there."

Engelbert Schucking suspects this is the work of Felix Pirani, and since I discovered (during a visit to his house) that Pirani wrote similar skits for several other conferences, I am almost certain he is correct: one can recognize the lion by its paw!

relativity and quantum gravity independently from concerns in particle physics, and this shift in fact coincides with a more general trend to treat gravitational physics as a worthwhile field in its own right (the so-called 'renaissance,' which is covered in the next and final chapter).

···

REFERENCES

Alkemade, A. (1994) *On Vortex Atoms and Vortons*. Ph.D. thesis, Technische Universiteit Delft.

Barrett, J. A. and P. Byrne (eds.) (2012) *The Everett Interpretation of Quantum Mechanics: Collected Works 1955–1980 with Commentary*. Princeton: Princeton University Press.

Bergmann, P. G. (1959) Summary of the Colloque International De Royaumont. In A. Lichnerowicz and M.-A. Tonnelat (eds.), *Les Théories Relativistes de la Gravitation* (pp. 463–72). Paris: Centre National de la Recherche Scientifique, 1962.

Blum, A. and T. Hartz (2017) The 1957 Quantum Gravity Meeting in Copenhagen: An Analysis of Bryce S. DeWitt's Report. *The European Physical Journal H* J42(2): 107–57.

Bromberg, J. (1982) *Fusion: Science, Politics, and the Invention of a New Energy Source*. Cambridge: MA: MIT Press.

Clifford, W. K. (1982) On the Space-theory of Matter. In R. Tucker (ed.), *Mathematical Papers*. London: Macmillan.

DeWitt, C. and D. Rickles (eds.) (2011) *The Role of Gravitation in Physics: Report from the 1957 Chapel Hill Conference*. Berlin: Edition Open Access. http://edition-open-access.de/sources/5/index.html.

DeWitt, B. (1967) Quantum Theory of Gravity. I. The Canonical Theory. *Physical Review* 160: 1113–48.

Einstein, A., L. Infeld, and B. Hoffmann (1938) The Gravitational Equations and the Problem of Motion. *Annals of Mathematics* 39 (1): 65–100.

Einstein, A. and N. Rosen (1935) The Particle Problem in the General Theory of Relativity. *Physical Review* 48(1):73–7.

Feynman, R. and J. Wheeler (1949) Classical Electrodynamics in Terms of Direct Interparticle Action. *Reviews of Modern Physics* 21(1): 425–33.

Finkelstein, D. and C. Misner (1959a) Some New Conservation Laws. *Annals of Physics* 6: 230–43.

Finkelstein, D. and C. Misner (1959b) Further Results in Topological Relativity. In A. Lichnerowicz and M.-A. Tonnelat (eds.), *Les Théories Relativistes de la Gravitation* (pp. 409–13). Paris: Centre National de la Recherche Scientifique, 1962.

Friedman, J. and R. Sorkin (1980) Spin 1/2 from Gravity. *Physical Review Letters* 44: 1100–3.

Gannon, D. (1975) Singularities in Non-simply Connected Space-times. *Journal of Mathematical Physics* 16(12): 2364–7.

Giulini, D. (2016) Aspects of 3-manifold Theory in Classical and Quantum General Relativity. *Abhandlungen aus dem Mathematischen Seminar der Universität Hamburg* **86**(2): 235–71.

Giulini, D. (2009) Matter from Space. arXiv: 0910.2574.

Grünbaum, A. (1973) The Ontology of the Curvature of Empty Space in the Geometro-dynamics of Clifford and Wheeler. In P. Suppes (ed.), *Space, Time and Geometry* (pp. 268–95). Dordrecht: D. Reidel Publishing Company.

Kaiser, D. (2000) *Making Theory: Producing Physics and Physicists in Postwar America.* Ph.D. dissertation, Harvard University.

Kragh, H. (2002) The Vortex Atom: A Victorian Theory of Everything. *Centaurus* **44**: 32–114.

Lehmkuhl, D. (forthcoming) The Genesis of Einstein's work on the Problem of Motion in General Relativity. *Studies In History and Philosophy of Modern Physics.*

Lehmkuhl, D. (2018) The Metaphysics of Super-substantivalism. *Noûs* **52**(1): 24–46.

Mehra, J. (1994) *The Beat of a Different Drum: The Life and Science of Richard Feynman.* Oxford: Oxford University Press.

Misner, C., K. Thorne, and W. Zurek (2009) John Wheeler, relativity, and quantum information. *Physics Today*, April 2009: 40–6.

Pauli, W. (1993) *Wolfgang Pauli: Scientific Correspondence with Bohr, Einstein, Heisenberg A.O.; Vol. III: 1940–1949* (K. v. Meyenn, ed.). New York: Springer.

Schucking, E. (1990) Views from a Distant Past. In N. Ashby, D. Bartlett, and W. Wyss (eds.), *General Relativity and Gravitation* (pp. 479–489). Cambridge University Press.

Silliman, R. H. (1963) William Thomson: Smoke Rings and Nineteenth-Century Atomism. *Isis* **54**(4): 461–74.

Thompson, W. (1867) On Vortex Atoms. *Proceedings of the Royal Society of Edinburgh* **VI**: 94–105.

Weyl, H. (1924) Was ist Materie? *Naturwissenschaften* **12**(30):604–11.

Wheeler, J. (1998) *Geons, Black Holes, and Quantum Foam.* New York: W. W. Norton and Company.

Wheeler, J. (1968) Superspace and Quantum Geometrodynamics. In C. DeWitt and J. Wheeler (eds.), *Battelle Rencontres: 1967 Lectures in Mathematics and Physics* (pp. 242–307). New York: W. A. Benjamin.

Wheeler, J. (1964) Geometrodynamics and the Issue of the Final State. In C. DeWitt and B. DeWitt (eds.), *Relativity. Groups and Topology, 1963 Les Houches Lectures* (pp. 317–520). New York: Gordon and Breach.

Wheeler, J. (1957) On the Nature of Quantum Geometrodynamics. *Annals of Physics* **2**: 604–14.

Wheeler, J. (1955) Geons. *Physical Review* **97**: 511–36.

Wheeler, J. (1946) Problems and Prospects in Elementary Particle Physics. *Proceedings of the American Philosophical Society* **90**(1): 36–47.

9

Institutionalizing Quantum Gravitational Research

Remind me not to come to any more gravity conferences!
Richard Feynman[1]

In order for good, sustained quantum gravitational research to be carried out, there needs to be an environment in which to do so. General relativity got off to a roaring start, with many of the top people investigating the theory, but it had very quickly stagnated by the 1920s.[2] By the late 1940s, universities were, by and large, against the subject. The journals were suspicious.[3] There were no dedicated conferences. Supervisors warned their students to avoid the subject, lest they render themselves unemployable. General relativity and gravitational physics became associated with crackpot research: Einstein's own search for a unified field theory was considered a dead-end, and Eddington's fundamental theory was considered the work of someone who has become thoroughly detached from reality. This all changed in a rather dramatic way in the 1950s, and indeed quantum gravity's subsequent rapid growth emerged from the more general growth in gravitational physics in the 1950s (combined with several fortuitous circumstances, as we will discuss in this chapter). But it can also be said that the changing fortunes of *classical* gravitational research also benefited from the pursuit of quantum gravity since the two were inextricably linked as a result of a host of contingent factors, many of which strain credulity.

---

[1] From a letter to his wife from what would be called the GR3 conference in Jablonna in 1962 (reproduced in *What do you Care what other People Think?* W. W. Norton, 1988, p. 91).

[2] Recall Pauli's remark from the previous chapter: "classical field theory is a completely squeezed out lemon that can't possibly produce anything new" (Pauli letter to Einstein, September 19th, 1946: Einstein archive: 19–182). See Chapter 11 of [Eisenstaedt (2006)] for a clear account of this stagnation, along with its reasons.

[3] As Bryce DeWitt noted, even in the mid-fifties, Samuel Goudsmidt, then Editor-in-Chief of *Physical Review*, indicated that he would release an editorial pointing out that "papers on gravitation or other fundamental theory" would no longer be accepted by the journal (nor by *Physical Review Letters*)—it was due to John Wheeler that this decision was reversed ([DeWitt (2009)], p. 414). One must suppose that by "fundamental theory" he had in mind *unified* theory (or something of the same sort Eddington produced in his final work), which was by then looked upon as a fool's errand.

Covered with Deep Mist: The Development of Quantum Gravity (1916–1956). Dean Rickles, Oxford University Press (2020). © Dean Rickles.
DOI: 10.1093/oso/9780199602957.001.0001

This state of affairs, and the transition, were well-known to various people in the 1950s, and the "neglect" of gravitational physics was often raised as an issue to be tackled, as was the reversal of this neglect. In other words, the participants (engaged in the transitionary phase) were conscious of the changes at the time. Of course, while classical general relativity had a renaissance, quantum gravity was, rather, taken out of its incubator—this chapter describes the establishment of such a quantum gravity-friendly environment that enabled it to go out into the world on its own, somewhat less dependent on other areas of physics. We are, then, more concerned with the development of basic *infrastructure* here, but let us briefly review alternative explanations of the sudden mid-century growth in gravitational research.

9.1   Explaining the Gravitational Renaissance

Despite its strong early public profile following its announcement, then, research on general relativity had degenerated to a trickle by the early 1950s. It didn't really generate a thriving research program until the second half of the twentieth century, with a pivot-point around the middle of the 1950s (see fig. 9.1).[4] Today, of course, it finds itself once again at the centre of attention and is intermingled with both cosmology and particle physics. Indeed, it is hard to believe that it was ever in trouble. Yet it was: it spent a sizeable portion of its early life in mathematics departments, inspired few PhD theses, and was more or less experimentally dormant. Many of the kinds of phenomenon we now naturally associate with general relativity—black holes, gravitational waves, etc.—were only established after this pivot-point. An important historical question is: what triggered this remarkable turnaround?[5]

We can divide the kinds of possible explanation into two categories: internal and external. By 'internal' I have in mind explanations that point to factors such as an increase in mathematical sophistication (such as the discovery of the Petrov classification of the Weyl tensor or the Einstein–Infeld–Hoffmann result). External explanations

---

[4] Of course, one has to be a little careful here since the rise might simply be due to a general overall rise in activity after the war—see [Goenner (2016)] for a critical analysis of the concepts of 'renaissance' (and 'golden age'), along these lines. However, even correcting for this (by examining publications on other fields of physics in the same way), one finds a significant increase in the publication rate beyond this general trend. We should also note that there are many significant papers missing from this analysis as it stands, with topics and titles somewhat at odds with our search terms. But it is enough to see the general trend of rise and fall and subsequent rise again. We need not take this as historically definitive, and its suggestiveness is enough. Moreover, there were enough remarks from mid-century physicists, pointing to the "neglect" of gravitational physics (some of which we will present in this chapter) to be fairly confident in locating real phenomena in the research landscape of physics corresponding to the rises and falls in this graph.

[5] There are really two components to the puzzle: (1) what caused the initial stagnation? and (2) what caused the subsequent awakening of interest. Jean Eisenstaedt [Eisenstaedt (1986)] has labeled (1) "the low water mark," while Clifford Will [Will (1989)] labels (2) "the renaissance" of general relativity. Naturally, we might expect that some of the reasons for the low water mark will, when 'put right,' influence the renaissance. Elements of this can certainly be seen; but not all of the renaissance can be thus explained—see [Blum, Lalli, and Renn (2017)], [Blum, Giulini, Lalli, and Renn (2017)], and [Lalli (2017)] for recent work on this issue that takes the complexity of the problem more seriously.

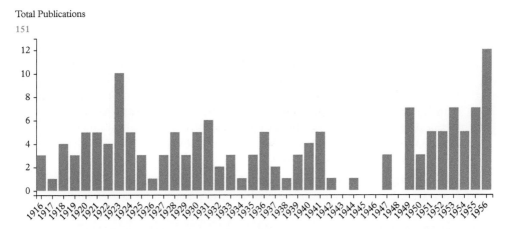

Fig. 9.1 *Graph of published works on general relativity and quantum gravity from 1916 to 1956. One can see quite clearly a peak after 1922, followed by a significant decline (Eisenstaedt's 'low-water mark period') in the 1940s. One also sees clear evidence of a regeneration after 1955 (the beginnings of the 'renaissance'). Source: Web of Science [keywords: 'general relativity,' 'quantum gravity,' 'quantengravitation,' 'allgemeine Relativitatstheorie,' or 'relativite generale.']*

point to factors outside of the subject itself that nonetheless influenced the progress of the theory. Here we can include developments in other areas of physics (such as the problems facing quantum field theory that hadn't found their own internal resolution, or the introduction of Yang–Mills theories); experimental and observational advances (including the development of atomic clocks and the discovery of quasars); and social factors, such as the launching of Sputnik, or the applicability of new funding sources (we can also include serendipitous factors such as particular physicists, with quantum gravitational interests, being selected to oversee such funding and being its earliest beneficiaries).

There have been various explanations offered in the literature, none of which strike me as complete (even when taken together):

- **Jean Eisenstaedt: "The Low Water Mark of General Relativity, 1925–1955"** [Eisenstaedt (1986)]. Eisenstaedt argues that the shift came about because of the potential for new experiments, as a result of developments in astrophysics, and precision tests of old results (often involving new technologies, such as atomic clocks). In particular, the Pound and Rebka experiment, Kruskal's solution, and the discovery of quasars and of strong gravitational fields.[6] However, these experiments

---

[6] Certainly, Bob Dicke had complained, in 1959, that "it is a serious lack of observational data that keeps one from drawing a clear portrait of gravitation" ([Dicke (1959)], p. 40). Incidentally, Dicke's doctoral research (1941) had been on inelastic proton and neutron scattering.

came somewhat later than the turnaround and amount to more of an *effect*. If we want to find the *cause*, we must look further back.

Eisenstaedt's periodization of the stagnant phase of general relativity, 1925–1955, is a good approximation, but ignores the incredibly important work done on the canonical formulation of the theory by several groups, including Peter Bergmann's Syracuse group and Alfred Schild's group (with Felix Pirani and Ray Skinner) at Carnegie Tech. This work laid the foundations for what are the key technical and conceptual hurdles in finding true observables for general relativity, and therefore for contemplating the resolution of the problem of quantizing the theory along canonical lines.

- **Clifford Will: "The Renaissance of General Relativity"** [Will (1989)]. Will also traces the renaissance of general relativity back to series of high-precision experimental tests (between 1960 and 1980) that confirmed general relativity to a greater degree of accuracy than the earlier 'classic tests' (Mercury's perihelion, light deflection, and gravitational red shift). Also of crucial importance for Will (the turning point) was the discovery of various high-energy astronomical objects such as quasars, in the early 1960s—these, of course, were essential for confirming predictions sufficiently distinct from Newtonian gravitation.

    But this is too simplistic. General relativity was already considered well confirmed. No one doubted its accuracy. The question mark over it was not one of truth value, but of value *simpliciter*. By showing how general relativity was intimately bound up with the central problems of elementary particle physics, and not some alien other, the theory was pursued with increased vigour. These changes in outlook were not the result of experiments, rather the fact that those experiments were funded and pursued in the first place was as a result of the already much improved status of general relativity.

- **Sam Schweber: "Einstein and Oppenheimer: Interactions and Intersections"** [Schweber (2006)]. Schweber traced the "reawakening of interest in the field [of GR]" ([Schweber (2006)], p. 526) to the Bern conference (celebrating the golden jubilee of special relativity), GR0, in 1955.[7] He also argues that interest in quantum gravity was made "respectable" as a result of Feynman's course at Caltech in 1962–3 (ibid., p. 527). The Bern conference *was* important. It was a distinctly European affair, however. In the United States, as we have seen, there were several converging lines of attack leading to a reawakening of interest. Indeed, I would argue that since the Bern conference consisted mostly of an older generation who had persistently (albeit often in their unpublished work) thought about general relativity and quantum gravity for decades, the phrase 'reawakening of interest' is not applicable in this case. Klein, Pauli, and Rosenfeld, for example, were veterans when it came to the study of both. Hence, I think that its pivotal role has been somewhat over-estimated: it marked more of an *end* of an era than a new beginning.

---

[7] André Lichnerowicz [Lichnerowicz (1992)] has also pointed to this conference as being of crucial importance.

Further, to trace the respectability of quantum gravity research to Feynman's course is over-stretching. Wheeler had been including material on quantum gravity from the time he began teaching his general relativity course at Princeton almost a decade earlier. In addition to this, there were, as Schweber himself notes, several strong 'schools' concentrating on gravitation research (including quantum gravity) by the end of the 1950s. That Feynman's course happened was as a result of the increased respectability already well in operation (as well as advances in Yang–Mills theory)—moreover, though he had been mulling over quantum gravity for some time, Feynman's interest was piqued largely as a result of his participation in the 1957 Chapel Hill conference and the discussions that happened there.

However, let us reflect a moment on the Bern conference, which was undoubtedly important. The Bern conference was certainly in the minds of the DeWitts during their preparations for the Chapel Hill conference in 1957. Indeed, it was their explicit aim to avoid as far as possible the "exact duplication of results" from the earlier conference (letter from Bryce DeWitt to Alfred Schild, dated May 29th, 1956: CDWA). Asked (by Abraham Pais) what caused the change of pace and growth of relativity theory post-1950, Dennis Sciama, while acknowledging the Bern conference, identifies the Chapel Hill conference as the pivotal one:

> The Bern conference was followed two years later by the Chapel Hill conference organized by Bryce DeWitt ... [8] This was the real beginning in one sense; that is, it brought together isolated people, showed that they had reached a common set of problems, and inspired them to continue working. The "relativity family" was born then.
>
> ([Pais (1982)], p. 270)

However, the Bern conference was certainly an inspiration, and achieved several important advances:

1. The experimental foundations of general relativity (then amounting to just two tests) were critically scrutinized. The potential to devise cosmological tests, that were easier to shield from disturbances, and making it easier to isolate the effects, was exposed.

2. Global problems were posed in a systematic way. Roger Penrose would soon pursue this line and quite radically alter the kinds of questions that were asked of general relativity.

3. The nature of gravitational waves was investigated; though still left uncertain, it was on the agenda as a central concern.

4. The problem of the quantization of the gravitational field received a review and a state-of-the-art assessment. This included several attempts that went

---

[8] I feel obliged to intervene here and note that, as can be seen from the archival material, the **vast** majority of the organization was in fact carried out by Cécile DeWitt-Morette!

beyond the standard *quantization* of ordinary classical GR. For example, Wigner considered the possibility that it is quantum theory that might be modified in order to produce a more general version. Both Klein and Bergmann discussed further issues.[9] Bergmann put centre stage the idea that in order to resolve the quantum problem one needs to first resolve the problem of the invariant description of classical theory. Here, Géhéniau's work was invoked as a way of resolving that preliminary problem using the values of the four scalars constructible from the curvature tensor.

5. The theory of general relativity was freed from its association with Einstein's own search for a classical unified field theory that was able to account for apparently quantum phenomena. Though there was one talk, given by Kaufman, on Einstein's progress, this was more of a review of what he had achieved rather than a suggestion that it is worth continuing. By detaching GR from what was considered to be a foolish enterprise, the theory gained some credibility back amongst serious researchers.

Within the 7 years following the jubilee there were several very important general relativity conferences, starting with the 1957 Chapel Hill conference, then the Royaumont conference in 1959, and a conference at Jablonna, Warsaw, in 1962.[10]

- **David Kaiser: "A ψ is just a ψ? Pedagogy, Practice, and the Reconstitution of General Relativity, 1942–1975"** [Kaiser (1998)]. Kaiser discussed the rise of general relativity from backwater to a central part of the physicist's toolbox. His view is that the infiltration of the non-geometrical, Lagrangian approach played a crucial role in the spread of the theory. Feynman, for example, presented GR as a quantum field theory of a massless spin-2 field *because his students would be more familiar with this approach*. That is, the theory was morphed to suit the pedagogical needs of the time. The original geometrical formulation demanded too specific a mathematical toolkit, and one absent from the education of most physicists in those days (see especially ibid., p. 329). In addition to this, one must also add the crucial motivation underpinning Feynman's *original* reason for pursuing general relativity, namely to produce a quantum theory of the gravitational field.[11]

---

[9] I would add to the Bern conference, in terms of importance to *quantum gravity*, the Bohr festschrift [Pauli (1955)] (featuring essays for his 70th birthday). This, appearing the same year as the Bern conference, was also of crucial importance for quantum gravity, introducing to a wide readership the notion, due to Landau, that gravitational interactions might be of relevance in providing a cutoff to cure the ultraviolet divergences of quantum field theory.

[10] These were known as the GRG conferences, with the Chapel Hill conference retrospectively gaining the honorific title GR1 and the Bern conference becoming GR0. We should note, however, that gravity continued to, and still does, lag behind much other physics research. For example, after the Chapel Hill conference, the next conference *solely devoted to relativity* to be held in the United States was not until 1969, in Ohio. This conference coincided with the first issue of what was then the only journal devoted to gravitation, namely *General Relativity and Gravitation*—this journal itself partly flowed from the work that was begun as a result of the external contributions described in this chapter.

[11] I note that Kaiser also draws attention to Roger Babson and the Gravity Research Foundation as playing a role in the renaissance (see, eg., [Kaiser (2000)], p. 591), and also mentions, albeit briefly, Agnew Bahnson's role

While acknowledging that there is some degree of truth to each of these stories, in this chapter I argue that they ignore an important contribution in the form of private patronage and other funding sources connected to the military-industrial complex. In this case the impact was unusually (for physics) significant and had much to do with the emergence of precisely the kind of environment that allowed the kinds of advances mentioned in Eisenstaedt's *et al.* responses. We aim also to reveal the remarkable contingencies (and curiosities) in the development of gravitational physics (focusing on quantum gravity) and the institutions in which quantum gravitational research was carried out.

9.2 The Role of Private Patronage

As hard to believe though it might seem, I argue here that quantum (and classical) gravity's growth, from its period of stagnation, was in large part spurred on, at least initially, by two American philanthropists: Roger Babson and Agnew Bahnson—the former more indirectly than the latter. They founded two institutions, The Gravity Research Foundation and the Institute for Field Physics respectively, that would provide a much-needed shot of funds and infrastructure at a time the discipline most needed it. Let us begin with Babson's contribution.

> Lots of people, including Einstein, talk about gravity, the restraining force which makes people walk on floors instead of floating in midair. What worries Roger Babson, 74, economic oracle and head of the Babson Institute, is that no one does anything about it.
> [*Time Magazine*, January 2nd, 1950]

Roger Babson (1875–1967) was a man obsessed with gravity. He wanted to control it; to exploit it to solve the world's energy problems; but most of all, *to screen it out*. As he saw it, gravity was a natural evil responsible for the death of his sister and his grandson, both of whom drowned. Writing of his sister's death he wrote:

> When I was a boy my oldest sister drowned while bathing in Annisquam River, Glouster, Mass. Yes, they say she was "drowned", but the fact is that, through temporary paralysis, or some other cause (she was a good swimmer) she was unable to fight gravity, which came up and seized her like a dragon and brought her to the bottom.
> ([Babson (1948)], p. 828)

The "dragon" Gravity would once again rear its ugly head in the summer of 1947, snatching his grandson Michael as he attempted to save someone's life. This event brought gravity back into Babson's mind with a sense of urgency. Gravity was a menace;

(ibid., pp. 592–3)—though Kaiser seems to uphold Babson as the decisive player; below, we will take Bahnson as the more decisive figure. In a joint paper [Kaiser and Rickles (2018)], Kaiser and I argue, as I do here, that it was in combination that Babson and Bahnson had the oversized effects they had.

something to be battled with and conquered. However, as Don Howard has noted, the real culprit was not deemed to be gravity *per se*, but our ignorance of its workings ([Howard (unpublished)], p. 2).[12] By this time, Babson had accumulated the financial resources to do something about it.

As Babson explains in his memoirs, *Actions and Reactions*, he began the Gravity Research Foundation [GRF] in January 1949 "as a non-profit institution to collect and distribute information in connection with gravitation" ([Babson (1950)], p. 340). He clearly modelled the GRF on the company that made him a millionaire, a financial analysis company called *Babson's Statistical Organization*.[13] This company provided the latest and best financial and business information in a simple digest that banks and investors could subscribe to.[14]

In a suitably symmetrical fashion, physics entered into his approach to finance— indeed, this is an interesting early example of econophysics. One of his tools, the so-called "Babsonchart" (of American Business Conditions), was based directly on Newton's third law, the principle of action and reaction. Babson used this dubious tool to predict the 1929 Wall Street Crash and the Great Depression that the best financial minds had failed to see—see *The Commercial and Financial Chronicle*, September 7th, 1929. This seems to have been pure luck on Babson's part: there is no necessity in the mechanism that Babson clung to in his prediction. Babson's contemporaries thought so to. In his book on the Great Crash, J. K. Galbraith writes of Babson's business acumen that:

> [He] was not a man who inspired confidence as a prophet in the manner of Irving Fisher or the Harvard Economic Society. As an educator, philosopher, theologian, statistician, forecaster, and friend of the law of gravity he had sometimes been thought to have spread himself too thin. The methods by which he reached his conclusions were a problem. They involved a hocus-pocus of lines and areas on a chart. Intuition, and possibly even mysticism, played a part. Those who employed rational, objective, and scientific methods were naturally uneasy about Babson, although their methods failed to foretell the crash. In these matters, as so often in our culture, it is far, far better to be wrong in a respectable way than to be right for the wrong reasons. Wall St. was not at a loss as what to do about Babson. It promptly and soundly denounced him.
>
> ([Galbraith (1997)], p. 85)

Babson was resolute: his successful prediction was nothing but the work of his hero, the great Isaac Newton, acting at a distance. His penchant for all things Newtonian, already

---

[12] Howard (ibid.) argues that Babson's quest can be understood in terms he labels "Physics as Theodicy": battling with the evils of nature by seeking to better understand its workings. This can clearly be seen in the section entitled 'Purposes, Objects and Powers' of the document of incorporation of the Gravity Research Foundation formed by Babson. Clause (1) states that the purpose of the corporation is to "Observe the phenomena of nature and encourage, promote and support investigations in search of underlying knowledge of these phenomena. Conduct theoretical and experimental studies to discover the laws which affect them and evolve new technological concepts for the improvement and welfare of mankind" (Document from the personal archive of Cecilé DeWitt-Morette: CDWA).

[13] The company still exists as *Babson-United Investment Reports*.

[14] Babson started his career in finance working on the trading floor, as it were. Apparently, his overly ethical attitude, involving him questioning his superiors' practices, saw him dismissed.

strong (from when he was first properly introduced to Newton's life and work in a course at M.I.T.) increased manyfold after this.[15] It is highly probable that this denouncement by the scientific establishment considerably influenced Babson's future interactions in which he displays a marked tendency to avoid academic scientists and 'experts' in favour of inventors and entrepreneurs. This impacts directly on the institutional model adopted by the GRF and ultimately explains its failure to thrive.

The GRF was based in New Boston, New Hampshire, located for maximal safety in the event of a nuclear attack in World War III! The site was set up to achieve self-sufficiency "free from industrial plants, able to feed, clothe and heat itself under any circumstances, and independent of uncertain water, light and power utilities" (ibid., p. 340). Babson had several goals for his GRF. The *primary* purpose of the GRF was to house the most complete library on gravity in existence (ibid., p. 341). In this he seems to have been successful.

Babson was certainly correct in writing that gravity in this period was a neglected area of research.[16] He notes that "[t]he mention of gravity too often brings a smile as if the enquiry were not taken seriously" (ibid., p. 341)—all too true, of course, but Babson hardly helped matters in this regard, as we will see. This provides the second purpose of the GRF: "to get in touch with interested people scattered throughout the world and let them know that they have a sympathetic friend" (ibid., p. 341). Crucially, Babson did not much care about the credentials of such interested people and his vision of 'canceling out' and controlling gravity meant a pre-selection bias that rejected experts (and favoured the more crackpot element).

A further task of the GRF had to do with the relationship between gravity and health, harking back to the dragon. He notes that for those over 60 years of age "Gravity may be called Enemy Number One" (ibid., p. 342). Credulity is stretched somewhat by some of the research foci that fall within this aim. For example, Babson states that some time will be devoted to probing a possible correlation between the phases of the moon and the incidence of accidents and disease:[17]

> [O]ne of the tasks of our Gravity Foundation is to collect from hospitals, insurance companies and physicians the day and, if possible, the hour, of a fracture and learn how this time correlates with the phases of the moon. Not only does the pull of the moon and sun counteract (or relieve) at certain times the downward pull of the earth on an individual, but this same gravity may affect the temporary judgment or awareness of the

[15] This included the establishment of one of the largest collections of "Newtonia" (as he called it: [Babson (1950)], p. 340) in the world, on the campus of the Babson Institute at Wellesley, Massachusets. One of the library's rooms is an actual room used by Newton while in his final years in London. This was purchased by Babson's wife when she discovered the building was being demolished, shipped over from the UK, and rebuilt on site as Newton would have used it "with the same walls, doors and even the identical shutters containing the hole through which he carried on his first experiments in connection with the diffusion of light" (ibid., p. 340).

[16] This harks back to my earlier remark about the 'low-water mark' period being a well-known phenomenon at the time, rather than a recent historical construct.

[17] This overestimation clearly shows Babson's profound lack of understanding of gravitational physics. Though we might look more charitably upon such studies as early examples of the effects of microgravity environments on such things as bone density, so essential for subsequent space missions.

individual. After ascertaining definite data on the above, it must further be recognized that the variation in this gravity pull of the sun and moon may affect the judgment of individuals differently according to their mental capacity and development.

([Babson (1950)], p. 342)

In addition to such 'gravitational astrology,' Babson was also deeply concerned with the *harnessing* of the power of gravity, to develop 'free' power:

The general impression is that gravity will not be harnessed until a partial insulator, reflector or absorber is discovered to develop a *differential*. It is further believed that this discovery will be accomplished through stumbling on some alloy which will give the desired results. Hence, the Foundation is encouraging all engineers and chemists who work with alloys to be on the watch for such a discovery. It surely would be a great blessing to mankind.

([Babson (1950)], p. 343)

This vision—amounting in Babson's mind to the creation of perpetual motion (a vision influenced by his conversations with Edison)—would guide the entry requirements for essays in the GRF's competition (which we turn to below), threatening to further bury the enterprise. Babson was profiled in *Time* magazine in 1950, just after he had established the GRF (and the same year he revised his autobiography). In this article the reporter clearly adopts a mocking tone. But though Babson was certainly viewed as a crank, by both economists and physicists, he had the connections to be able to attract external funds for the GRF. As the *Time* article notes:

A leading shoe manufacturer offered Babson $100,000 for "something that can be put into the sole of a shoe to insulate against gravity". Floor-covering manufacturers showed a lively interest in the possibility of "flying" carpets.

We strongly suspect that the GRF's patrons had a similar distrust of 'scientific experts,' being mostly self-made men. A GRF conference was held in 1951, at which attendees included Clarence Birdseye of frozen food fame who was also one of the original trustees of the GRF, himself having a very keen interest in anti-gravity. The conference employed specially designed 'anti-gravity' chairs that reclined their occupants in such a way as to aid circulation. A bulletin read that "It is the hope that New Boston will gradually become the center where physicists, engineers, metallurgists and others especially interested in the causes and the possibilities of gravitation will come as a mecca in the summer" (quoted in Gardner [Gardner (1957)], p. 94).

But the GRF could never really gain the prestige it desired. Babson held too much control over which lines of research were investigated. Since he was no scientist, these tended to be at odds with scientific orthodoxy. For example, in one GRF bulletin, the biblical miracle of Jesus walking on water was offered up as *evidence* in the possibility of anti-gravity shields, as was the ability of angels to defy gravity! The GRF stood no chance (at least not in this form). Martin Gardner famously mocked Babson's own gravity ideas in an article entitled "Sir Isaac Babson," calling the GRF "perhaps the most useless scientific project of the twentieth century" ([Gardner (1957)], p. 93). He was referring to the stated aims of the GRF *and* its annual essay competition; namely to discover some kind of gravity screen ('the right kind of alloy'). Gardner rightly points out that the concept of a material that is opaque to gravitational interactions was made obsolete by the shift to the general theory of relativity. He clinically demolishes other views of Babson and his associates. In Gardner's opinion, however strong Babson's love of Newton was, "he...failed to emulate the great scientist" for "he has failed to acquire more than an elementary knowledge of physics" ([Gardner (1957)], p. 100). He goes on to note: "there is surely a touch of pride in his refusal to accept advice from competent physicists on how money could best be spent for the good of science and humanity" (ibid.). That rather gets to the psychological heart of the matter, and is the crucial contrast between Babson and Bahnson.[18]

Agnew Bahnson was a close friend of George Rideout (the president of Babson's GRF, and the person who had initially suggested the idea of a GRF to Babson). As we shall see in a moment, on the basis of how the initial organizational meetings went, Gardner's critique offered up a recipe for a more successful venture, primarily bankrolled by Bahnson, in which expert advice on how to spend money is well-taken. To get to that story we must first consider an important transition in the GRF's essay competition which led, curiously, to a positive snowballing effect for the fortunes of quantum gravity and directly influenced the more positive venture.

9.3   The GRF Essay Contest

The GRF would have probably suffered a well-justified early demise had it not been for the essay competition it established "for the best two thousand word essays on the possibilities of discovering some partial insulator, reflector or absorber of gravity waves" ([Babson (1950)], p. 344). Even this competition, had it not been hijacked by quality people who tore apart or simply ignored the basic theme of the competition, would have quickly been consigned to the dustbin of history. This competition is all that remains of the foundation, at least in the public eye,[19] and it has, for some time now, basked in the mantle of scientific respectability as a result of the respectability of its entrants (which now include virtually all of the great general relativists >1960).

The winnings for 1st prize in the early years was $1000, a not inconsiderable sum in those days (roughly equivalent to a postdoctoral fellow's annual salary) and pretty good

---

[18] I would add that ego (and, indeed, more than a dash of selfishness) was at the root of many of the GRF's chosen topics. For example, in his autobiography he notes that the "Foundation is interested in the ocean tides at Gloucester, Massachusetts" ([Babson (1950)], p. 344), which is where Babson just happened to spend his summers—not tides *in general*, but where the tides might affect *him*. We should, however, not be quite so hard on Babson, and perhaps Gardner is a little too harsh. Babson's concerns were outwardly directed as well as inwardly. He seems to have genuinely wanted to have a large positive impact on the world, and was simply misguided in his methods—my thanks to David Kaiser for this impressing this point on me.

[19] Aside, that is, from the curious GRF monuments scattered around various US campuses, installed to "remind students of the blessings forthcoming when science determines what gravity is, how it works, and how it can be controlled," as the engravings point out.

even today! David Wittry (from the University of Wisconsin) won the first ever 1st prize, in 1949, for his historical essay describing (failed) attempts to produce an anti-gravity screen. The second competition's 1st prize in 1950 was awarded to a graduate student at Princeton University. This was covered by *Time* magazine, which suggests that at least in the public eye gravity still held some fascination—likely through its association with Einstein, who was then, and perhaps still is now, the epitome of a scientist for most people and would be mentioned periodically in the popular press as he pursued his unified field theory. There was evidently seen to be something mysterious and difficult about gravity that had the power to hook a general audience. For our purposes, 1953 and 1954 were important years for the essay competition, though for very different reasons.

Not long after Babson had been profiled in *Time* magazine, at the end of 1950, a young postdoc named Bryce DeWitt, who had just completed his PhD on the then very unfashionable topic of quantum gravity (under the rather minimal supervision of Julian Schwinger, at Harvard), was making his way from the ETH in Zürich to Bombay (now Mumbai), where he was taking up a Fulbright scholarship at the Tata Institute—where he would write his first published paper on quantum gravity (involving spinors in a gravitational field: [DeWitt and Morette DeWitt (1952)]), co-authored with his new wife Cécile Morette. The two would soon connect, albeit indirectly, through the essay competition.

Bryce DeWitt was born Carl Bryce Seligman, on the 8th January 1923, in Dinuba, California. He was the eldest of four boys. His father was a country doctor, of German Jewish descent.[20] His mother was a teacher of latin and mathematics of French Huguenot ancestry—she had written an undergraduate thesis on the history of the calculus (a copy of which remains in the DeWitt archive in Texas). DeWitt was something of a maverick. He wrote (at Harvard, where he had also been an undergraduate) what is arguably the first ever PhD (in the west: Bronstein's being the only other such PhD) devoted to the subject of quantum gravity.[21] He had just gotten engaged, to Cécile Morette (fig. 9.2), when he set off for the ETH to work with Pauli. Morette had previously worked on meson physics under Heitler and de Broglie, but when she met Bryce at the Institute

---

[20] Bryce dropped the Seligman part of his name as a result of anti-semitic incidents surrounding him and his three brothers—all of the them changed their names on the instructions of their father, Emil (see [Weinberg (2008)]: Weinberg claims that Felix Bloch vetoed DeWitt's professorship at Stanford in 1972 as a result of his change of name, which Bloch viewed as an anti-semitic decision (ibid., p. 10). Bryce was raised in the Presbytarian church; the only Jewish remnants were, as he put it himself, the matzos his grandfather brought over at Passover—see DeWitt's own rendition of his obituary (Bryce DeWitt Papers: Box 4RM174).

[21] DeWitt's Pre-Doctoral thesis had absolutely nothing to do with gravity, quantum or classical. It was entitled: "An outline of the theory of specific heats (20 May 1946 - a term paper)" 106pp - with a bibliography of 6 items! This includes a quantum mechanical theory of specific heats. It is, however, brilliantly written, and shows his penchant for extreme number crunching. But certainly no sense of the QG work to come—though it will have set him up well for his work on hydrodynamical detonation (also at Harvard)! Curiously, the latter was heavily computational, and the skills he built up here would later be used in the service of quantum gravity (though primarily colliding black hole and gravitational 2-body problem) simulations. Also, one might wonder whether DeWitt's decision to pursue hydrodynamics was influenced by the work that was recently being undertaken at Harvard on underwater ballistics (for the Navy)—see [Rees (1980)], p. 612. DeWitt later began work on (Lagrangian) hydrodynamics at the Lawrence Livermore Laboratory—he discusses his time at Livermore (the 'Rad Lab') in [DeWitt (1985)].

Fig. 9.2 *Cécile and Bryce DeWitt, who would together direct the Institute for Field Physics at Chapel Hill. [My thanks to Chris DeWitt for allowing me to use this photograph of their wedding in 1951.]*

for Advanced Study at Princeton, 1949, she had fallen under the spell of Feynman's path integral approach to the formulation of quantum theory and was busy mastering its mathematical details, initially with Freeman Dyson, and then directly from Feynman himself, in Santa Fe. Bryce was struggling somewhat career-wise, and was having serious trouble getting employment to work on quantum gravity—it is evident that supervisors' warnings to their students to avoid gravitation were not misplaced at this time. He was, in the years following the completion of his doctoral thesis, on the verge of giving up quantum gravitation, but was encouraged by Freeman Dyson not to throw in the towel. As he explains in a letter to various department heads:

> Owing to the difficult and tedious nature of research in gravitational theory, and also owing to the apparent complete lack of any immediate practical application of its results, I was, until recently, strongly resolved to discontinue further work along these lines and to turn my attention elsewhere. A conversation I had with F. J. Dyson this summer, however, has left me with somewhat altered views. He stressed to me the urgent need for workers in field theory who have a thorough understanding of gravitational theory and its problems.
> (Bryce DeWitt, letter to Raymond Thayer Birge, November 11th, 1951: CDWA)

But the letter fell on deaf ears, and the situation in gravitation remained as dire as before for several more years. The best DeWitt managed, after the Tata Institute trip, was to get a position at Livermore working on the hydrogen bomb under Edward Teller's general direction.

Bryce entered Babson's radar (or, at least, George Rideout's) in 1953 when he submitted an essay for the GRF's essay competition (fig. 9.3). Though not considered a respectable institution, given Bryce's frustration with the lack of interest in gravitation, it must nonetheless have pleased him to find anything at all on the subject, even if ill-conceived in its aims. Recall that the condition was in place that submissions must deal in some way with the notion of anti-gravity and a gravity shield. DeWitt's article attended to this point, but tore apart the very possibility of such an anti-gravity shield, "essentially giving them hell for such a stupid…way it had been phrased in those early years".[22] Surprisingly, the essay won first prize. DeWitt wrote the essay in a single evening calling it "the quickest $1000 I ever earned" (ibid.). (Note that in this essay, DeWitt views the problem of gravity's quantization as all but completed, using Hamiltonian methods, save for a few remaining details.)

Rideout clearly saw DeWitt as a person who could lift the respectability of the GRF, working as he was in the Rad Lab at Berkeley. Indeed, several people have remarked that this seemed to be the case, arguing that whereas the prize was previously avoided by 'serious' physicists, after DeWitt won, the floodgates opened. Cécile herself made just this point: "until Bryce, no real physicist wanted to touch it, because it was considered as a total crackpot project". It is indeed true that the next year saw Richard Arnowitt and Stanley Deser (both students of Julian Schwinger, as was DeWitt) take first prize,[23] but Bryce DeWitt is less sanguine about his role in shifting the competition's reputation, writing that "it took probably five or six years for the atmosphere to change" (ibid.). It is something of a myth that the GRF attracted no notable people until DeWitt won the competition. In fact, there were people *more* distinguished than DeWitt (at that time) that were entering *before* 1953, including J. M. Luttinger (4th prize in 1951[24]) and Martin Perl (in 1952; later winning the Nobel prize for his discovery of the tau lepton)—E. T. Jaynes

---

[22] Interview of Bryce DeWitt and Cécile DeWitt-Morette by Kenneth W. Ford on February 28th, 1995 Niels Bohr Library & Archives, American Institute of Physics: www.aip.org/history-programs/niels-bohr-library/oral-histories/23199. DeWitt would reiterate this point in a reassessment essay he submitted to the 1960 round of the essay competition, writing: "any frontal attack on the problem of 'harnessing' the power of gravity along lines based on analogy with our experiences with electromagnetism is a waste of time" ([DeWitt (1960)], p. 1). As David Kaiser has noted, Rideout was very serious about respecting this gravity-shield criterion writing (in his second annual report to the trustees, from August 1st, 1950) that "We will not accept any essays simply on the subject of Gravity. Some of them sound just like a text book. We are insisting on adherence to the subject, namely, the objective of discovering some partial insulator, reflector or absorber of gravity" (GRF archives, Box 1, Folder 5—as cited in [Kaiser (2000)], p. 575).

[23] Much to the chagrin of Robert Oppenheimer who was supervising both, as postdocs, at the Princeton Institute for Advanced Study. Oppenheimer thought that entering the competition and accepting the prize brought the IAS into disrepute. He believed that Arnowitt and Deser exploited their positions at the institute—more on this below.

[24] Curiously, Luttinger was Pauli's own pick for the potential husband of Cécile Morette, and would try to engineer such a coupling—against Luttinger's wishes. (Interview of Cécile DeWitt, by Dean Rickles and Donald Salisbury: March 4th, 2011.)

New Directions for Research in the Theory of Gravitation

by Prof. Bryce S. DeWitt

Radiation Laboratory
University of California
Berkley 4, Cal.

1953

Before anyone can have the audacity to formulate even the most rudimentary plan of attack on the problem of harnessing the force of gravitation, he must understand the nature of his adversary. I take it as almost axiomatic that the phenomenon of gravitation is poorly understood even by the best of minds, and that the last word on it is very far indeed from having been spoken.

Nevertheless, the theoretical investigation of gravitation has received relatively little attention during the last three decades. There are several reasons for this. First, the subject is peculiarly difficult; the existing body of theory on it involves rather recondite mathematics, and the fundamental equations are almost hopeless of solution in all but a very few special cases. Although the accepted theory is motivated by two or three beautifully simple yet profound principles, these guiding principles have so far been of little help in predicting the general features of the solutions of the equations to which they give rise. And, as any researcher in the field knows, one can develop a serious case of "writer's cramp" in that manipulation of tensor indices which is usually necessary in order to prove only a single tediously trivial point.

Secondly, modern gravitational theory has few consequences which are even remotely susceptible of experimental verification. The old Newtonian theory, involving action-at-a-distance, has, for practical purposes, been far too adequate. Consequently, stimuli for the theoretical investigation of gravitation are virtually non-existent, and gravitational research is almost totally unrewarding. It is a field which had its brief brilliant hour, but which has since fallen into a state of near disrepute.

In spite of all this, it is very probable that the phenomenon of gravitation will eventually have to be reckoned with again in re-spectable circles, and it may well happen that this reckoning will present itself in a rather acute form. It is one of the purposes of this note to suggest that we may be already in the first phases of such a new development, and to point out some new directions into which we are likely to be led as a result.

Fig. 9.3 *Bryce DeWitt's winning entry for the GRF essay competition in 1953.*

applied the same year as DeWitt. It is true that there was an emerging overall increase in the rate of quality submissions (for example, Felix Pirani submitted a version of his paper demonstrating definitively the transport of energy by gravitational radiation by 1957, presented as a means of controlling gravity to a certain extent), but remember that there was an overall increase in the production of work on general relativity *simpliciter*.

In any case, Richard Arnowitt and Stanley Deser's 1954 entry was supposedly something of a physicist's joke, playing with several then trendy topics in a way that sounded like a somewhat plausible method of gravity manipulation linking gravity and nuclear physics. Deser recalls the events as follows:[25]

> DESER: What happened was this thing [the GRF essay competition] had never existed before. And let me tell you, being a postdoc at the Institute put a lot of pressure on one also, so you need to do something to break it up. So we thought it would be funny to write a paper, to submit a thing that we never even thought that we would win because it was so manifestly—
> SALISBURY: By that time, you were convinced that it was a reputable organization that was awarding the prize.
> DESER: No! No, I wasn't!
> RICKLES: So DeWitt must have mentioned [it], because he won it the year before in 1953, and you cite DeWitt's paper ...
> DESER: Oh, we may have cited it, but no, we never talked to Bryce about it, no. We just got the idea of saying, "Wouldn't it be funny if" and we didn't know. Actually, that outfit was pretty crackpot! So no, the answer is no. We did it just as a lark. And of course, it turned out Oppenheimer—because we won, which was a major mistake. We were ready to give them back the dough, but it turned out that that would have been even worse.
> RICKLES: Yes, more publicity.
> DESER: So we told Oppenheimer, and we went on our knees to his office and so on, but he really got very unhappy and refused. He felt that we were using the name of the Institute in vain, which perhaps we were, but it was certainly no evil intention. To put it mildly, if you take that paper, it compares I would say favorably with Sokal at NYU, the sociology paper ...
> SALISBURY: So it was meant as satirical?
> DESER: Yes.

In a letter to Oppenheimer, following a mention of their paper in the *New York Herald Tribune* and elsewhere, Deser made light of their ill gotten gains:

> Such little experience as I have had with publicity inclines me to the view that it might be wisest, since there has apparently been little echo of the articles, to forget the whole thing; scientists would either laugh at the joke (as people at Princeton did when they heard we had won with that essay) or dismiss it as another example of garbled science reporting. The non-scientific public, I would imagine, skim all rocket-to-the-moon stuff and then forget it. (Stanley Deser, letter to Robert Oppenheimer, December 13th, 1955: Oppenheimer Papers, Box 30)

---

[25] Interview of Stanley Deser, by Donald Salisbury and Dean Rickles, March 12th, 2011.

Deser was likely right that the "non-scientific public" would pay no heed; but there were other eyes on the article, and the 'echoes' would be heard by what would become important funding agencies for gravitation. We will see below that what was nothing more than a silly joke to Arnowitt and Deser had an unexpected and rather fruitful punchline. The idea expressed in the competition essay, that gravitational phenomena could be converted to nuclear energy,[26] suggested to some less knowledgeable folk that particle accelerators could be useful for 'gravity manipulation' (i.e. for anti-gravity). For example, Agnew Bahnson, who we will meet in the next section, appears to have assumed that UNC's new accelerator might be useful for the study of gravity, and it seems highly likely that he was motivated by this paper through his association with George Rideout— he also explicitly refers to the paper in several 'memoranda' from the DeWitts' Institute of Field Physics he helped found and bankroll.

In 1954, the GRF and its essay competition, despite having been won by Bryce DeWitt the year prior, was still considered something of a fringe endeavour, notable only for the lucrative prize money it afforded young scholars. However, in addition to being taken over by a higher caliber of researcher in the mid-1950s, the GRF also triggered, via the personal network of its vice-president George Rideout, several new directions which would be taken far more seriously and lead to a far more significant regeneration of gravitational physics. It would turn out that the source of the fringe-nature of the GRF (namely, the promise of anti-gravity: gravitation's answer to the atom bomb), would trigger the interest of a whole range of sponsors who, as mentioned above, *inadvertently* (with the promise of anti-gravity) funded pure, basic research in classical and quantum gravity. We consider these sponsors in subsequent sections. In the next section we see how several examples of institutional design and funding were indirectly spawned by the GRF, via George Rideout. This all happened at roughly the same time, in the middle part of 1955—it may or may not be a coincidence that Einstein died shortly before...

9.4   Institute of Field Physics, Inc.: Bahnson Contra Babson

The Institute of Field Physics set much of the quantum gravity agenda during the late 1950s and 60s—this is one of the strokes of good fortune that quantum gravity had

---

[26] I confess, I'm not 100% convinced that the paper *was* intended to be a hoax. We saw in Chapter 4 that W. H. McCrea had submitted a paper on 'Relativity Theory and the Creation of Matter' in 1951 [McCrea (1951)], in which he postulated a very similar link between "zero-point stress" (introduced to allow the use of general relativity in the context of continuous creation models) and quantum field theory. This paper in fact marks a curious stage in the life of the continuous creation models, since it reveals a potential, though curtailed, quantum gravity model. Recall that McCrea writes: "The creation process, if it actually occurs, must be the spontaneous appearance of elementary particles, that is, an atomic quantum process. One of the attractive features of the hypothesis of such creation is that it therefore relates the large-scale properties of the universe to atomic phenomena" (ibid., p. 573). The quantum gravity link comes about since "the creation process must follow from the existence of a zero-point stress in space" (ibid.). Felix Pirani [Pirani (1955)] returned to this idea in 1954 (I believe marking the end of this line of thinking) by postulating a zero rest-mass entity with negative energy, called a 'gravitino,' that is created—this is connected to quantum theory by assuming that the gravitino is simply a negative energy neutrino.

in this early phase (namely that it was Bryce DeWitt—certainly the most quantum gravity-oriented physicist at this time—that was approached by Bahnson). It was formed to "provide a place where a number of physicists can work quietly, in financial and professional security, in a presently neglected field". Even in 1955, when the institute was being set up, general relativity and quantum gravity were treated as a province of less reputable folk. As Bryce DeWitt put it in the opening passages of one of the founding statements of the Institute of Field Physics, "Remarks on a Presently Neglected Area of Physical Research":

> The modern theory of gravitation, as formulated by Einstein in 1915, represented the high point of a profound revolution in human ideas as to the nature of the physical universe. The fruits of that revolution and of simultaneous upheaval occasioned by the advent of the quantum theory are today everywhere to be found—**except**, strangely enough, at the summit itself. The general theory of relativity (i.e. Einstein's gravitational theory) remains almost totally barren, its only applications so far being cosmological theory and in the interpretation of certain minute astronomical effects.
>
> This situation did not come about through any lack of interest in gravitation on the part of physicists immediately following Einstein's formulation. Indeed the foremost physicists of the older generation entered the arena of general relativity theory with enthusiasm, hoping to bind the phenomenon of gravitation to the rest of physics (or vice versa) in an intimate, fundamental way. However, these men failed, and those who followed them, being thus discouraged from making similar attempts, gradually left the field to the cranks and crackpots.
>
> (Document dated October 5th, 1955: CDWA, p. 1)

Amongst the reasons DeWitt presents for the lack of interest was, as he had claimed in his GRF essay, the lack of *incentives*. Without some hope of reward (financial or otherwise) and esteem, to pursue GR is rendered a lonely journey. In addition to such psychological factors impeding progress, he also mentions previous failures to extend GR; the lack of experimental input; and the difficulty of the mathematics employed.[27] Curiously, in his motivation for the study of GR, DeWitt mentions that the 'Golden Rule of science' (to base theory on empirical input) might not be as inflexible as is often supposed—"it is not clear how tight a restriction it imposes on the theorist" (ibid., p. 4). Citing the history of the Dirac equation, he argues that in order to bridge "the gap between gravitation and the rest of Nature we will be forced to go out on a limb in order to attain success." It is certainly true that DeWitt went out on a limb in working together with Bahnson, who himself fell into the "cranks and crackpots" category.

Despite DeWitt's clear-headed proclamations, the Institute of Field Physics was formed in rather strange circumstances. The story will require that we regress through some back-stories. As mentioned, Rideout is the point of origination of some of these through his connections to others interested in spearheading gravitational projects of their own, including Bahnson. We begin with the aviation company, the Glenn L. Martin

---

[27] Note how closely these align with the various recent resolutions of the puzzle of general relativity's renaissance mentioned earlier.

Company[28] (now Lockheed Martin). Almost simultaneously with the running of the Bern conference, George Trimble, vice-president of the Glenn Martin Company, wrote Bryce DeWitt, then still a Rad Lab postdoc, that:

> During a recent conversation with Mr. George Rideout, president of Roger Babson's Gravity Research Foundation, we were commiserating on the unfortunate state of the affairs that knowledgeable folks do not wish to get "mixed up" in the field of gravity research. During the course of the conversation he reviewed with me your suggestion that perhaps his Gravity Research Foundation might be transformed from its present function into an active center of research concentrating on the field of gravity. He also told me that the foundation was not able to undertake such an expansion. (Letter from G. S. Trimble to Bryce DeWitt, dated, June 10th, 1955 - [CDWA])

It seems that DeWitt had suggested to George Rideout something along the lines of the Institute for Advanced Study—a model he was very familiar with since both he and Cécile both spent time in several of them.[29] Trimble himself describes the proposed activity as an "modest version of the Institute for Advanced Study".[30] This model better aligned with The Glenn Martin Company's plans. However, their ultimate goal was not pure research for its own sake, but something more applied (with military applications). The letter goes on:

> It occurred to us sometime ago that our industry was vitally concerned with gravity. As time goes on we become more and more concerned because we feel certain that sooner

---

[28] Note that Agnew Bahnson's later anti-gravity collaborator Thomas Townsend Brown (T. T. Brown) worked for the Glenn Martin Company in the late 1930s, as a materials engineer. Bahnson formed a laboratory, not long after the Institute of Field Physics received its certificate of incorporation, in which he and Brown attempted to build flying saucers based on Brown's idea that anti-gravity can be generated through strong electric fields (that is, that strong electric fields had gravitational effects so that the latter could be controlled by manipulating the former)—he was guided by a simplistic electrodynamical analogy (of much the same kind that guided the early theoretical work on quantum gravity, in fact; though, following the likes of W. F. G. Swann, a regular of Babson's 'Gravity Days,' Brown questioned the veracity of Einstein's theory of relativity, and considered the dielectric material paramount in testing gravitational forces). Brown had attempted to establish his own gravity research institute, through a joint grant (for 'Project Winterhaven') with The Franklin Foundation via his own 'Townsend Brown Foundation,' which, in addition to a general study of gravitation and its relation to electromagnetism, proposed to redo the Troton–Noble experiment to detect the Earth's motion through the ether. The Franklin Institute Laboratories for Research and Development discussion documents for this venture (available at: http://www.thomastownsendbrown.com/hydro/winterhaven.pdf) mention quantum gravity, noting that the "smallness of magnitude of the interrelationships, as pertaining to terrestrial experiments, would continue to exist and that any practical bearing which the interrelationships might have upon us would lie in their effects in some large-scale cosmological situation which they control" (ibid., A-25). For this reason, the institute was not keen on funding practical work.

[29] Cécile was sent to the Dublin Institute for Advanced Study when she worked in Joliot's lab in the mid-40s, having earlier been charged, by Joliot, with the task of reviewing the famous Bohr–Wheeler paper on fission (Interview of Bryce DeWitt and Cecile DeWitt-Morette by Kenneth W. Ford, February 28th, 1995, Niels Bohr Library & Archives, American Institute of Physics: www.aip.org/history-programs/niels-bohr-library/oral-histories/23199).

[30] Trimble makes an interesting remark concerning the tight relationship between scientific research and society: "we feel morally obligated to push forward in the basic sciences and we believe as a dynamic industry we can provide the motivation for advances that can be obtained in no other way." In other words, for better or for worse, the pursuit of certain areas of basic research demand some kind of motivation beyond the search for deeper knowledge. Practical applications are one way to motivate such study.

or later man will invade space and we see it as our job to do everything possible to speed this event. At least one category of the things one must study, when he desires to bring space flight to a reality, is the laws of nature surrounding the force of gravity.

(ibid.)

Trimble bemoans the fact that most of those working on gravitation are "mad men and quacks"—perhaps he has those connected with Babson's own endeavour in mind here? The relevant work that they had done on space flight, notes Trimble, had been contracted out to German scientists working within Germany. But, the construction of a space vehicle was the ultimate goal of the Glenn Martin gravitational project, and a small research group focusing on the theoretical principles of gravitation was one of their routes.

The Glenn Martin Company established their institute, known as RIAS [Research Institute for Advanced Study, directed by a man named Welcome Bender], with Louis Witten[31] placed in charge of the hiring of gravitational personnel (the first hire was Bob Bass, who would go on to present at the Institute for Field Physics' 1957 conference), and in fact Witten was the first hired person. DeWitt was not interested, and was simultaneously being approached by another industrialist, Agnew Bahnson, a very wealthy North Carolinian gravity enthusiast, offering more lucrative terms (both for him and, eventually, his wife, Cécile): the promise of a position in a more traditionally academic environment, coupled with the freedom of an externally funded position, would ultimately win out.[32] Though, without the intervention of John Wheeler, DeWitt would have also turned down Bahnson's proposal, viewing him as yet another crackpot.[33]

There was much truth to DeWitt's diagnosis, but, fortunately, Bahnson was no Babson. Bahnson contacted DeWitt via letter, May 30th, 1955 (see fig. 9.5), and it can clearly be seen that Bahnson is captivated by the spectacular technological possibilities gravity might afford, just as much (if not more) than Babson was. Bahnson liked to get his hands dirty too, with amateur experiments, and had his own quirky theories of gravity and unlimited energy extraction from the gravitational field. Yet, despite this, the research institute Bahnson backed adopted a far more sober approach than the GRF, at least in

---

[31] Louis Witten, born in 1921, had previously worked for the Glenn Martin Company on pilotless aircraft, more as an electronics engineer than a theorist, after he left the army in 1946, but before he did his PhD in physics. Despite this late entrance into physics, Witten tenaciously pushed himself through 12 courses per semester to catch up and learn sufficient mathematics to do research. Following his PhD work, Witten worked on Spitzer's Stellerator as a postdoc at Princeton, together with Schwarzschild and Kruskal. He shifted back to the Martin Company after just a year, due to financial circumstances. (Interview of Louis Witten by Dean Rickles and Donald Salisbury on March 17th, 2011, Niels Bohr Library & Archives, American Institute of Physics: www.aip.org/history-programs/niels-bohr-library/oral-histories/36985.)

[32] Trimble and The Glenn Martin Company played a role in helping Bahnson and the DeWitts' institute get off the ground. Not only did they purchase a Founder's Membership for the Institute of Field Physics, for the considerable sum of $5000, they also offered their support to solicit further funding.

[33] See Interview of Bryce DeWitt and Cecile DeWitt-Morette by Kenneth W. Ford, February 28th, 1995, Niels Bohr Library & Archives, American Institute of Physics: www.aip.org/history-programs/niels-bohr-library/oral-histories/23199. In this interview, Bryce notes how he had already encountered Wheeler before this when Wheeler came to visit him in Danville (halfway between Livermore and Berkeley, where Cécile was teaching at the time) to discuss quantum gravity.

public life.[34] For example, in the foreword of an early draft for the Institute of Field Physics' promotional brochure, entitled 'The Glorious Quest,' Bahnson wrote:

> In the minds of the public the subject of gravity is often associated with fantastic possibilities. From the standpoint of the institute no specific practical results of the studies can be foreseen at this time.
>
> (Document dated November 17th, 1955: CDWA)

There is no mention of flying carpets, anti-gravity soles for shoes, or any such whacky Babsonesque ideas. In many ways Babson's GRF was used as a test-case, or foil, highlighting things to copy but (more importantly) things to avoid. A lengthy exchange of letters between Bahnson and several senior physicists (especially John Wheeler, whom Bahnson clearly admired a great deal) set to work on eradicating any aspects that might lead to claims that the institute was for crackpot research. Wheeler did much behind the scenes sculpting of the Institute of Field Physics, and was perhaps most responsible for the bringing about of the Institute. For example, writing to the acting president of the University of North Carolina, Harris Purks, November 25th, 1955, Wheeler writes of:

> the absolute necessity to avoid identification with so-called "anti-gravity research" that may be today's version of the last century's search for a perpetual motion machine. [...]
>
> Unfortunately, there are sensationalists only too willing to confuse in the public mind the distinction between so-called "anti-gravity research"...and responsible, well informed attempts to understand field physics and gravitational theory at the level where it really is mysterious, on the scale of the universe and in the elementary particle domain. (CDWA)

He goes on to applaud the step (in fact suggested by Wheeler himself, earlier) of attaching to every piece of Institute of Field Physics publicity a 'disclaimer' to the effect that the Institute of Field Physics is in no way connected to such anti-gravity research. This, Wheeler, says, is necessary to avoid discouraging both sponsors and scientists. The message that anti-gravity connotations must be avoided at all costs runs through much of the correspondence and foundation documents like a mantra. It clearly played a vital role (in the minds of physicists) in establishing the legitimacy of the enterprise. The following "Protection Clause" would therefore be attached to each Institute of Field Physics statement (CDWA):

> The work in field physics and gravitation theory carried on at the University of North Carolina at Chapel Hill, and financed by the Institute of Field Physics, as fund raising agency, has no connection with so-called "anti-gravity research" of whatever kind and for whatever purposes. Its scientists, basing their investigations upon verifiable data, accept

---

[34] Bahnson would repeatedly overstep the mark in terms of using the Institute's name to further his own speculative theories. Just a few months before his death he visited Max Born, using Wheeler's friendship and introducing himself as the president of the institute, to push his views about utilizing zero point energy (see fig. 9.4).

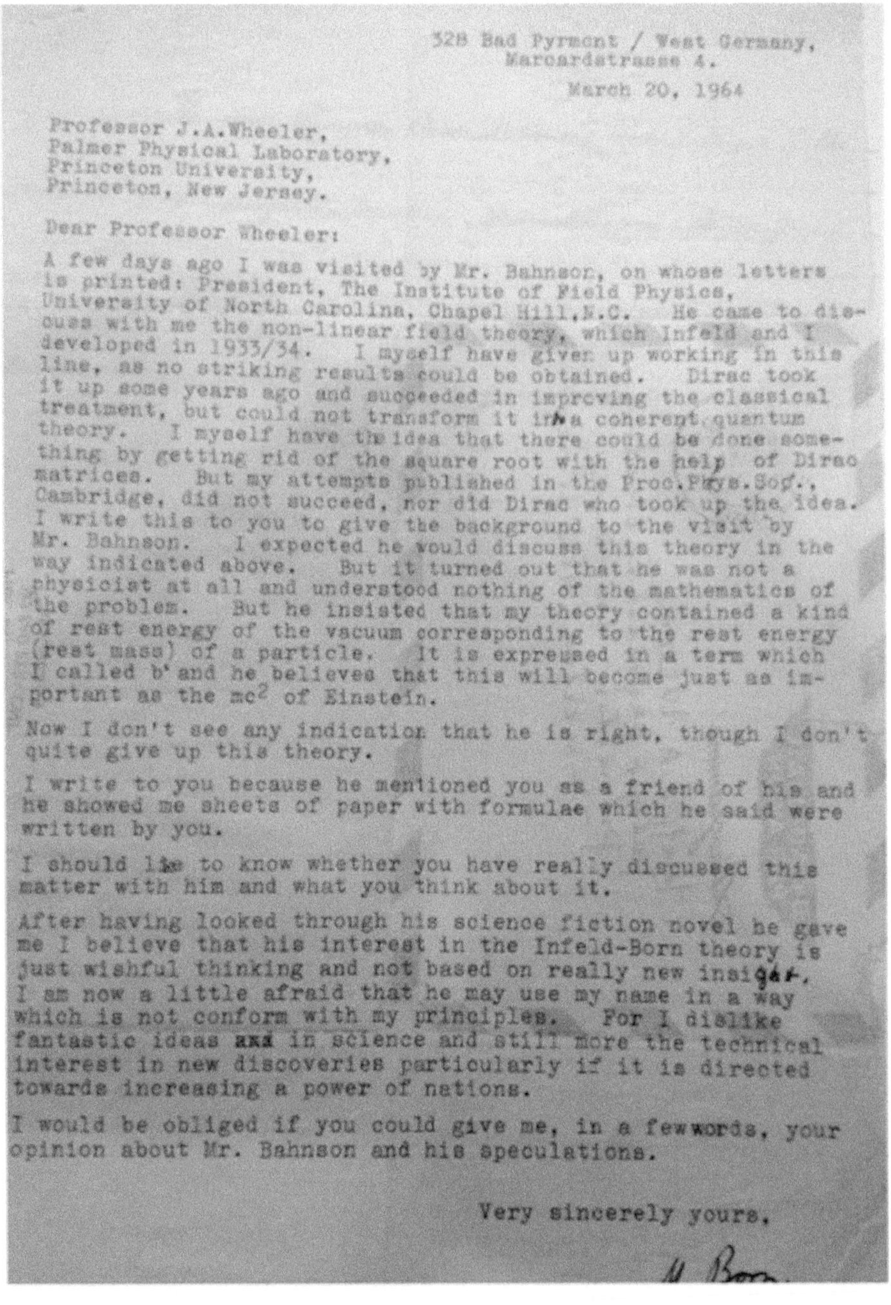

328 Bad Pyrmont / West Germany,
Marcardstrasse 4.

March 20, 1964

Professor J.A.Wheeler,
Palmer Physical Laboratory,
Princeton University,
Princeton, New Jersey.

Dear Professor Wheeler:

A few days ago I was visited by Mr. Bahnson, on whose letters
is printed: President, The Institute of Field Physics,
University of North Carolina, Chapel Hill,N.C.   He came to discuss with me the non-linear field theory, which Infeld and I
developed in 1933/34.   I myself have given up working in this
line, as no striking results could be obtained.   Dirac took
it up some years ago and succeeded in improving the classical
treatment, but could not transform it into a coherent quantum
theory.   I myself have the idea that there could be done something by getting rid of the square root with the help of Dirac
matrices.   But my attempts published in the Proc.Phys.Soc.,
Cambridge, did not succeed, nor did Dirac who took up the idea.
I write this to you to give the background to the visit by
Mr. Bahnson.   I expected he would discuss this theory in the
way indicated above.   But it turned out that he was not a
physicist at all and understood nothing of the mathematics of
the problem.   But he insisted that my theory contained a kind
of rest energy of the vacuum corresponding to the rest energy
(rest mass) of a particle.   It is expressed in a term which
I called b' and he believes that this will become just as important as the mc² of Einstein.

Now I don't see any indication that he is right, though I don't
quite give up this theory.

I write to you because he mentioned you as a friend of his and
he showed me sheets of paper with formulae which he said were
written by you.

I should like to know whether you have really discussed this
matter with him and what you think about it.

After having looked through his science fiction novel he gave
me I believe that his interest in the Infeld-Born theory is
just wishful thinking and not based on really new insight.
I am now a little afraid that he may use my name in a way
which is not conform with my principles.   For I dislike
fantastic ideas and in science and still more the technical
interest in new discoveries particularly if it is directed
towards increasing a power of nations.

I would be obliged if you could give me, in a few words, your
opinion about Mr. Bahnson and his speculations.

Very sincerely yours,

M Born.

Fig. 9.4 *Letter from Max Born to John Wheeler, concerning Agnew Bahnson's visit to him. [Image source: CDWA]*

THE BAHNSON COMPANY
Complete Industrial Air Conditioning

1001 SO MARSHALL ST
WINSTON-SALEM, N C

May 30, 1955

Mr. Bryce S. DeWitt
University of California
Theoretical Division
Radiation Laboratory
Berkeley 4, California

My dear Mr. DeWitt:

For several years I have had correspondence with Mr. George Rideout, President of the Gravity Research Foundation in New Boston, New Hampshire. Recently, I wrote him about the fact that the Burlington Mills Company in Greensboro, North Carolina, with whom we have done a considerable amount of business over a period of thirty years, has given a two hundred thousand dollar grant to the State College of the University of North Carolina, at Raleigh, North Carolina, for the building of a nuclear reactor and other laboratory facilities. That laboratory was dedicated about a week ago. I had hoped to attend the dedication but was unable to do so. I did talk to Mr. Spencer Love, Chairman of the Board of Burlington Mills, who is a good friend of mine, and mentioned to him the thing that has been of interest to me for over twenty years. He seemed quite willing to investigate the possibility further in connection with the work in this nuclear laboratory at State College.

You may have recently heard of the division of Glenn L. Martin Aircraft company and I believe of the Convair Division of General Dynamics, that has set up research in anti-gravity as a new method of supporting heavier than air machines above the surface of the earth. You may have also read the article by Mr. William Lear of the Lear Radio Company in the last October issue of FLYING MAGAZINE predicting that fifty years from now the air plane would be a horse and buggy and that anti-gravitational reaction would support aircraft at any desired height above the earth. This may sound a little like the flying saucer deal but I believe it has a very practical opportunity of being worked out during our normal lifetime. Twenty years ago such ideas were not received with much hope of practical consummation. I recall discussing such things with Dr. David Griggs, who is now at the University of Southern California in the Geological Department. He lives in Brentwood which is on the outskirts of Los Angeles. I doubt if you have ever crossed his path but I am sending him a copy of this letter in the hope that you may have had some contact or may have such contact if you are ever in his vicinity or he is in yours.

I agree with your letter that the field of gravitation is quite unexplored. It seems to be one of the most important pioneer frontiers in science today. The

Bahnson Company

HUMIDIFYING · HEATING · VENTILATING · COOLING · AIR FILTERING · DEHUMIDIFYING

Fig. 9.5 *The initial letter from Agnew Bahnson to Bryce DeWitt that led to the establishment of The Institute for Field Physics, the first research center devoted to quantum gravity. [Image source: CDWA]*

the Newton–Einstein analysis of gravity as free of a single established exception, and as the most comprehensive physical description wc have today. They seek the implications of gravity and other fields of force at the level of the elementary particles. More generally, the Chapel Hill project is a modest attempt to learn more about the nature of matter and energy.

Wheeler did not hold back on the need for the Institute of Field Physics, though his claims were moderated somewhat by a knowledge that progress might well be very slow. Writing to Bahnson he states:

> It is hard to see how one can get to the bottom of the elementary particle problem—the central issue of modern physics–without coming to the very foundations of our physical world and the structure of space and time. Gravity, fields, and particles must in the end be all one unity. The absence of any paradox or discrepancy in gravitation theory at the human and astronomical levels creates an obligation to apply Einstein's ideas down to smaller and smaller distances. One must check as one goes, until one has either a successful extension to the very smallest distances, or a definite contradiction or paradox that will demand revision. ... The challenge cannot be evaded. Exactly how to proceed is a matter of wisdom, skill, judgement, and a good idea. Nobody guarantees to have a good idea, but the DeWitts, fortunately, have a very sound plan of what to do while searching for a good idea. They propose to do something that has long needed doing – help make clear the fundamental facts and principles of general relativity so clearly and inescapably that every competent worker knows what is right and what is wrong. They can do much to clear away the debris of ruined theories from the rocklike solidity of Einstein's gravitation theory so its meaning and consequences will be clear to all. This is a great enterprise. Einstein's theory of the space-time-gravitation field is even richer than Maxwell's theory of the electromagnetic field. That field has been investigated for many years, and now forms the foundation for a great science. One cannot feel physics has done its job until a similarly complete investigation has been made for the gravitational field. (John Wheeler, letter to Bahnson, November 25th, 1955—CDWA)

Though there is, of course, a fair amount of rhetoric in this passage, it nonetheless shows the importance in Wheeler's mind of the role that the Institute of Field Physics (and the DeWitts) would play. It also reflects Wheeler's own personal hopes for the study of gravitation in providing answers to the particle problem (issues not so much on DeWitt's radar).

The plan was to house an institute within an academic institution, so as to avoid the conflict that physicists felt working in an industrial setting. The chosen location was the physics department at the University of North Carolina, Chapel Hill. In order to lend prestige to the Institute of Field Physics, Bahnson secured letters of comment from several of the most prominent physicists of the day, including Oppenheimer, Dyson, Teller, Feynman, and Wheeler. The various letters of support (dating from between October 1955 and January 1956), for which the preceding letter from Wheeler to Purks provides a cover letter, highlight the recognition that general relativity and gravitational research had been unfairly neglected, and the need for a renewal of interest. Oppenheimer writes that he "shares with most physicists the impression that this field has been rather neglected by us". Dyson seconds this (as does Nordheim), but adds some conditions for success, more or less reiterating what Wheeler had already said: that immediate results should not be expected, and that ("to avoid becoming isolated and sterile") the institute should be settled as firmly as possible in "the framework of normal university life". Edward Teller remarks that "a comprehensive examination of general relativity and high-energy physics, together with an investigation of the interaction between these two fields, may very well lead to the essential advance for which we are all looking."

Feynman too voiced the opinion that "the problem of the relation of gravitation to the rest of physics is one of the outstanding theoretical problems of our age." However, he was less positive about the chances of the proposed institute in its original form. Feynman was not convinced that an industrially funded institute, detached from a university, could possibly deliver the requisite flexibility to develop new fundamental knowledge: that required absolute freedom to bounce around between topics, as one chose (see fig. 9.6).[35] On learning that the institute was to be housed in a university, Feynman was unreservedly positive about the proposal (letter to Wheeler, dated December 2nd, 1955: CDWA).

John Toll, head of physics at the University of Maryland, writes, directly discussing the other letters:

> Most of my colleagues have pointed out in their comments that the field of general relativity has not received the attention which it deserves and that it is particularly important to attempt to obtain some synthesis of the methods and concepts used in general relativity with the ideas now employed to discuss elementary particles. One reason for the neglect of general relativity has been the great difficulty of work in this field which challenges even the best theoretical physicists; solution of the major problems involved will probably require a determined program which may extend over many years. *A second and related reason has been the difficulty of obtaining adequate support for this field; the problems are not of the type which are supported by federal agencies which finance so much of the research in physics in the United States by short term contracts, mostly in fields which appear to have more immediate applicability to defence problems.* (Letter from J. S. Toll to John Wheeler, dated December 28th, 1955: CDWA—emphasis mine)

This was all written towards the end of 1955. By the same period of 1957, just two years later, the picture looked remarkably rosier, and Toll's second reason had been virtually nullified.

An initial meet and greet session of the institute was held June 8th–10th, 1956, at Roaring Gap in North Carolina, where Bahnson had a summer house. This was open to all members of the institute and a few select others, including Freeman Dyson and Lothar Nordheim, potential funders, and a reporter from the Winston-Salem Journal & Sentinel. As Bahnson put it, in his "4th Memorandum" (of June 20th, 1956) the purpose

---

[35] These sound remarkably similar to DeWitt's own concerns on being approached by Glenn Martin. It seems likely that there was an on-going debate about the best environment for generating research following Vannevar Bush's 1954 report on "The independent research institution" (*Physics Today*, 7: 19).

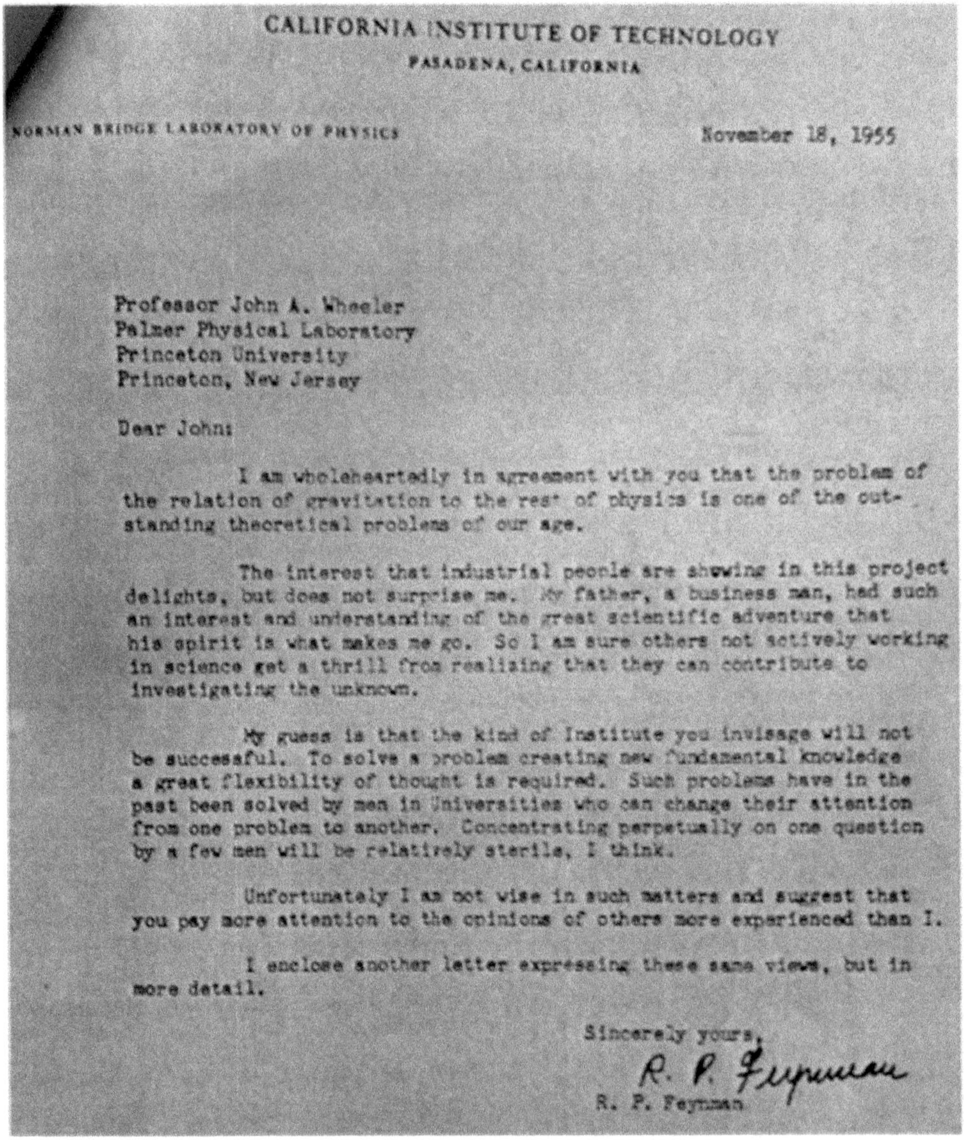

CALIFORNIA INSTITUTE OF TECHNOLOGY
PASADENA, CALIFORNIA

NORMAN BRIDGE LABORATORY OF PHYSICS   November 18, 1955

Professor John A. Wheeler
Palmer Physical Laboratory
Princeton University
Princeton, New Jersey

Dear John:

I am wholeheartedly in agreement with you that the problem of the relation of gravitation to the rest of physics is one of the outstanding theoretical problems of our age.

The interest that industrial people are showing in this project delights, but does not surprise me. My father, a business man, had such an interest and understanding of the great scientific adventure that his spirit is what makes me go. So I am sure others not actively working in science get a thrill from realizing that they can contribute to investigating the unknown.

My guess is that the kind of Institute you invisage will not be successful. To solve a problem creating new fundamental knowledge a great flexibility of thought is required. Such problems have in the past been solved by men in Universities who can change their attention from one problem to another. Concentrating perpetually on one question by a few men will be relatively sterile, I think.

Unfortunately I am not wise in such matters and suggest that you pay more attention to the opinions of others more experienced than I.

I enclose another letter expressing these same views, but in more detail.

Sincerely yours,

R. P. Feynman

R. P. Feynman

Fig. 9.6 *Letter from Richard Feynman to John Wheeler, endorsing the need for further research on the area of gravitation, but not able to endorse the proposed institute. [Image source: CDWA]*

of the meeting was "to introduce members of the Institute and their guests to Mr and Mrs DeWitt and 'to define more clearly' the problems to be dealt with at the institute" (with gravity as "the focal point of interest"—CDWA).

December 7th, 1956 saw DeWitt deliver a paper focusing on current research in gravitational physics to the American Astronautical Society (published as [DeWitt (1957)]). By this time he was able to give his position as the Director of the Institute of Field Physics. The talk was clearly intended as a piece of propaganda and advertising. DeWitt opened by distancing his work from any foreseeable practical applications, in the manner of the protection clause (we might call this 'de-Babsonification'). He then notes the lack of serious research being carried out; counting just seven institutions with gravitation research projects: Syracuse, Princeton, Purdue, UNC, Cambridge, Paris, and Stockholm; with RIAS, Inc, on the private/industrial side.[36] The Institute of Field Physics received its certificate of incorporation on September 7th, 1955, becoming one of the most important research centers for gravitation. This was enough of an event to attract the front page (and a considerable chunk of real estate elsewhere) in North Carolina's preeminent newspaper the *Salem Sunday Journal and Sentinel*, calling it "one of the most significant developments to have occurred in North Carolina in recent years" (June 24th, 1956).

In his APS report, DeWitt mentions even at this early stage of quantum gravity history the problems that would plague the quantum geometrodynamical approach throughout its existence (until it transformed into loop quantum gravity): these are the problem of defining the energy and the quantities that are conserved with respect to it (i.e. the observables), and the factor ordering problem.[37] The former was studied by Bergmann's group at Syracuse, while the latter problem was studied by DeWitt's own group at the Institute of Field Physics. The agenda for the Institute was heavily biased towards quantum gravitational research (see fig. 9.7).

Bahnson maintained his own exuberant interests in gravitation, and was sufficiently well connected to have these (often naive, sometimes crackpot) ideas examined by the likes of Edward Teller, and also Bryce DeWitt, with the assistance of the head of physics at UNC, Everett Palmatier.[38] Bahnson's interest, as with Babson's, was with

---

[36] In terms of personnel, we have Peter Bergmann and his students at Syracuse; Wheeler and his students at Princeton; Suraj Gupta and Frederik Belinfante at Purdue (with Vaclav Hlavatý nearby at the University of Indiana); the DeWitts at UNC; Hermann Bondi and his students, including Felix Pirani, at Cambridge; André Lichnerowicz and his students at Paris; Oskar Klein and his students, including Bertel Laurent, at Stockholm; and Louis Witten at RIAS. This list of 'schools' curiously misses Alfred Schild at Carnegie Tech, despite the fact that Schild had supervised Pirani and written one of the first papers on the canonical quantization of gravity just five years earlier.

[37] This problem refers to an issue caused by the straightforward canonical quantization of general relativity according to which, when one meets a momentum term, one substitutes a derivative. However, when this procedure is applied in general relativity, one faces situations were there exist products, and so one has to multiply as well as differentiate. The order in which one does this matters for the form of the final wave equation.

[38] Neither DeWitt nor Palmatier took Bahnson's ideas seriously, and would frequently joke around in their analyses of Bahnson's experiments, marking them "top secret!" There was one stage, however, where DeWitt seriously wondered whether Bahnson would drop him as a result of his repeated rejection of the experiments (letter from DeWitt to Palmatier, August 1958: CDWA).

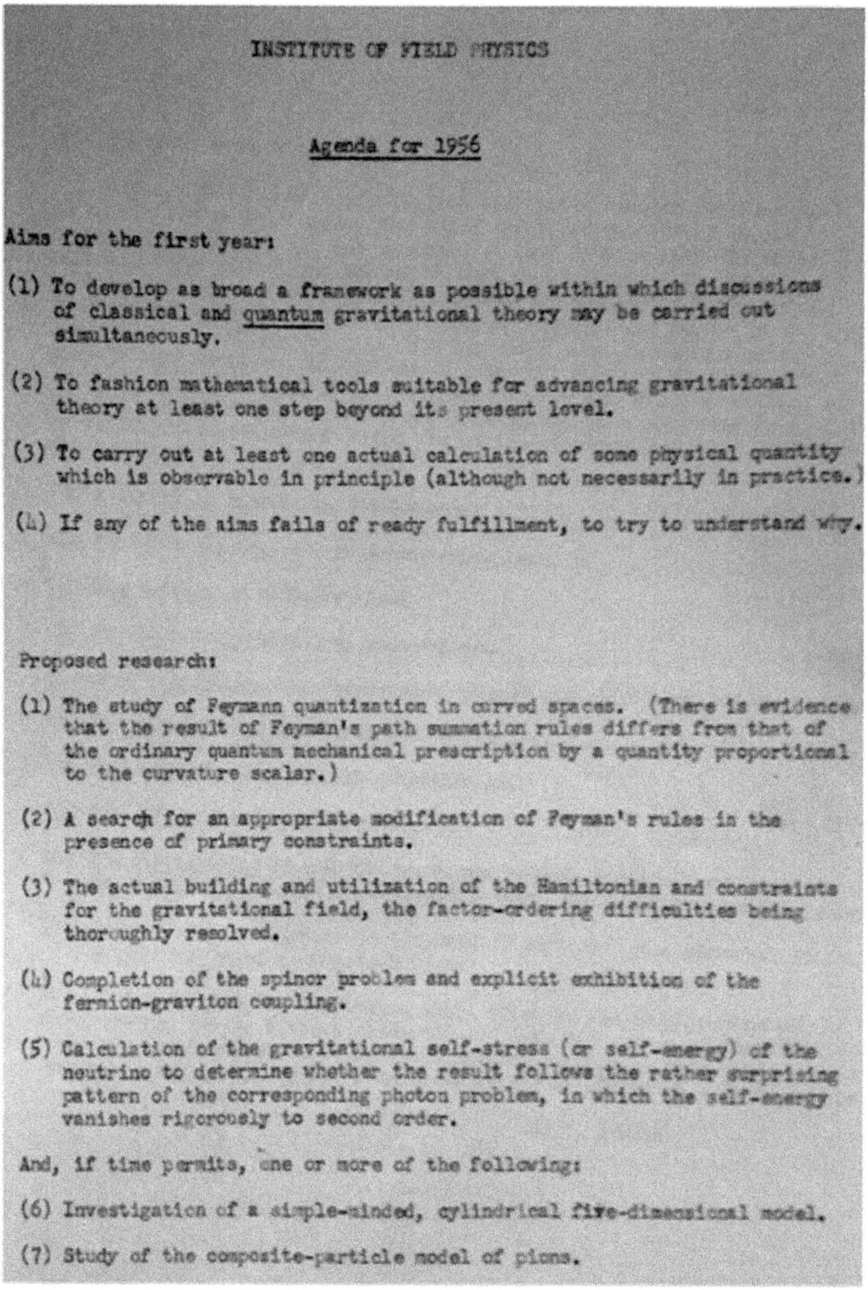

INSTITUTE OF FIELD PHYSICS

Agenda for 1956

Aims for the first year:

(1) To develop as broad a framework as possible within which discussions
of classical and quantum gravitational theory may be carried out
simultaneously.

(2) To fashion mathematical tools suitable for advancing gravitational
theory at least one step beyond its present level.

(3) To carry out at least one actual calculation of some physical quantity
which is observable in principle (although not necessarily in practice.)

(4) If any of the aims fails of ready fulfillment, to try to understand why.

Proposed research:

(1) The study of Feynmann quantization in curved spaces. (There is evidence
that the result of Feyman's path summation rules differs from that of
the ordinary quantum mechanical prescription by a quantity proportional
to the curvature scalar.)

(2) A search for an appropriate modification of Feyman's rules in the
presence of primary constraints.

(3) The actual building and utilization of the Hamiltonian and constraints
for the gravitational field, the factor-ordering difficulties being
thoroughly resolved.

(4) Completion of the spinor problem and explicit exhibition of the
fermion-graviton coupling.

(5) Calculation of the gravitational self-stress (or self-energy) of the
neutrino to determine whether the result follows the rather surprising
pattern of the corresponding photon problem, in which the self-energy
vanishes rigorously to second order.

And, if time permits, one or more of the following:

(6) Investigation of a simple-minded, cylindrical five-dimensional model.

(7) Study of the composite-particle model of pions.

Fig. 9.7 *The first year's agenda for the Institute of Field Physics. [Image source: CDWA]*

electrically induced anti-gravity (or, to use Brown's terminology, 'electrogravitics'). But while Babson wanted to shield it for its evils, Bahnson wanted to master it to create spaceships. In 1957 he hired T. T. Brown to assist him with a series of experiments, ultimately to build an anti-gravity powered flying saucer.[39] In other words, while the public face of Bahnson's institute involved an explicit blanket dismissal of anti-gravity research, 'The Bahnson Lab' was busily experimenting in precisely this area, indeed, one might call it "the world-center of anti-gravity research."[40] Louis Witten recalls visiting this lab when he was at RIAS:[41]

> Witten: I was already doing quite a bit of relativity by the time Bahnson came along. But I don't know what you want to know about Bahnson. He claimed that he had an anti-gravity thing, and Welcome Bender sent me down to see what he had.
>
> Rickles: Did he claim to have built something?
>
> Witten: Yes, he had a laboratory. I went to visit him and visited his laboratory, and the basic idea of his laboratory was he had a strong electrostatic field, which was about 150,000 volts over a distance of about like that, about a meter. And he had an operator operating this thing. But I knew enough about experiments to know that this was not a very happy place, because I knew that for strong electrostatic fields, there shouldn't be any sharp points around. Everything should be curved, and nothing was curved.
>
> Salisbury: Oh my goodness.
>
> Witten: And the operator was working on it, his hair was standing up. [Chuckling] And then Bahnson took a long cylindrical pipe, and he smoked a cigarette, and he blew through the pipe into this central place where the electrostatic field was, and low and behold, the smoke rose. Explain it. [Laughter] Just at that moment, there's a table with a sharp corner, and I was standing with my back to it, about a foot away, and there was a spark from my backside to the corner of the table. So I said, "Let's go down into the hall." [Laughter] And I said it's nothing worth explaining. It's completely understandable. You all explain it. It's not worth explaining. You had an electric field, you got ionization, you get motion. I said look at the operator, his hair is standing on end! [Laughter]…
>
> Witten: And I didn't talk to him much; I just left.

While this sounds like nothing but an amusing anecdote, the influence of anti-gravity on the development of legitimate areas of gravitational physics (including, especially, quantum gravity) should not be underestimated. The bulk of funding largely responsible

---

[39] See [Yost (1991)] for a discussion of the notebooks from this collaboration.

[40] Old 8mm footage of Bahnson and Brown, and an assistant, J. Frank King Jr. (Bahnson's brother-in-law) has been publicly released: https://www.youtube.com/watch?v=vWuUJt7iSAo—a youthful, and slightly bemused-looking, Bryce DeWitt makes an appearance in the film at around 14 minutes (this visit occurred on December 19th, 1957); it must have been somewhat disconcerting for DeWitt to see Bahnson dabbling in such ventures, especially knowing full well that the 'anti-gravity' phenomena were simple 'ionization' effects.

[41] Interview of Louis Witten by Dean Rickles and Donald Salisbury on March 17th, 2011, Niels Bohr Library & Archives, American Institute of Physics: www.aip.org/history-programs/niels-bohr-library/oral-histories/36985.

for the significant infrastructure shift that had to occur before the fruits of the renaissance could be produced was grounded in the (mistaken) belief that anti-gravity (or at the very least some practical applications) were forthcoming. In other words, the lack of knowledge of gravitation in the low-water mark period was both a curse and a blessing.

9.5   Quantum Gravity and the Military-Industrial-Academic Complex

The year 1957 was notable for another reason beyond the Chapel Hill conference: Sputnik was launched, much to the surprise of most Americans. This had a marked effect in the funding of scientific projects, including those involving gravitation, since gravitation was linked to aviation and spaceflight. The Cold War element cannot be underestimated, and key gravitational players were behind the Iron Curtain. Fock was in the USSR, Infeld was in Poland, Papapetrou was in East Berlin, and Rosenfeld and Géhéniau were considered a security risk. There was some speculation that Fock had managed to develop a new 'graviplane' flight technology (based on a 'lift anomaly'). This was nonsense, and presumably propaganda.[42]

1955 was a good year for gravitation. The Aeronautical Research Laboratories[43] (at the Wright-Patterson Air Force Base, outside Dayton, Ohio) decided to expand their support to projects involving general relativity. Dr. Max Scherberg, who was in charge of these decisions (as 'Chief of Applied Mathematics'), provided the first ever military funding for gravity to Vaclav Hlavatý, a Czech researcher then based at Indiana University. Scherberg himself was trained as an applied mathematician, and had written his doctoral thesis, in 1931, on the degree of convergence of a series of Bessel functions, with Dunham Jackson at the University of Minnesota. However, he had worked on flight modeling (especially modeling flight spins), and produced a report[44] for NACA [the National Advisory Committee for Aeronautics, before it became NASA] before his doctoral work. But it is clear that he didn't have knowledge of general relativity, and it is most likely that he approached Hlavatý, hardly the best general relativity scholar at the time, because of a bit of a media frenzy that had occurred in 1953 on account of Hlavatý solving the equations of Einstein's then-latest unified field theory. In a popular TV program 'Johns Hopkins science review,' Hlavatý was presented as a great man of

---

[42] Bahnson, Memorandum No. 1, Feb 3, 1958 (CDWA). The source of the story was *American Aviation* magazine, which appears to have been the mouthpiece of Gravity Rand Ltd.—it is difficult to probe the origins of this curious venture (I suspect that Bahnson and T. T. Brown were involved in some way, especially since Brown established a company called 'Rand International Limited').

[43] The name was changed to the Aerospace Research Laboratory soon after adding general relativity to its portfolio, and that should give some information about the expectations of the Air Force, in terms of applications, when they added relativity. However, there is also a case to be made that the direct Air Force support, with a knowledgeable researcher guiding funding decisions, saved the Air Force the trouble of sifting through many proposals that were scientifically ungrounded (see [Kennefick (2007)], pp. 116–7, for more on this 'protective function' of ARL's gravity funding).

[44] "Mass distribution and performance of free flight models" (NACA Technical Note 268, October 1st, 1927): https://ntrs.nasa.gov/archive/nasa/casi.ntrs.nasa.gov/19930081026.pdf.

science.[45] More importantly, Hlavatý was also mentioned in a revealing story, "Conquest of Gravity Aim of Top Scientists in U.S.", featured in the *New York Herald-Tribune* (Sunday, November 20th, 1955: pp. 1 and 36—the author was Ansel E. Talbert, the military and aviation editor for the paper). Talbert writes that Hlavatý "believes that gravity simply is one aspect of electromagnetism—the basis of all cosmic forces—and eventually may be controlled like light and radio waves".[46] This idea (and the story in which it was expressed) was undoubtedly a driving force behind many anti-gravity speculations and funding opportunities: if gravity is just an aspect of electromagnetism, and electromagnetism can be controlled and shielded, then it surely stands to reason that gravity can likewise be controlled and shielded!

Scherberg was asked to find someone to oversee Air Force funding in gravitation. He eventually settled on Joshua Goldberg, a student of Peter Bergmann's, who was then at the Armor Research Foundation.[47] Goldberg joined the staff at the Aeronautical Research Laboratories (as part of the General Physics Laboratory) in September, 1956, just as the Chapel Hill conference was being organized. In addition to overseeing funding, it was also a genuine research group, and would later include Roy Kerr. It ran until 1972, when the 'Mansfield amendment' put an end to the funding of basic research by the Defence Department.

One of the many elements of serendipity in the history of this period of quantum gravity was the fact that the DeWitts had just started the Institute of Field Physics venture around the same time that funding for gravitational projects started at the ARL, transforming what would have been a more local affair into a truly international event (not least through the military air transport [MATS] flights offered for non-US participants). As mentioned, Wheeler was consulted by Goldberg and mentioned the Institute of Field Physics as a potential recipient of support. This new funding was also known to Bahnson right away, because of his connections. Indeed, the Institute of Field Physics was the very first recipient of ARL support (for the Chapel Hill conference), and would lead to the earliest subsequent recipients since Goldberg was able to interact with the leading relativists at the conference. Goldberg was also offered a rare opportunity to

---

[45] One can watch this program by visiting Johns Hopkins University: https://catalyst.library.jhu.edu/catalog/bib_2405701. Hlavatý's solution of Einstein's unified field equations was a *tour de force* (involving 64 unknowns), and it is clear that the fact that Einstein himself deemed their solution impossible played a role in the media frenzy—physicists were not so taken, simply because they distrusted Einstein's unified field theory approach.

[46] We note that Arnowitt and Deser's GRF essay is also picked up by this article, which quotes the following passage: "One of the most hopeful aspects of the problem is that until recently gravitation could be observed but not experimented on in any controlled fashion, while now with the advent in the past two years of the new high-energy accelerators (the Cosmotron and the even more recent Berkeley Bevatron) the new particles which have been linked with the gravitational field can be examined and worked with at will"—we shall return to this in a moment. The article is full of 'gee-whiz' speculation and false claims (such as the claim that "[t]here is no scientific knowledge or generally accepted theory about the speed with which it [gravity] travels across interplanetary space." But this, and the Hlavatý mention, would have obviously stuck in the minds of those looking for anti-gravity projects to fund.

[47] Goldberg has given an account of the history of this episode in [Goldberg (1993)]. He notes that Scherberg first asked several others, including Bergmann himself, but also Jordan, Géhéniau, Hlavatý, and Rosen. Bergmann had suggested Goldberg to Scherberg. Goldberg himself went on to consult with John Wheeler and Peter Bergmann soon after he was hired in order to determine where support was needed.

travel by the ARL, in the states and overseas, in order to investigate potential recipients of grants. Thus, this position came with a built-in method for linking the various researchers together. He interacted, in a short period of time, with Géhéniau and Debever in Belgium; Lichnerowicz in Paris; and Jordan in Hamburg. Goldberg was a hub in an emerging network.[48]

There was naturally a condition that funding could not be given to support fundamental research: there had to be some discernible military purpose, however remote. Anti-gravity was the answer. As Goldberg puts it:

> There were people, and I don't know who—this is one of those hearsay things that nobody can verify, so I will say it, but it's totally unverified—that some officer in the Air Force, thinking about the next big thing that the Air Force needed, was an anti-gravity device. And so they needed somebody to work on general relativity ...

There are some very revealing reports from the United States Air Force that tell us much about what was in their minds when they decided to fund gravitational projects.[49] The primary target seems to have been 'electrogravitics.' When we look at these reports, we quickly realize why Hlavatý was the first person to be supported by the Air Force, for it was a naive understanding of Einstein's unified field theory that motivated the decision. Again, this simply indicates that the level of knowledge of gravitation (even such basic aspects as the impossibility of constructing gravity shields or absorbers) was in a dire state. Even experts on physics had only a basic grasp of the possibilities. It is little wonder that funding agencies were confused, especially given the power of physics demonstrated in the bomb projects. Nuclear power must have seemed almost magical and capable of many things and inspired almost blind faith in the power of physics and physicists.[50]

An important example of such a report is entitled "Electrogravitic Systems: An Examination of Elecrostatic Motion, Dynamic Counterbary and Barycentric Control" [TL 565 A9: AF Wright Aeronautical Laboratories Technical Library Wright-Patterson Air Force Base, Ohio].[51] This report views gravitation as aviation's enemy, arguing

---

[48] In fact, political factors constrained Goldberg in various ways. Géhéniau, for example, could not receive funding despite Goldberg's desire to make it happen since the Air Force forbade it on account of his political leanings—this despite, as Goldberg notes ([Goldberg (1993)], p. 95), the fact that they were happy to offer support to Jordan who had been associated with the Nazis (of course, 'Operation Paperclip' would overlook many such cases).

[49] As Daniel Keffenick has nicely summed it up: "The lack of comprehension of relativity that the air force labored under was well matched by the relativists' inability to fathom what it was that the air force wanted from them" [Kennefick (2007), p. 116].

[50] Robert Serber had already written on the possibility of using nuclear power for rockets as early as 1946 [Research Memorandum No. 1. Project RAND, July 5th, 1946].

[51] The report claims to have been prepared (and finalized on February 25th, 1956) by the 'Gravity Research Group, Aviation Studies (International) Limited,' which gives as its address "London 29–31 Cheval Place, Knightsbridge, London SW7, England" (with internal report number: GRG 013/56 February 1956. Curiously, this address corresponded to residential flats, and seems to have been such for some time. It is likely that the 'group' in question was a single individual, namely, the director, a Mr. R. G. Worcester, a fairly well-known expert on air policy and aviation). So far as I can tell, no other reports from this 'group' exist. A declassified Douglas Aircraft Company Inc. report on "Unconventional Propulsion Systems" (from W. B. Klemperer to

that fundamental research "to discover the nature of gravity from cosmic or quantum theory...would [if successful] change the concept of sustentation, and confer upon a vehicle qualities that would now be regarded as the ultimate in aviation" (pp. 1–2). The report then summarizes past and current work in the field, including, primarily, 'electrogravitics.' However, also central is Hlavatý, who is heralded as "the most authoritative voice in micro-physics," and is presented as having discovered a "concept of gravity as an electromagnetic force that may be controlled like a light wave." Later on, mention is made of finding and solving the equations of the unified field theory which "Einstein hopes to find a way of doing this before he dies."

Also presented is a list of interested industrial parties interested in "counterbary" (1950s aviation-speak for anti-gravity), including Douglas, Hiller, Glenn Martin (quoted as saying that "gravity control could be achieved in six years, but they add that it would entail a Manhattan District type of effort to bring it about"), Sikorsky (quoted as saying that "gravity is tangible and formidable, but there must be a physical carrier for this immense trans-spatial force"). General Electric are claimed to be working on a way "to make adjustments to gravity;" Bell Labs claimed to have hardware that is able "to cancel out gravity," and Lawrence Bell himself is said to be "convinced that practical hardware will emerge from current programs." There are also companies more connected with navigating the challenges of gravity, such as Lear, Inc., Convair, Sperry (who went on to develop traffic control systems and other digital aviation technologies). In each case, the idea is that if "a physical manifestation exists, a physical device can be developed for creating a similar force moving in the opposite direction to cancel it." This lack of understanding of the modern conception of gravitation thus motivated a surge of investment in gravitational research.

The article goes on to mention research institutes with gravity foci, including the GRF, RIAS (at which the new gravitational wing is mentioned), and the Institute for Field Physics (incorrectly written as the Institute for *Pure* Physics), which also is described as under proposal at this stage.

There is clearly an electrogravitics bias in the article, with T. T. Brown hailed as "the equivalent of Frank Whittle in gas turbines." The author also expresses bemusement as to how the Germans could have possibly overlooked electrogravitics, given how close they were to the Americans with respect to the nuclear program. Once again, however, it seems that Hlavatý's conception of the unified field theory is playing a major role:

E. P. Wheaton, dated March 1st, 1955: MTM-622, Part 2) indicates correspondence with the research group who are described as "Management Consultants...who prepare and distribute the Aviation Reports discussing technical, commercial and political developments in the world of aviation." The report, however, is something of a favourite amongst UFO conspiracy theorists on the world wide web. While their interpretations are somewhat left-field, the report is genuine (it appears in the Library of Congress in an issue of *Aviation Studies*, for example), and appears to have been taken seriously. Of course, it is perfectly true that the US Air Force were interested in all manner of propulsion methods, and this included 'spaceships' and methods of space travel— even Bryce DeWitt wrote an early report on 'The Scientific Uses of Large Spaceships' (*General Atomic Report GAMD* **965**, 1958) and Lyman Spitzer, of the Matterhorn Project, wrote on 'Interplanetary Travel' as early as 1954. But this is simply a natural part of the development of spaceflight, leading, amongst other things, to the creation of NASA and the JPL. Any secrecy is due to the fact that the military wanted to keep its funded research under wraps from its enemies.

If Dr. Vaclav Hlavatý thinks gravity is potentially controllable that surely should be justification enough, and indeed inspiration, for physicists to apply their minds and for management to take a risk. Hlavatý is the only man who thinks he can see a way of doing the mathematics to demonstrate Einstein's unified field theory—something that Einstein himself said was beyond him. Relativity and the unified field theory go to the root of electrogravitics and the shifts in thinking, the hopes and fears, and a measure of progress is to be obtained only in the last resort from men of this stature. Major theoretical breakthroughs to discover the sources of gravity will be made by the most advanced intellects using the most advanced research tools.

(ibid., p. 12)

The logic seems to be clear, at least: Because Einstein was a great genius and Hlavatý did something Einstein claimed to be impossible, Hlavatý must also be a great genius too, whose ideas must be taken as authoritative.

Finally, we see a clear reference made to the 'hoax' essay of Arnowitt and Deser (neither of whom was an expert on gravitational physics when they wrote it), with the remark:

Though gravity research, such as there has been of it, has been unclassified, new principles and information gained from the nuclear research facilities that have a vehicle application is expected to be withheld. The heart of the problem to understanding gravity is likely to prove to be the way in which the very high energy sub-nuclear particles convert something, whatever it is, continuously and automatically into the tremendous nuclear and electromagnetic forces.

(ibid., p. 13)

However, the report sounds the cautious note of the scientific establishment that "nothing can be reasonably expected from the science for yet awhile," but ignores this in favour of the engineering positions, pointing out that "NACA is active, and nearly all of the Universities are doing work that borders close to what is involved here, and something fruitful is likely to turn up before very long." There is certainly truth in the view that the various aviation institutions believed they were close to stumbling on something big. But there was also explicit recognition that the levels of uncertainty were very large indeed. For example, in a NACA Aviation Report (October 19th, 1954), it is reported that:

Glenn Martin now feels ready to say in public that they are examining the unified field theory to see what can be done. It would probably be truer to say that Martin and other companies are now looking for men who can make some kind of sense out of Einstein's equations. There's nobody in the air industry at present with the faintest idea of what it is all about. Also, just as necessary, companies have somehow to find administrators who know enough of the mathematics to be able to guess what kind of industrial investment is likely to be necessary for the company to secure the most rewarding prime contracts in the new science. This again is not so easy since much of the mathematics just cannot be translated into words. You either understand the figures, or you cannot ever have it explained to you. This is rather new because even things like indeterminacy in quantum mechanics can be more or less put into words.

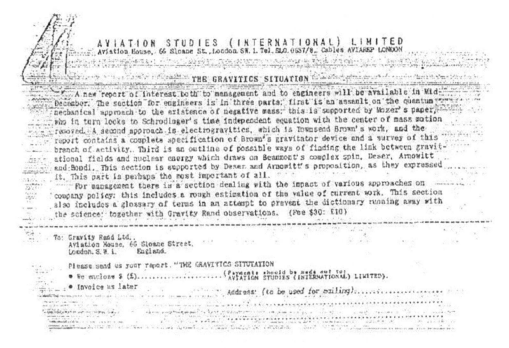

Fig. 9.8 *Order form for Gravity Rand's (aka Aviation Studies [International] Limited's) then-forthcoming report on the "Gravitics Situation". [Source: [Valone (2004)], p. 8]*

That so many ideas that we would consider foolish today were taken seriously by the military and their industrial partners, we should simply see as an indication of the poor state of knowledge in gravitation at the time. When the understanding of gravitational physics is unfavourably compared with the understanding of indeterminacy in quantum mechanics, then that has to be trouble. This explains the rather disconcerting fact that Arnowitt and Deser's spoof GRF entry, relating gravity and nuclear particles, was taken up so strongly in this anti-gravity frenzy. A gravity group (again: members unknown) known as 'Gravity Rand Ltd.' was formed[52] to consider what the author/s view as potentially plausible anti-gravity schemes. To this end, a report entitled "The Gravitics Situation" was published in December, 1956 (see fig. 9.8). The report states:

> [T]he time is fast approaching when for the first time, it will be within the capability of engineers with tevatrons to work directly with particles that—it is increasingly accepted—

[52] This group, like the Gravity Research Group, was based in London (based at address: 66 Sloane Street, London, SW1—now housing an exclusive estate agents): it seems obvious that they are one and the same company: Aviation Studies (International) Ltd. (It might be no coincidence that T. T. Brown formed a company known as 'Rand International, Ltd.', this just after a spell at a Bahnson-owned company, part of Bahnson Labs, known as the 'Whitehall-Rand Project'—it seems plausible that Bahnson was connected to Gravity Rand, though I have been unable to verify this).

contribute to the source of gravitation. And while that in itself may not lead an absorber of gravity, it will at least throw more light on the sources of the power. Another task is solution [see Appendix IV] of outstanding equations to convert gravitational phenomena to nuclear energy. The problem—still not solved—may support the Bondi–Hoyle theory that expansion of the Universe represents energy continually annihilated instead of being carried to the boundaries of the Universe. This energy loss manifests itself in the behavior of the hyperon and K-particles which would—or might—form the link between the microcosm and Macrocosm. Indeed, Deser and Arnowitt propose that the new particles are a direct link between gravitationally-produced energy and nuclear energy. If this were so, it would be the place to begin in the search for practical methods of gravity manipulation. It would be realistic to assume that the K-particles…are such a link. Then a possible approach might be to disregard objections which cannot be explained at this juncture until further Unified Field links are established. As in the case of the spin and orbital theories which were naive in the beginning, the technique might have to accept the apparent forces and make theory fit observation until more is known. Some people feel that the chances of finding such a Unified Field Theory to link gravity and electrodynamics are high. Yet still think that the finding of a gravity shield is slight because of the size of the energy source and because the chances of seeing unnoticed effect seem slender. Others feel the opposite and believe that a link between nuclear energy and gravitational energy may precede the link between the Einstein General Relativistic and Quantum Theory disciplines. Some hope that both discoveries may come together while a few believe that a partial explanation of both may come about the same time. Which will afford sufficient knowledge of gravitational fields to perfect an interim type of absorber using field links that are available. The latter seems the more likely since it is already beginning to happen. There is not likely to be any sudden full explanation of the microcosm and Macrocosm but one strand after another joining them will be fashioned as progress is made towards quantizing the Einstein theory.[53]

Again, we see how two strands reassert themselves: on the one hand there is the Arnowitt–Deser idea, now explicitly mentioned; and on the other hand there is Hlavatý's idea. Both do indeed suggest the possibility of anti-gravity since they suggest a deeper unity between gravity and other forces that are controllable. It is clear that these ideas led to gravitational project funding from a range of sources with the, perhaps not unreasonable given the state of knowledge even amongst experts, expectation of technological advances.[54]

The report includes the following interesting aside on funding in this period, and on the role of 'the nuclear experience':

> It is a common thought in industry to look upon the nuclear experience as a precedent for gravity and to argue that gravitics will similarly depend on the use of giant tools beyond

---

[53] The report cites the works of Bryce DeWitt, including his PhD Thesis. This, it would seem, must be someone who was aware of DeWitt himself. My sense is that it was likely written by Bahnson himself, or else by T. T. Brown with whom he was working at the time—or perhaps the pair of them together.

[54] Indeed, DeWitt speculates that the termination of his Air Force grant in 1966 was related to the fact that by this time "the military realized that they weren't going to get magical results from gravity research" [DeWitt (2009), p. 415]. The evidence seems to bear this opinion out.

the capabilities of the air industry. And that companies will edge into the gravitational age on the coat tails of the Government as industry has done—or is doing—in nuclear physics. But this overlooks the point that the 2 sciences are likely to be different to their investment. It will not need a place like Hanford or Savannah River to produce a gravity shield or insulator once the know-how has been established. As a piece of conceptual engineering, the project is probably likely to be much more like a repetition of the turbine engine. It will be simple in its essence. But the detailed componentry will become progressively more complex to interpret in the form of a stable flying platform. And even more intricate when it comes to applying the underlying principles to a flexibility of operating altitude ranging from low present flight speeds at one extreme to flight in a vacuum at the other. This latter will be the extreme of its powers.

The author speaks as if the production of a gravity shield or insulator (of the kind Babson envisaged and was so humiliatingly vilified by Gardner for proposing) is a foregone conclusion given enough effort and support. As mentioned above, however, we shouldn't be too hard on the folks who bought this line of thinking given that the anti-gravity suggestions were coming from reputable sources (or so they believed).[55]

There's serendipity at several levels in this story, but ultimately the lesson is that poor understanding of classical and quantum gravity led to the funding opportunities that transformed it. As this transformation continued, improving knowledge, the unusual funding opportunities were pruned. Beyond serendipity, however, it was vital that people with good sense and training (such as Wheeler and Goldberg) were in control of major decisions at key moments.

One aspect that we have not delved into here, since it is not so relevant from the perspective of quantum gravity, is the *commercial* involvement in gravity through its impact on geology and geological features such as oil deposits: oil is often to be found in areas of anomalous density which are associated with variations in the strength of gravity on the Earth's surface above them (itself a result of the oblate structure of the Earth). One can, in this way, map subsurface geology. High precision gravitational measurements were used to prospect for oil (the first being conducted in Czechoslovakia in 1915), and were rather successful at a time when few other technologies were available. This work in itself led to advances in technologies used to measure gravity (such as better pendulums, torsion balances, and the 'gravimeter')—see [Eckhardt (1940)] for a historical review of this work. Taken together with the other industrial and military work on gravity, it reveals that despite the lack of knowledge of gravitation in the mid-1950s and earlier, it was nonetheless entangled with a great variety of endeavours, practical and theoretical, each of which contributed to the renaissance.

---

[55] We should note that there was at least some scope for a discussion of gravity shields even in the context of general relativity. For example, in 1956, Peter Bergmann [Bergmann (1957)] (on an Air Force research grant) considered the possibility of gravity shields via negative mass particles (which, though not likely to exist, were not ruled out entirely). Bergmann showed that only in the case where the negative mass particle possessed an electric charge would such a particle be polarized gravitationally, by applying an electric field. But even in this case, the shielding effect would be so small as to offer no benefit.

9.6 Conclusion: Quantum Gravity on the Rise

For those readers who know something about the current quantum gravity research landscape, many questions and features that are now deemed to be of crucial importance are nowhere to be found in the earliest work we have looked at. Supersymmetry is nowhere to be seen. The notion of black hole entropy is absent. Effective field theory is unknown, though its initial roots are forming. Such notions radically alter the way quantum gravitational research is conceived and pursued, offering up a whole new set of constraints to guide the construction and evaluation of proposals. However, the work we have discussed in this book shows how the groundwork was laid for such constraints to come about. The earliest days of quantum gravity research were marked by a kind of nonchalance about the magnitude of the task. There were just a handful of particle types and two (and then three) forces binding them together, which we treated as more or less similar. In this final concluding section, let us simply say a few brief words about the state of quantum gravity and the explosion that occurred in 1957, at which point some of the special features of gravity were apparent.

In the years following 1956, quantum gravity ascended fairly rapidly, assuming the mantle of respectability of which it had long been deprived.[56] In his 1960 'stock taking' submission to the Gravity Research Foundation's essay competition, Bryce DeWitt usefully sums up the improved state of affairs at the time:

> [T]he researcher in gravitational theory is no longer so lonely as he was then [1953— DR], and his gamble seems less one-sided than it did just a few years ago. Papers in small but increasing numbers, concerned with detailed, down-to-earth, and non-speculative aspects of general relativity are appearing in many quarters. Two books containing survey articles as well as fundamental contributions are in course of publication. The relativist is now able to speak without apologizing for himself. The circle of full- or part-time relativists is itself growing by leaps and bound, both in number of individuals and in diversity of activities. Within the space of five years three major international conferences of general relativity have been held, a series of events without precedent throughout the entire preceding forty years. And this appears to be only the beginning, both in respect of numbers and in respect of the problems under attack. The problems of gravitation today are in clearer focus. The "big" theoretical problem is still that of quantization.
>
> [DeWitt (1960), p. 2]

DeWitt was himself, of course, responsible for a large part of this shift, not least through the Chapel Hill conference.[57] In a *Physics Today* (July, 1957, p. 53: no author specified) report on this conference, we find that the necessity of quantization is centre stage:

> The first question raised in the sessions of quantum field theory was: Why quantize the gravitational field? The theory of measurement in the context of gravitation can tell us whether there is a logical necessity for quantizing the gravitational field. It seems

---

[56] We might note that an enormous impetus came with the discovery of high energy gravitating sources that led to the probing of gravity beyond the weak-field limit (cf. [Bonolis (2017)]).

[57] In fact, as the archives amply reveal, the conference organization was primarily due to the efforts of Cecilé DeWitt.

at present that there is a logical need to quantize the gravitational field or else to introduce a new basic principle in physics. Even without settling this question, physicists are interested in tackling the technical problems of quantization of the gravitational field, partly because of their present difficulties in understanding existing quantum field theories and elementary particles, and partly because it is a challenging problem in itself.

Note also, crucially, that there is the statement that quantum gravity might be pursued as "a challenging problem in itself," independently of its role in other areas of physics. This is a milestone and, I believe, the first statement of its kind in a public arena.

Similar sentiments were expressed in January 1957, in a piece entitled "Relativity Re-explored", in which a correspondent for *Science News* re-states the old quip about relativity, finding it no longer true (see fig. 9.9):

> Four or five years ago such a meeting would not have been possible. If held, it would have consisted of "three friends talking together," Einstein and his two close associates, Drs. Peter Bergmann of Syracuse University and Valentin Bargmann of Princeton University.
> (*Science News Letters* 71(5), p. 68.)

However, the problem of quantum gravity is once again center-stage, and presented in the same terms as turn of the century physicists facing the problems of atomic physics.

Half a year after the Chapel Hill conference, DeWitt organized a smaller, exclusive month-long 'exploratory research session' (fig. 9.10) in Copenhagen (at the Universitets Institut for Teoretisk Fysik), now devoted specifically to quantum gravity. In addition to DeWitt, there was Stanley Deser, Charles Misner, Christian Møller, Oskar Klein, and Bertel Laurent.[58] This session solidified the various approaches, though firm canonical versus covariant divisions had still not yet fully crystallized, and the atmosphere was more one of openness to ideas.[59] More crucially, perhaps, it solidified the idea that gravity is not like other fields. Quoting from the workshop report we find the statement: "Dr. Misner also expresses a similar view, calling attention to the fact that if gravitation is to occupy a significant place in modern physics, it can do so only by being qualitatively different from other fields. As soon as we assume the gravitational field to behave qualitatively like other fields we find that it is quantitatively insignificant. It is in its qualitative difference that its very special importance lies."[60]

The Stevens Meetings (fig. 9.11), also originating in 1957 (coming almost another half year after the exploratory research session) and organized by James Anderson,[61] marked an important phase in the building of a gravity community (see fig. 9.11). A thorough

---

[58] This would not have been possible without the establishment of infrastructure described earlier in this chapter, and was funded by the U.S. Air Force.

[59] As Agnew Bahnson put it in one of his 'Memoranda' for the Institute for Field Physics, of which this was a part, the idea was that each participant "should endeavor to expound his own point of view and accomplishments to the others" (Memorandum No. 10, September 25th, 1957: CDWA).

[60] I refer the reader to the excellent, detailed review of this meeting (including the publication of the report itself: [DeWitt (2017)]) by Alexander Blum and Thiago Hartz [Blum and Hartz (2017)].

[61] As Anderson put it, "the ground rule was you could talk for ten minutes, and then you could have as much discussion as you wanted". He goes on: "the whole point of the meetings was to get discussion and sort of cook ideas and see what comes out of it, and then people went back and came back and talked some more. So it was like a revival meeting, almost. Whoever wanted to talk came. There was not a set agenda. I don't

68   Science News Letter *for February 2, 1957*

PHYSICS

Relativity Re-explored

Progress is being made toward a better understanding of relativity after years of neglect. The "quantization" of general relativity is one approach now being studied.

➤ IF ALBERT EINSTEIN were alive, he would be astounded to know that 45 of the world's top physicists, mathematicians and astronomers would meet for six days in the United States to discuss his theories on gravitation, as they have recently.

Four or five years ago such a meeting would not have been possible. If held, it would have consisted of "three friends talking together," Einstein and his two close associates, Drs. Peter Bergmann of Syracuse University and Valentin Bargmann of Princeton University.

The renewed interest in gravitational theory has developed after many years of neglect due to the lack of experimental guideposts and mathematical difficulties. Both remain troublesome, but much progress has been made toward better understanding of general relativity by considering approximate solutions of the equations.

One approach now being thoroughly studied is the "quantization" of general relativity. This involves treating gravitational radiation not as being continuous but as consisting of tiny packets, much as light, which was once thought continuous radiation, is now known to consist of tiny light packets, or photons.

The scientists struggling with general relativity theory are in much the same position now as were those earlier this century when quantum mechanics was being devised to account for the tiny world of the atom.

Gist of the problem is that one set of laws seems to apply to nuclear particles such as electrons, protons, etc., while another is needed for large objects such as the earth, stars and galaxies.

So far there are only three known proofs for Einstein's general relativity. One is changes in the orbit of the planet Mercury, which observations have shown come very close to what the theory predicts. Another is the bending of light from far-away stars by the sun's gravitational field, which also comes very close to the amount predicted. The third is the displacement, or reddening, of spectral lines from certain very massive stars.

There are, on the other hand, literally thousands of proofs, with more accumulating every day, that the laws of quantum mechanics hold true for minute particles of matter.

Some day scientists may be aided in bridging this gap by using giant computers. That will not be possible, however, until they have figured out the necessary equations.

As one scientist attending the International Conference on the Role of Gravita-

tion in Physics at the University of North Carolina in Chapel Hill said: "If it were tried now, either the mathematician preparing the instructions for the computer would blow out his brains, or the machine itself would blow up."

Science News Letter, February 2, 1957

INVENTION

Air Traffic Control Provides Safer Flying

➤ SAFER FLYING for the nation's countless air travelers is the promise of an automatic air traffic control system that keeps planes on the straight and narrow at all times.

Key to the device's actions is a magnetic memory tape coupled with an electronic computer. They are designed to sit in the cockpit along with a human pilot and an automatic pilot.

This is how the system works. Before flight, a path through the skies is plotted for the pilot. He is also given specific times for being at a specific spot on his flight path. This information is fed into the memory device. The computer then checks the actual flying time and flight path while the air liner is moving and correlates it with the predetermined information given the pilot by the control tower.

If the plane is off its time or path, the computer automatically feeds signals to the autopilot and throttle control, thereby correcting the flight to maintain the plane on course and schedule.

One of the big advantages of the system, its inventor, Thomas M. Ferrill Jr. of Garden City, N. Y., says, is that the position and movement of each aircraft at any given instant can be radio controlled by a tower operator to prevent possible collision.

Science News Letter, February 2, 1957

MINING

Discarded Diggings Found Valuable

➤ DIGGINGS, thrown away as worthless by disgruntled gold miners 20 years ago, are proving to be a storehouse of valuable minerals today.

The water-deposited storehouse of strategic materials has been found in the Pacific Northwest, a report by A. J. Kauffman Jr. and K. D. Baber of the U. S. Bureau of Mines shows.

Chromite is being recovered from alluvial deposits in Oregon. Columbite-tantalite and "radioactive blacks," which are dark sands

Fig. 9.9 *Note from* Science News Letters *indicating the importance of the Chapel Hill Conference for quantum gravity, and relativity research more generally.*

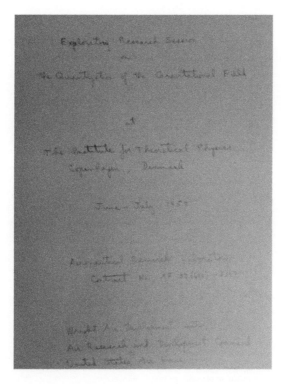

Fig. 9.10 *Handwritten report from the first ever workshop devoted solely to quantum gravity, June–July 1957, in Bryce DeWitt's handwriting. Image source: CDWA.*

account of these Stevens Institute conferences remains to be carried out. However, it is clear that they were another vital piece of the puzzle of how quantum gravity came to be viewed as a valid subject of study in its own right. These took the form of both seminars and conferences, with several running each year. The first took place on November 19th, 1957. The titles of the talks for the first three years were as follows:[62]

Relativity Seminar 1: November 19th, 1957:

- Peter Bergmann, Relations between canonical and Landau and Lifschitz energy-momentum; invariant properties of field singularities.

think I ever had a list of speakers. If people wanted to speak, they came and they spoke, and then there was the discussion, and people talked"—interview with Don Salisbury and the author, March 19th, 2011.

[62] My thanks to Dieter Brill for making the contexts of these talks available to me—though, as he notes, the 'titles' of the talks were often his interpretations of the talks, since it was rare for speakers to submit titles in advance, rather instead simply turning up and presenting in an impromptu fashion. Also, it is possible that what are labeled 'relativity seminars' might not always be Stevens Institute events, but Princeton events. Regardless of this, we nonetheless see a clear record of increased relativity activity, along with a sense of what were considered the most important problems. Indeed, the absence of a serious historical investigation of these seminars and conferences constitutes a serious gap in the literature on quantum gravity and gravitational physics in general.

Fig. 9.11 *Stevens Institute Relativity Conference. Amongst those present are James Anderson (standing at the front, in his capacity as host), Dieter Brill, Felix Pirani, Bob Dicke, David Finklestein, Josh Goldberg, Wolfgang Kundt, Arthur Komar, Frederik Belinfante, Paul Dirac, John Stachel, Ezra Newmann, Peter Bergmann (standing up in the centre of the audience), Hermann Bondi, John Boardman, Stanley Deser, Joe Weber, Louis Witten, Bryce DeWitt, Alfred Schild, and John Wheeler. Image courtesy of James Anderson.*

- David Finkelstein, on Spin without Spin—already unified theory simplified by spinors
- Arthur Komar, on Scalar Invariants.

Relativity Conference 1: January 28th, 1958:

- David Finkelstein, Schwarzschild solution in the new coordinates.[63]
- Richard Lindquist, 2-body problem.
- James Anderson, True observables.
- Huzihiro Araki, Amount of freedom in time-symmetric initial value problem.
- Bryce DeWitt, Constraints.
- Bertel Laurent, Representation of group of general coordinate transformations.
- H. Dehnen (?), Work of Jordan's group.

[63] Note that this was the first statement of the 'event horizon' concept, leading to the modern understanding of black holes.

Relativity Seminar 2: March 18th, 1958:

- Stanley Deser, Schwinger's action principle to quantize gravity.
- Arthur Komar, Invariants.
- Oskar Klein, Eddington Relations.[64]

Relativity Conference 2: October 28th, 1958:

- Felix Pirani, Definition of energy.
- Baldwin, Spherical waves in linear approximation.
- Paul Dirac, Hamiltonian form of gravitational field equations.[65]
- Charles Misner, Variational principle and coordinate conditions.

Relativity Seminar 3: December 16th, 1958:

- Plebanski, New derivation of equations of motion.
- Bryce DeWitt, Remarks on space-like description.

Relativity Seminar 4: January 22nd, 1959:

- Bryce DeWitt, Commutation rules for Dirac's quantized gravitation.
- James Anderson, Elimination of non-physical variables.

Relativity Conference 3: January 27th, 1959:

- James Anderson, Solution of initial value equations.
- Paul Dirac, Hamiltonian formulation.
- Joe Weber, Experiment on gravitational waves.
- Felix Pirani, Plane waves.
- Wolfgang Kundt, Invariant characterization of solutions.

Hence, a steady flow of talks and interactions, involving quantum gravity, continued after the Chapel Hill conference, and included a workshop devoted solely to quantum gravity. All in one year: 1957. A remarkable turnaround.

***

[64] Note that at this time, Klein gave a series of talks at Princeton on quantum mechanics and general relativity, discussing the revival of the 5-dimensional model—this was a course, which ran for several weeks.

[65] Dirac then spoke on the quantization of general relativity at Princeton, on November 10th, and for several weeks thereafter.

The end point of this book really only marks the very beginnings of quantum gravity as a research field in its own right. It was, at that point, and still *is*, covered with a deep mist. Perhaps more so, the more we have learned about the problem. Hence, the reader should not come away from this book feeling like all was well for quantum gravity after 1957. Indeed, following the 1974 Oxford Symposium, John Wheeler, when asked by a reporter at the conference whether a young student wanting to work on quantum gravity "should be a relativist or a field theorist," somewhat surprisingly perhaps, replied that "such a person would be better advised to find something other than quantum gravity to work on!" [Nature correspondent (1974), p. 283]. Worse still, at that same conference, John Taylor was reported to have said that quantum gravity, as a result of its unrenormalizability, "is dead" (ibid, p. 282.).[66] Hence, while this struggle represents a magnificent human achievement in many ways—indeed, one cannot fail to be impressed how far these pioneers got, like explorers, not realizing the magnitude of their journey—it would not at all surprise me if in another century we find the problem still unresolved and as misty as ever.

··

References

Babson, R. W. (1950) *Actions and Reactions: An Autobiography of Roger W. Babson.* (2nd revised edition). New York: Harper & Brothers Publishers.

Babson, R. W. (1948) Gravity—Our Enemy Number One. Reprinted in H. Collins *Gravity's Shadow: The Search for Gravitational Waves* (pp. 828–31). Chicago: University of Chicago Press, 2004.

Bergmann, P. G. (1957) Gravity Shields. *Science* 125(3246): 498–9.

Blum, A., D. Giulini, R. Lalli, and J. Renn (eds.) (2017) *The Renaissance of Einstein's Theory of Gravitation* (special issue). *The European Physical Journal H* 42(2): 95–393.

Blum, A. and T. Hartz (2017) The 1957 Quantum Gravity Meeting in Copenhagen: An Analysis of Bryce S. DeWitt's Report. *The European Physical Journal H* J42(2): 107–57.

Blum, A., R. Lalli, and J. Renn (2017) The Renaissance of General Relativity: How and why it Happened. *Annalen der Physik*. DOI: 10.1002/andp.201600105

Bonolis, L. (2017) Stellar Structure and Compact Objects before 1940: Towards Relativistic Astrophysics. *The European Physical Journal H* 42(2): 311–93.

DeWitt, B. and C. Morette DeWitt (1952) The Quantum Theory of Interacting Gravitational and Spinor Fields. *Physical Review* 87: 116–22.

DeWitt, B. S. (2017) Exploratory Research Session on the Quantization of the Gravitational Field. *The European Physical Journal H* J42(2): 159–76.

DeWitt, B. S. (2009) Quantum Gravity: Yesterday and Today. *General Relativity and Gravitation* 41: 413–19.

---

[66] This talk does not seem to have made it into the proceedings volume.

DeWitt, B. S. (1985) The Early Days of Lagrangian Hydrodynamics at Lawrence Livermore Laboratory. In J. M. Centrella, J. M. LeBlanc, and R. L. Bowers (eds.), *Numerical Astrophysics: Proceedings of the Symposium* (pp. 474–81). Boston: Jones and Bartlett.

DeWitt, B. S. (1960) Gravitational Research: The Coming Decade. GRF essay entry, 1960. GRF archives.

DeWitt, B. S. (1957) Principal Directions of Current Research Activity in the Theory of Gravitation. *Journal of Astronautics* 4: 23–8.

Dicke, R. H. (1959) Gravitation–An Enigma. *American Scientist* 47(1): 25–40.

Eckhardt, E. A. (1940) A Brief History of the Gravity Method of Prospecting for Oil. Geophysics 5(3): 231–42.

Eisenstaedt, J. (2006) *The Curious History of Relativity.* Princeton: Princeton University Press.

Eisenstaedt, J. (1986) The Low Water Mark of General Relativity, 1925–1955. In D. Howard and J. Stachel (eds.), *Einstein and the History of General Relativity* (pp. 277–92). Boston: Birkhaüser.

Galbraith, J. K. (1997) *The Great Crash of 1929.* New York: Mariner Books.

Gardner, M. (1957) *Fads & Fallacies In The Name Of Science.* New York: Dover.

Goenner, H. (2016) A Golden Age of General Relativity? Some Remarks on the History of General Relativity. arXiv: 1607.03319.

Goldberg, J. (1993) US Air Force Support of General Relativity: 1956–1972. In J. Eisenstaedt and A. J. Kox (eds.), *Studies in the History of General Relativity* (pp. 89–102). Boston: Birkhaüser.

Howard, D. (unpublished). Physics as Theodicy. Talk delivered at the workshop on *Scientific Perspectives on Natural Evil,* Vatican Observatory, September 15th, 2005 http://www.nd.edu/~dhoward1/Physics%20as%20Theodicy.pdf.

Kaiser, D. (2000) *Making Theory: Producing Physics and Physicists in Postwar America.* Ph.D. dissertation, Harvard University.

Kaiser, D. (1998) A ψ is just a ψ? Pedagogy, Practice, and the Reconstitution of General Relativity, 1942–1975. *Studies in the History and Philosophy of Modern Physics* 29(3): 321–38.

Kaiser, D. and D. Rickles (2018) The Price of Gravity: Private Patronage and the Transformation of Gravitational Physics after World War II. *Historical Studies in the Natural Sciences* 48(3): 338–79.

Kennefick, D. (2007) *Traveling at the Speed of Thought.* Princeton: Princeton University Press.

Lalli, R. (2017) *Building the General Relativity and Gravitation Community During the Cold War.* New York: Springer.

Lichnerowicz, A. (1992) Mathematics and General Relativity: A Recollection. In J. Eisenstaedt and A. J. Kox (eds.), *Studies in the History of General Relativity* (pp. 102–108). Boston: Birkhaüser.

McCrea, W. H. (1951) Relativity Theory and the Creation of Matter. *Proceedings of the Royal Society of London. Series A* 206(1087): 562–75.

Nature correspondent (1974) Limited Progress with Quantum Gravity. *Nature* **248**: 282–3.

Pais, A. (1982) *Subtle is the Lord: The Science and the Life of Albert Einstein*. Oxford: Oxford University Press.

Pauli, W. (ed.) (1955) *Niels Bohr and the Development of Physics*. New York: Pergamon Press.

Pirani, F. (1955) On the Energy-Momentum Tensor and the Creation of Matter in Relativistic Cosmology. *Proceedings of the Royal Society of London. Series A* **228**(1175): 455–62.

Rees, M. (1980) The Mathematical Sciences and World War II. *The American Mathematical Monthly* 87(8): 607–21.

Schweber, S. (2006) Einstein and Oppenheimer: Interactions and Intersections. *Science in Context* 19(4): 513–59.

Unidentified author (1950) Science: The Trouble with Gravity. *Time* January 1950.

Valone, T. (ed.) (2004) *Electrogravitics Systems*. Washington, DC: Integrity Research Institute.

Weinberg, S. (2008) *Bryce Seligman DeWitt, Biographical Memoir*. Washington, DC: National Academy of Sciences.

Will, C. (1989) The Renaissance of General Relativity. In P. C. W. Davies (ed.), *The New Physics* (pp. 7–33). Cambridge: Cambridge University Press.

Yost, C. A. (1991) T. T. Brown and the Bahnson Lab Experiments. *Electric Spacecraft Journal* Apr/May/June: 6–12.

Appendix A
PhDs on Classical and Quantum Gravity ≤ 1957

Below is a list of doctoral dissertations written in the period covered by this book. Since these involve the submission dates, I have extended the date to 1959 to allow for the fact that the dissertations will likely have been commenced within my chosen period. I have also included works on classical gravitation in order to reveal the status of gravitational physics in general. Though I have of course aimed for completeness, I make no claims that this is an exhaustive list.

- Amar, Henri. *A General-Relativistic Approach to Short-Range Forces.* PhD, Ohio State University: 1952.
- Anderson, James LeRoy. *On the Quantization of Covariant Field Theories.* PhD, Syracuse University: 1954.
- Barkas, Walter H. *Cosmological Models and Intergalactic Free Electrons.* PhD, University of Washington: 1936.
- Bastin, E. W. *A Development of Eddington's Fundamental Theory from New Axioms, and its use as a Critique of Eddington's Scientific Philosophy.* PhD, University of London, King's College: 1952.
- Bastin, E. W. *A Theory of the Process of Extrapolation in Physics, with Special Reference to the Investigation of the Nature of the Limits of Observation in Quantum Theory and in Cosmology, and to Connections between Them.* PhD, University of Cambridge: 1958.
- Blankfield, Judith R. *Approximations to Einstein's Equations of General Relativity.* PhD, University of Illinois: 1958.
- Brandner, Fred August. *The Motion of a Planet Under Einstein's Law of Gravitation.* MSc, University of Illinois: 1923.
- Brehme, Robert Woodruff. *The Stationary Charged Particle in a Static Gravitational Field.* PhD, The University of North Carolina at Chapel Hill: 1959.
- Bronstein, Matvei Petrovich. *Quantization of Gravitational Waves.* PhD, Leningrad Physico-Technical Institute: 1935.
- Bryan, Russell Byrne. *The Introduction of the Fundamental Length Concept in the Quanization of the Electromagnetic Field.* PhD, North Carolina State College: 1954.
- Camp, Leon Walton. *Analysis of a Method Using Electrostatic Forces in Measuring Variations in the Earth's Gravitational Field.* PhD, The University of Texas at Austin: 1941.
- Caplan, Davis Isaac. *On the Quantization of Interacting Einstein, Maxwell, and Dirac Fields.* PhD, Purdue University: 1957.

- Casson, Benedict. *Theory of Gravitational Spin.* PhD, California Institute of Technology: 1930.
- Chou, Pei Yuan. *The Gravitational Field of a Body with Rotational Symmetry in Einstein's Theory of Gravitation.* PhD, California Institute of Technology: 1928.
- Coleman, Albert John. *A Study in Relativistic Quantum Mechanics based on Sir A. S. Eddington's Relativity Theory of Protons and Electrons.* PhD, University of Toronto: 1944.
- Cornish, F. H. J. *The Mathematical Form and Physical Content of Unified Field Theories Derived from a Variational Principle.* DPhil, University of Oxford: 1955.
- Clark, G. L. *The Relativity Theory of Gravitation.* PhD, University of Cambridge: 1940.
- Crosby, Douglas Richard. *Tensor Analysis in Finsler Space and the Problem of a Unified Theory of Gravitation and Electromagnetism.* PhD, Princeton University: 1946.
- Davidson, W. *Problems in Relativity and Cosmology* PhD, University of London, Royal Holloway College: 1958.
- Davis, William Robert. *Kritik einheitlicher Feldtheorien: Beiträge zum allgemeinen Formalismus, eine Interpretations möglichkeit und einige Quantisierungsprobleme.* PhD, Technical University of Hannover: 1956.
- Davy, N. *The Effect of Temperature on Gravitative Attraction.* MSc, University of London: 1929.
- Deser, Stanley. *Relativistic Two-Body Problems.* PhD, Harvard University: 1953.
- Drill, Harry T. *A Search for an Electrostatic Analogy to the Gravitational Red Shift.* PhD, University of Washington: 1939.
- Etherington, I. M. H. *On Relativistic Cosmology, and on the Definition of Distance in General Relativity.* PhD, The University of Edinburgh: 1932.
- Everett III, Hugh. *On the Foundations of Quantum Mechanics.* PhD, Princeton University: 1957.
- Fletcher, John George. *Local Conservation Laws in Generally Covariant Theories; their Relationships and their Uses.* PhD, Princeton University: 1959.
- Freistadt, Hans. *Approximate Solutions of the Field Equations and Motion of Particles in Einstein's Generalized Theory of Gravitation.* PhD, University of North Carolina: 1950.
- Goldberg, Irwin. *Transformation Groups and Lagrangian Quantization.* PhD, Syracuse University: 1957.
- Goldberg, Joshua. *Equations of Motion in a Covariant Field Theory.* PhD, Syracuse University: 1952.
- Goody, A. J. *Gravity Waves of Finite Amplitude. I: Stationary Waves. II: Progressive Waves of Finite Depth.* MSc, University of London, King's College: 1955.
- Gutzwiller, Martin C. *Quantum Theory of Wave Fields in a Space of Constant Curvature.* PhD, University of Kansas: 1953.
- Harding, D. *Einstein's Unified Field Theory.* MSc, University of London, Sir John Cass College: 1959.
- Harrison, Bertrand Kent. *Exact Three-Variable Solutions of the Field Equations of General Relativity.* PhD, Princeton University: 1959.

- Haskey, H. W. *Some Unitary Theories of Gravitation and Electricity.* PhD, University of London, External Degree: 1936.
- Haywood, J. H. *The Equations of Motion and Co-Ordinate Conditions in General Relativity.* PhD, The University of Manchester: 1951.
- Heller, Jack. *The Canonical Formulation of a General Field Theory with Spin Properties.* PhD, Polytechnic Institute of Brooklyn: 1950.
- Hide, T. G. S. *Static Models in Relativistic Cosmology.* PhD, University of London, King's College: 1943.
- Holland, Dan Howard. *Lattice Space Quantization of Coupled Meson and Nucleon Fields.* PhD, Stanford University: 1955.
- Huke, Aline. *Some Relativity Fields in N Dimensions (Gravitation).* AM, The University of Chicago: 1927.
- Hunt, J. N. *Mathematical Investigations of Gravity Wave Problems, Including Waves of Finite Height.* PhD, University of London, Imperial College of Science and Technology: 1952.
- Ingraham, Richard Lee. *Conformal Relativity.* PhD, Harvard University: 1952.
- Janis, Allen Ira. *Coordinate Conditions and True Observables.* PhD, Syracuse University: 1957.
- Johnson, Clare P, Jr. *I. The Extraction of Aluminum From Clay. II. A Unified Theory of Gravitation and Electromagnetism.* PhD, Harvard University: 1952.
- Kennedy, Warren Leslie. *The Constraints and Commutators of the Interacting Field of Electrons, Electromagnetism and Gravity.* PhD, Purdue University: 1957.
- Kirkwood, Robert Lord. *On the Theory of Relativity.* PhD, Stanford University: 1951.
- Komar, Arthur Baraway. *Some Consequences of Mach's Principle for the General Theory of Relativity.* PhD, Princeton University: 1956.
- Kursunoglu, B. *Gravitation and Electrodynamics.* PhD, University of Cambridge: 1952.
- Leichter, Michael. *Quantization in 'The Irreducible Volume Character of Events'.* PhD, Ohio State University: 1952.
- Lemaitre, G. H. *The Gravitational Field in a Fluid Sphere of Uniform Invariant Density, According to the Theory of Relativity.* PhD, Massachusets Institute of Technology: 1927.
- Laurikainen, Kalervo Vihtori. *Über die Gravitationsenergie des materiefreien Elektromagnetis-chen Feldes.* PhD, Universitat Helsinki: 1950.
- Levinson, Horace Clifford. *The Gravitational Field of Masses Relatively at Rest According To Einstein's Theory Of Gravitation.* PhD, The University of Chicago: 1922.
- Leybourne, G. G. *The General Theory Of Relativity.* PhD, The University of Wales: 1932.
- Littman, Walter. *On the Existence of Gravity Waves Near Critical Speed.* PhD, New York University: 1956.
- Liverhant, S. E. *The Unification of Gravitation, Electro-Magnetism and Quantum Mechanics.* MSc, University of London, King's College and Queen Mary College: 1945.
- Marder, L. *On The Existence Of Cylindrical Gravitational Waves.* PhD, University of London, King's College: 1958.
- Mauger, F. E. *A Discussion of Four-dimensional Unified Field Theories.* MSc, University of London, King's College: 1956.

- McVittie, G. C. *Electromagnetism and Gravitational Fields*. PhD, University of Cambridge: 1930.
- Mikhail, F. I. *Relativity-Solutions of Einstein's Gravitational Equations For Empty Space $G_{\mu\nu} = 0$*. MSc, The University of Wales: 1949.
- Mikhail, F. I. *Relativistic Cosmology and Some Related Problems in General Relativity Theory*. PhD, University of London, Royal Holloway College: 1952.
- Misner, Charles William. *Outline of Feynman Quantization of General Relativity; Derivation of Field Equations; Vanishing of the Hamiltonian*. PhD, Princeton University: 1957.
- Moffat, J. W. *The Foundations of a Generalisation of Gravitation Theory*. PhD, University of Cambridge, Trinity College: 1958.
- Mould, Richard A. *An Axiomatization of General Relativity*. PhD, Yale University: 1957.
- Mulhern, John E. Jr. *Analysis and Extension of Eddington's Unified Theory of Physics*. PhD, Boston University: 1954.
- Müller, Paul. *Über die Eigenschwingungen der Zylindrischen und Sphärischen Welt*. PhD, Graz Universität: 1939.
- Newman, Ezra Theodore. *Observables in Singular Theories by Systematic Approximation*. PhD, Syracuse University: 1956.
- Ogden, Edwin B. *The Consequences of the Two Underlying Assumptions in Sulaiman's Theory of Relativity as Applied to the Orbit of Mercury*. PhD, Boston University: 1936.
- Omer, Guy C., Jr. *Studies of Nonhomogeneous Cosmological Models*. PhD, California Institute of Technology, 1957.
- Perry, Byrne. *Methods for Calculating the Effect of Gravity on Two-Dimensional Free Surface Flows*. PhD, Stanford University: 1957.
- Penfield, Robert H. *Hamiltonians without Parameterization*. PhD, Syracuse University: 1954.
- Pirani, Felix. *On the Quantization of the Gravitational Field of General Relativity*. PhD, Carnegie Mellon University: 1951.
- Ratcliffe, J. F. *Unified Field Theories: The Theories of Kaluza and Klein, Weyl, Eddington and Einstein*. MSc, The University of Wales: 1934.
- Regge, Tullio. *On the Properties of Spin 2 Particles*. PhD, University of Rochester: 1957.
- Reimer, Edward Hawke. *Part I. The Relation of Weyl's Geometry to Eddington's. Part II. A New Eddingtonian Geometry with Applications to Differential Geometry*. PhD, University of California, Berkeley: 1927.
- Rindler, W. *Problems in Relativistic Cosmology*. PhD, University of London, Imperial College of Science and Technology: 1956.
- Roope, Percy M. *A Method of Measuring the Velocity of Gravitation*. PhD, Clark University: 1927.
- Rosen, Gerald Harris. *On the Quantum Theory of General Relativity*. PhD, Princeton University: 1959.
- Sachs, Rainer K. *Structure of Particles in Linearized Gravitational Theory*. PhD, Syracuse University: 1958.
- Salzman, George. *Born-Type Rigid Motion in Relativity*. PhD, University of Illinois: 1953.

- Schaffhauser-Graf, Edith. *Versuch einer 4-dimensionalen einheitlichen Feldtheorie der Gravitation und des Elektromagnetismus* Dr, Universite de Fribourg: 1953.

- Scheidegger, Adrian E. *Gravitational Radiation and Equations of Motion.* PhD, University of Toronto: 1950.

- Schild, Alfred. *A New Approach to Kinematic Cosmology.* PhD, University of Toronto: 1946.

- Schiller, Ralph. *Constraints and Equations of Motion in Covariant Field Theories.* PhD, Syracuse University: 1954.

- Scorer, R S. *Gravity Waves in the Atmosphere.* PhD, University of Cambridge: 1950.

- Seligman, Carl B. (DeWitt, Bryce Seligman). *Part I. The Theory of Gravitational Interactions. Part II. The Interaction of Gravitation with Light.* PhD, Harvard University: 1950.

- Shaw, H. *Interpretation of Gravitational Anomalies.* PhD, University of London, External Degree: 1931.

- Skinner, Ray O. *On the Quantization of Combined Gravitational, Electromagnetic and Electron Fields.* PhD, Carnegie Mellon University: 1953.

- Snyder, Hartland S. *Five Problems: Cascade Theory; Quadratic Zeeman Effect; Gravitational Collapse; Mesotron Collisions; Energy Levels of Fields.* PhD, University of California, Berkeley: 1940.

- Smith, Samuel S. *Motion of a Spheroid and a Sphere under Mutual Gravitation.* PhD, The University of Chicago: 1941.

- Stockum, W. J. van. *Axially Symmetric Gravitational Fields.* PhD, The University of Edinburgh: 1937.

- Swihart, James Calvin. *A Theory of Gravitation and its Quantization.* PhD, Purdue University: 1955.

- Taub, Abe H. *Quantum Equations in Cosmological Spaces.* PhD, Princeton University: 1935.

- Tauber, G. E. *A Study in Rotation and Gravitation.* PhD, University of Minnesota: 1951.

- Thomson, Robb. *Spin and Angular Momentum in the General Theory of Relativity.* PhD, Syracuse University: 1954.

- Tong, Kin-Nee. *Two-Dimensional Potential Flow in a Gravitational Field with a Known Free Surface.* PhD, University of Illinois at Urbana-Champaign: 1951.

- Topping, Alanson D. *Elastic, Elasticplastic, and Viscoelastic States of Stress and Strain around a Vertical Cylindrical Hole in a Semi-Infinite Gravitational Body.* PhD, University of Illinois at Urbana-Champaign: 1951.

- Tupper, B. O. J. *Five-dimensional Unified Field Theories and the Constants of Nature.* PhD, University of London, King's College: 1959.

- Vallarta, M. S. *Bohr's Atomic Theory from the Standpoint of the General Theory of Relativity and of the Calculus of Perturbations.* ScD, Massachusetts Institute of Technology: 1924.

- Vitousek, Martin Judy. *Some Flows in a Gravity Field Satisfying The Exact Free Surface Condition.* PhD, Stanford University: 1955.

- Watts, Henry Millard. *An Action-at-a-distance Theory of Gravitation.* PhD, John Hopkins University: 1953.

- Wilby, J. R. W. *On the Einstein Equations of the Gravitational Field in Orthogonal Coordinates with Two Explicit Independent Variables*. MSc, University of Leeds: 1928.

- Willmore, T. J. *Clock Regraduations and General Relativity*. PhD, University of London, External Degree: 1943.

- Wrede, Robert C. *n'' Dimensional Considerations of Basic Principles A and B of the Unified Theory of Relativity*. PhD, Indiana University: 1956.

- Zatzkis, Henry. *Conservation Laws in Nonlinear Field Theories*. PhD, Syracuse University: 1954.

Index